# MECHANICS AND MATERIALS FOR DESIGN

# MECHANICS AND MATERIALS FOR DESIGN

**Nathan H. Cook**

*Professor of Mechanical Engineering*
*Massachusetts Institute of Technology*

**McGraw-Hill Book Company**

New York   St. Louis   San Francisco   Auckland   Bogotá   Hamburg
Johannesburg   London   Madrid   Mexico   Montreal   New Delhi
Panama   Paris   São Paulo   Singapore   Sydney   Tokyo   Toronto

This book was set in Times Roman by Intercontinental Photocomposition Limited.
The editors were T. Michael Slaughter and David A. Damstra;
the production supervisor was Dennis J. Conroy.
The drawings were done by Danmark & Michaels, Inc.
The cover was designed by Jane Moorman.
Halliday Lithograph Corporation was printer and binder.

## MECHANICS AND MATERIALS FOR DESIGN

234567890HALHAL8987654

ISBN 0-07-012486-8

**Library of Congress Cataloging in Publication Data**

Cook, Nathan H.
    Mechanics and materials for design.

    Includes index.
    1. Mechanics, Applied.    2. Strength of materials.
3. Engineering design.    I. Title.
TA350.C745    1984        620.1        83-11337
ISBN 0-07-012486-8

# CONTENTS

# PREFACE

This text has been written as part of an effort at M.I.T. to modernize the well-established teaching of elementary applied mechanics. It is directed towards courses that are typically called "The Mechanics of Solids," "Strength of Materials," "Engineering Mechanics," "Mechanics of Materials," and the like.

Traditionally, (for the past half century) students have been introduced to mechanics in physics and/or statics courses. Given an elementary understanding of rigid-body mechanics, they then move on to the mechanics of deformable bodies. Many universities follow the tradition of a two-term sequence of Statics and Strength of Materials; at others, the material has been integrated into a single course in the Mechanics of Solids: first statics dealing with forces and moments in rigid, static bodies and structures then, strength of materials dealing with the effects of the forces and moments on real, deformable bodies and structures.

During the past 25 years, we have witnessed tremendous developments in computer application. In the field of applied mechanics, we now can solve routinely problems that were considered "unsolvable" just a few years ago. The tools of the trade now include a large array of sophisticated programs that are readily available to the practitioner of engineering.

By and large, these tools are too complex for full comprehension by more than a few specialists, yet they will be used by more and more people; some will use them correctly while others will not. There will be "good" answers and "bad" answers, "good" designs and "catastrophic" designs.

How should we develop our courses in applied mechanics to provide, at one and the same time, a solid grounding in the fundamentals of mechanics per se, and in the fundamentals of computer applications in mechanics?

The educational goal of this book is to provide *both* a firm, rigorous introduction to the mechanics of solids, *and* a rigorous introduction to computer methods in applied mechanics.

Generally, when we add material to an existing course, either the learning process must be improved, or some existing material must be removed. By and large, we are not very effective at the former, and are very hesitant about the latter.

The contents of this book are arranged to permit a great deal of flexibility in presentation, ranging from the older tradition of a two-term sequence to a considerably revised course in which some of the traditional content has been eliminated. Consequently, the book contains considerably more material than would be used in any given course, but will enable an evolutionary shift from the traditional to a more contemporary presentation.

Regarding the revised course content, the subjective decision on what should be eliminated, and what should be added, was based largely on three decades of experience and observation as an engineer-teacher-researcher-consultant. Underlying these decisions is a high expectation for the future availability of personal computers.

The text, in softbound version, has been used at M.I.T. for five semesters. Our experience with both the revised subject matter and the text has been most encouraging and rewarding.

The table below shows how the text could be used for the range of courses considered. The more traditional course is on the left, and the newer on the right.

**Possible Course Content**

| | Second term of Statics and Strength of Materials sequence | One term Mechanics of Solids | One-term "New Course" |
|---|---|---|---|
| 1. Fundamentals | Review | × | × |
| 2. Equilibrium | Review | × | × |
| 3. Axial Members | × | × | × |
| 4. Structural Analysis | — | — | × |
| 5. Stress & Strain | × | × | × |
| 6. Stress-Strain Relations | × | × | × |
| 7. Stress Limits | × | × | × |
| 8. Bending I | × | × | × |
| 9. Structural Analysis | — | — | * |
| 10. Torsion | × | × | × |
| 11. Structural Analysis | — | — | * |
| 12. Bending II | × | × | — |
| 13. Computer Applications | — | — | ** |

× Study this chapter.

— Omit this chapter.

* Problems only, subject matter for reference.

** Study one or two examples.

Note: The presentation of Chapters 8 and 9 is such that they can be inserted between Chapters 4 and 5 to give more time for working computer problems in bending.

Each chapter contains description and analysis, examples and problems. The examples are intended to be an *integral* part of the subject matter exposition, not simply an add-on for those who might be interested. Accordingly, students are expected to study the examples.

The problems are intended for use in homework assignments. Appendix 5 contains the answers to almost all of the numerical problems. As thorough as our checking of the problems has been, there may be a few answers that are still not correct; I would appreciate learning of same.

Naturally, much of the subject matter is traditional to this field. I list below some of the significant deviations from tradition as I perceive it.

- The Mohr's circle treatment for tensor transformation is introduced in Chapter 3 for the compliance of simple spring systems. Here the concepts are easily understood and student intuition is helpful. The same treatment, with the same subscripts and the same sign convention is used later for stress, strain, and the properties of beam cross-sections.

- Matrix structural analysis, with computer solution, is developed in Chapter 4 for the simplest case of axially loaded members (trusses). The introduction is complete, including:

    Structure definition
    Local stiffness matrices
    Global stiffness matrices
    Equilibrium equations in terms of structure displacements (the displacement method)
    Solution of equilibrium equations using the gaussian elimination method
    Effects of misfit and temperature change are included
    Determinate or indeterminate structures

    In Chapters 9 and 11 the structural analyses are revised to include the effects of bending and twisting (through revised local stiffness matrices). The subject matter in these chapters is intended primarily for reference, but the problems are intended for student use.

- The traditional treatment of beam bending has been split into two parts, I and II. Together, the coverage is as complete as in most conventional texts.

    Part I, which considers the bending of simple beams with end loading only, is a sufficient background for the development and use of the structural analysis programs, and for solving many real problems.

- Part II can be included in a more traditional presentation, can be excluded and used for reference only, or can be included section-by-section.

    Chapters 8 and 9 are written such that they can readily be inserted between Chapters 4 and 5. Such a presentation will permit use of the structural analysis programs for bending throughout a goodly portion of the course. Because students can easily relate to "bending," and because bending problems are so fundamental to this field, this arrangement may be attractive.

- There is a brief introduction to the Basic language which has become the de facto standard for personal computers. There are homework problems that ask the students to develop simple programs and to run them.

  Initially I had no intention of introducing a language or requiring any programming. However, the students, most of whom have had some computer experience, have requested more computer interaction.

- Most chapters contain an example of computer application to problems in that area. The examples range from the trivial (Mohr's circle, etc.) to methods for handling rather complex problems. These should indicate how easy it is to develop short programs for specific problems.

  In Chapter 13, the programs for tapered beams, the ring analysis, and shaft whip can be altered, by changing a few lines, to handle other problems.

  These examples are intended to be used, to be altered, or to be studied.

The title *Mechanics and Materials for Design* is intended to convey the sense that this book contains the essential mechanics analyses and material behavior concepts required for rational design decisions. No book can ever be complete in this regard, but this book is intended to be more complete than most.

There are a number of computer programs appreearing throughout the text. These are all written in NorthStarBasic* to run on NorthStar micro-computers (which our students at M.I.T. have been using). Please feel free to copy, use, and translate these programs. The programs were developed for this book, and may well contain "bugs" or errors. I would appreciate being informed of any bugs that are uncovered, and I would appreciate listings of any translations (to be made available to others).

Obviously, if these programs are used for any "real" problems, I will not be responsible for any errors they contain.

Many have contributed in diverse ways to the development of this book. With apologies for omissions, I wish to acknowledge a few:

First, my thanks to some hundreds of M.I.T. students who have suffered the inconvenience deriving from preliminary editions, dot-matrix print, poor figures, poor binding, errors, typos, incorrect answers, etc. I thank them for comments and criticism, ideas, enthusiasm, and patience.

My colleagues at M.I.T. who have been active in the teaching of the Mechanics of Solids and related subjects have been most forbearing in the presence of my enthusiasm, and constructive in their comments and criticism. I single out Ernest Rabinowicz, my friend and colleague for many years, whose wisdom regarding "applied academia" never ceases to amaze me.

My colleagues who have taught from the preliminary editions have, without exception, provided insight, thought, ideas, and enthusiasm. My thanks to Professors Lew Erwin, Tim Gutowski, Bruce Kramer, and Ming Tse.

---

* NorthStar Basic © NorthStar Computer, Inc.

Herb Richardson, head of the mechanical engineering department during the text development, has been particularly supportive of my attempt to modernize a basic mechanical engineering subject. He has provided moral support, administrative support, and the required computational facility.

I wish to acknowledge "Gloria," the name given to the NorthStar computer used in the development of this text. She has traveled with us, and has produced thousands of pages of text and thousands of lines of computer code. She has never failed!

Finally, I thank my wife, Collie, for enduring my sometimes overabundant dedication to duty and my long-term infatuation with Gloria I thank her for hours of proofreading, and her ever-present optimism, enthusiasm, love, and support.

*Nathan H. Cook*

# ONE

## FUNDAMENTALS

## 1.1 OVERVIEW

As we look around us we see a world full of "things": machines, devices, gadgets; things that we have designed, built, and used; things made of wood, metals, ceramics, and plastics. We know from experience that some things are better than others; they last longer, cost less, are quieter, look better, or are easier to use.

Ideally, however, every such item has been designed according to some set of "functional requirements" as perceived by the designers—that is, it has been designed so as to answer the question, "Exactly what function should it perform?" In the world of engineering, the major function frequently is to support some type of loading due to weight, inertia, pressure, etc. From the beams in our homes to the wings of an airplane, there must be an appropriate melding of materials, dimensions, and fastenings to produce structures that will perform their functions reliably for a reasonable cost over a reasonable lifetime.

The goal of this text is to provide the background, analyses, methods, and data required to consider many important *quantitative* aspects of the mechanics of structures. In practice, these quantitative methods are used in two quite different ways:

1. The development of any new device requires an interactive, iterative consideration of form, size, materials, loads, durability, safety, and cost. This text provides the analytic framework and methods fundamental to this process.

2. When a device fails (unexpectedly) it is often necessary to carry out a detectivelike study to pinpoint the cause of failure and to identify potential corrective measures. Our best designs often evolve through a successive elimination of weak points. Again, this text provides the analytic substance required for such a study.

To many engineers, both of the above processes can prove to be absolutely fascinating and enjoyable, not to mention (at times) lucrative.

In any "real" problem there is never sufficient good, useful information; we seldom know the actual loads and operating conditions with any precision, and the analyses are seldom exact. While our mathematics may be precise, the overall analysis is generally only approximate, and different skilled people can obtain different solutions. In this book most of the problems will be sufficiently "idealized" to permit unique solutions, but it should be clear that the "real world" is far less idealized, and that *you* usually will have to perform some idealization in order to obtain a solution.

The technical areas we will consider are frequently called "statics" and "strength of materials," "statics" referring to the study of forces acting on stationary devices, and "strength of materials" referring to the effects of those forces on the structure (deformations, load limits, etc.).

While a great many devices are not, in fact, static, the methods developed here are perfectly applicable to dynamic situations if the extra loadings associated with the dynamics are taken into account (we shall briefly mention how this is done). Whenever the dynamic forces are small relative to the static loadings, the system is usually considered to be static.

As we proceed, you will begin to appreciate the various types of approximations that are inherent in any real problem:

Primarily, we will be discussing things which are in "equilibrium," i.e., not accelerating. However, if we look closely enough, everything is accelerating.

We will consider many structural members to be "weightless"—but they never are.

We will deal with forces that act at a "point"—but all forces act over an area.

We will consider some parts to be "rigid"—but all bodies will deform under load.

We will make many assumptions that clearly are false. But these assumptions should always render the problem easier, more tractable. You will discover that the goal is to make as many simplifying assumptions as possible without seriously degrading the result.

Generally there is no clear method to determine how completely, or how precisely, to treat a problem: If our analysis is too simple, we may not get a pertinent answer; if our analysis is too detailed, we may not be able to obtain *any* answer. It is usually preferable to start with a relatively simple analysis and then add more detail as required to obtain a practical solution.

During the past two decades, there has been a tremendous growth in the availability of computerized methods for solving problems that previously were beyond solution because the time required to solve them would have been prohibitive. At the same time the cost of computer capability and use has decreased by orders of magnitude. In addition, we are beginning to experience an influx of "personal computers" on campus, in the home, and in business. Accordingly, we will begin to introduce computer methods in this text.

## 1.2 SCALE

A very vague, yet necessary, concept we must develop is that of "scale," both scale of size and scale of time. For instance, a cosmologist might, for some purposes, consider the universe to consist of uniformly distributed mass, whereas we "know" that we see discrete heavenly bodies. For some purposes, an automobile can be thought of as a solid, rigid entity, while for others it must be considered as an assembly of parts.

We can ask, "How much does a person weigh?" When he or she steps on a normal bathroom scale, the value given is single and unique. However, if that individual were to stand on a scale having greater sensitivity and "frequency-response," his or her "weight" would vary continuously because of the action of the heart, the lungs, and stabilizing muscles. For some purposes, knowledge of the "mean value" is sufficient; for others, we require the full dynamic range of information.

Whenever we start to analyze a mechanical system, we must in some intuitive way determine the scales of length and time that are (hopefully) appropriate. Unfortunately, there is no magic way; we must stumble along as best we can gaining confidence from experience, and experience from failure!

Our intuition can be guided to some degree by the observations of a brilliant French engineer, Adhemar Jean Claude Barre de Saint-Venant (1797–1886). While difficult to articulate, we can easily demonstrate "St. Venant's principle":

*Whenever a load is "applied" to a body it **must** be distributed over some finite area as shown in Fig. 1.1a. The loading will always have a "characteristic dimension" d as shown.*

*However, the effects of the load, at distances far removed from the load, are insensitive to the actual load distribution. For practical purposes, the distributed load can be replaced by an equivalent "point load" as shown in Fig. 1.1b. How far from the load must we be to justify this approximation? St. Venant teaches that if we are perhaps five characteristic dimensions away, the approximation will not greatly compromise our answer (outside the gray area in Fig. 1.1b).*

The concepts of St. Venant can be carried over to determination of time scales for temporal distributions, and, in fact, can serve as a general guide whenever we simplify a problem.

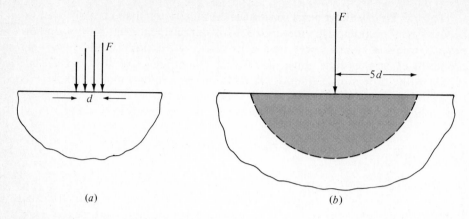

$(a)$ $(b)$

**Figure 1.1** St. Venant's principle: a load $F$, distributed over a characteristic length $d$ can be replaced by a point load $F$ for effects perhaps $5d$ away.

## 1.3 DYNAMICS

Although this text is concerned primarily with static structures and devices, we will be able to study many simple "dynamic" systems by applying a principle due to another Frenchman. Jean le Rond d'Alembert (1717–1783) was a student of law, medicine, mathematics, and the sciences. "D'Alembert's principle" states that if a particle of mass $m$ is accelerating with acceleration $a$, the particle can be treated as being in static equilibrium (not accelerating) if a fictitious "d'Alembert's force" of magnitude $F = ma$ is applied to the particle in a direction opposite to the acceleration (Fig. 1.2$a$). When the force $F = -ma$ is applied, the particle is said to be in "dynamic equilibrium" (Fig. 1.2$b$) and all the laws of statics apply.

By logical extension, a large mass $M$ undergoing acceleration $A$ can be placed in dynamic equilibrium by applying a force $F = -MA$ at the center of mass of the body. Also, a body undergoing rotational acceleration can be placed in dynamic equilibrium by applying an appropriate moment.

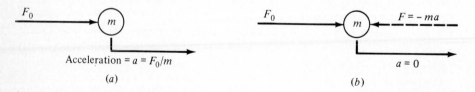

$(a)$ $(b)$

**Figure 1.2** D'Alembert's principle: ($a$) A particle of mass $m$, acted on by a force $F_0$, accelerates with $a = F_0/m$. The particle can be placed in dynamic equilibrium by applying "d'Alembert's force" ($F = -ma$) as shown in ($b$).

Thus if we can describe the acceleration of a body, we can apply fictitious d'Alembert's forces and treat the body as if it were in equilibrium. Clearly, if the d'Alembert forces cause deformations large enough to alter the accelerations significantly, then the problem becomes considerably more difficult.

## 1.4 FORCES AND MOMENTS

Thus far we have talked about forces and loads without defining them because we all have an intuitive concept of force. If we place a 100-lb weight on a set of scales, the weight pushes down on the scales with a force of 100 lb, and the scales push up on the weight with the same (but oppositely directed) force (Newton's third law).

We all "know" that a force has a magnitude (100 lb), a direction (down), and a point or area of application (on the scales). We all "know" that forces can be applied by direct contact (weight on scales) or at a distance by field effects (gravity, electrostatics, magnetism). Of course, if we look closely enough, we see that the "contact" forces are really electrostatic interatomic forces.

For our purposes we will represent a "point force" by a vector (Fig. 1.3a) showing magnitude, direction, and point of application or action. Clearly the total force acting on a body can also be "distributed" over the body in various ways: It can be distributed along the length of a slender member (beam) as shown in Fig. 1.3b, or over an area as in Fig. 1.3c. Forces can also be distributed within the volume of a body (body forces) through gravity (weight density), or inertia (d'Alembert's forces), etc.

While we all seem to have a good experience-based concept of force, we are much less familiar with the concept of "moment" or "torque" (which are synonymous). When we twist a screwdriver, we are applying a moment (usually, we also apply a force). When we drill a hole, we apply a torque to the drill. The drive shaft of a car transmits a moment from the transmission to the differential.

A "moment" implies the capability of twisting something. Because the

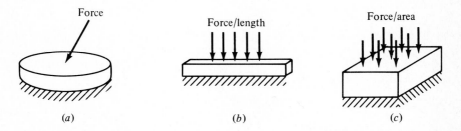

**Figure 1.3** Force applied to a body: (a) force applied at a point; (b) force distributed along a length; (c) force distributed over an area.

magnitude of a moment is the amount we are trying to twist about an axis, a moment (like a force) can be represented as a vector along that axis. Figure 1.4 shows the sign convention generally used for moment vectors (the right-hand rule). If the moment tends to produce twisting or rotation in the direction of the finger tips of the right hand, then the moment vector is in the direction of the thumb (Fig. 1.4c).

To distinguish a moment vector from a force vector, various conventions have been used; here we shall use a small curved arrow as shown in Fig. 1.4. To avoid ambiguity, and reliance on our artistic ability, always use the direction of the vector and the right-hand rule to determine the rotational sense of a moment rather than the direction of the curved arrow.

The magnitude of a moment is expressed in terms of distance times force [inch-pounds (in·lb), newton-meters (N·m)]. Figure 1.5a shows two equal and opposite forces $F$ separated by the distance $2r$. These forces, which act along the $x$ axis, produce a moment about the $z$ axis of magnitude $2rF$. This particular moment, produced by a *couple* of equal and opposite forces, is called a "couple." If you check, you will find that the moment due to these forces is the same about *any* axis parallel to the $z$ axis.

In Fig. 1.5b we see the moment about the $z$ axis caused by the force acting

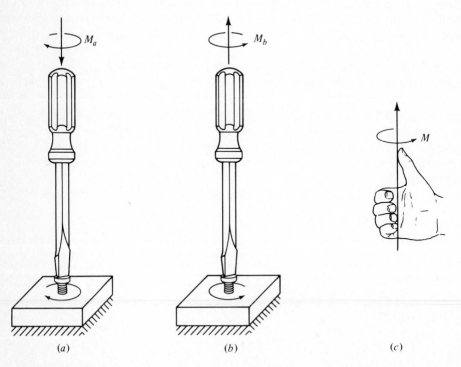

**Figure 1.4** (a) The moment $M_a$ to *tighten* a normal right-hand screw; (b) the moment $M_b$ to *loosen* the screw; (c) the right-hand rule.

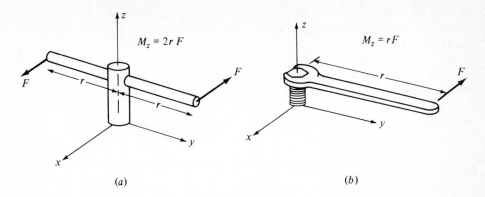

**Figure 1.5** Moments: (a) due to a "couple" of equal and opposite forces; (b) due to a single force.

on an open-end wrench. Here the magnitude is simply the force $F$ times the shortest distance from the force vector to the $z$ axis. In both examples in Fig. 1.5 the moment is positive according to the right-hand rule, i.e., the moment vector would be in the $+z$ direction.

Figure 1.6 illustrates a definition of a moment due to a force in terms of vector algebra; here a force $\mathbf{F}$, lying in the $xy$ plane, acts at a point $A$. The location of $A$ is defined by the position vector $\mathbf{r}$ relative to some point $O$. The moment of $F$ about $O$ is simply the vector cross-product $\mathbf{r} \times \mathbf{F}$.

1. The moment vector will always be perpendicular to the plane of $rF$.
2. The right-hand rule gives a vector sense that is consistent with cross-product rules.

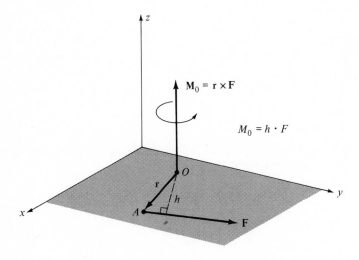

**Figure 1.6** The moment $\mathbf{M}_0$ of the force $\mathbf{F}$ about point $O$ is given by $\mathbf{M}_0 = \mathbf{r} \times \mathbf{F}$.

3. The magnitude of the moment will always be the product of the force times the shortest (perpendicular) distance $h$ from $O$ to $F$.

$$\mathbf{M} = \mathbf{r} \times \mathbf{F} \qquad \text{(vector cross-product)} \qquad (1.1)$$

The moment of $\mathbf{F}$ about any other axis through $O$ is simply the component of $\mathbf{M}$ along that axis. It should be clear that:

1. A force gives no moment about an axis parallel to the force.
2. If a force passes through an axis, it produces no moment about that axis.

A more complicated case is shown in Fig. 1.7. A right-angle structure is fixed (firmly supported) at the origin and loaded with a force $\mathbf{F}_z$ in the $+z$ direction as shown. The total (as opposed to some component) moment of $\mathbf{F}_z$ about $O$ is the vector $\mathbf{M}_0$ in the $xy$ plane. The total moment $\mathbf{M}_0$ can be "resolved" into components along the $x$ and $y$ axes as shown. The individual moment components are easy to calculate directly:

$$\mathbf{M}_x = y_A F_z \qquad \mathbf{M}_y = -x_A F_z \qquad (1.2)$$

That is, the moment about the $x$ axis is simply $F_z$ times $y_A$ (the perpendicular distance from $F_z$ to the $x$ axis). Likewise, $\mathbf{M}_y$ is $F_z$ times the shortest distance from $F_z$ to the $y$ axis. Study Fig. 1.7 until Eq. (1.2) makes sense to you.

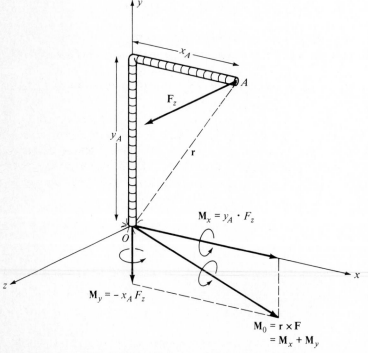

**Figure 1.7** The moment of $\mathbf{F}_z$ about $O$ is the vector sum of the moments of $\mathbf{F}_z$ about the $x$, $y$, and $z$ axes (because $\mathbf{F}_z$ is parallel to the $z$ axis, $M_z = 0$).

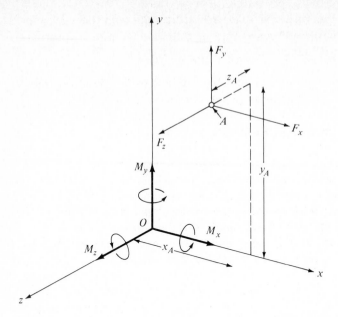

**Figure 1.8** General case of moments about $O$ due to three force components at $A$.

$$M_x = F_x \times 0 - F_y \times z_A + F_z \times y_A$$

$$M_y = F_x \times z_A + F_y \times 0 - F_z \times x_A$$

$$M_z = -F_x \times y_A + F_y \times x_A + F_z \times 0$$

If you prefer to treat moments as cross-products using vector algebra, fine. However, at this stage we get far more insight into the problems by calculating the individual moment components $(\mathbf{M}_x, \mathbf{M}_y, \mathbf{M}_z)$ directly. This seems to give a better "feel" for moments and what they do to a structure.

In the completely general case, where the force $\mathbf{F}$ has components in the $x$, $y$, and $z$ directions, and where the force location vector $\mathbf{r}$ has components in all three directions, we simply take the moment of each force component about each axis and sum moments about identical axes. You will have to keep your wits about you to keep the signs straight! Figure 1.8 shows the general case. See if you can agree with the equations shown. If you cannot, try, try again!

## 1.5 UNITS

The world of engineering is more than a world of concepts, theories, and things; it is a world of numbers. And except for pure, dimensionless numbers (radians and ratios), the numbers must have units. And the units are *arbitrary*. And that is unfortunate!

Any engineer (or engineering student) whose theory is correct but whose numbers are wrong will not get much applause. We each must come to grips with units at some time; if you have not yet, now is the time.

Anyone who reads this book will have some difficulty with the variety of different units used in different places and at different times. For practical work, two sets of units will suffice for your work in mechanics. These are:

The International System of units (called "SI units" from "Système International d'Unités"), which is predominant *outside* the United States
The U.S. Customary System (called "USCS units"), which is predominant *in* the United States at the current time

In both cases, arbitrary values have been defined for the units of mass, length, and time. In most mechanics problems, force is used much more than mass, and it becomes more convenient to discuss mechanics problems in terms of force, length, and time rather than in terms of mass, length, and time.

Table 1.1 gives the appropriate unit names and conversion factors.

Table 1.2 is an assembly of conversion factors and constants that the writer has found useful in a variety of problems. It is too bad that we have such a variety of units to contend with. While most of the world officially uses the SI system, many

**Table 1.1  Units for use in mechanics**

|  | SI | USCS | Conversion factor |
|---|---|---|---|
| Force | Newton (N) | Pound (lb) | 1 lb = 4.448 N |
| Length | Meter (m) | Inch (in) | 1 in = 0.0254 m |
| Time | Second (s) | Second (s) | |
| Mass | Kilogram (kg) | Pounds/386 | $1 \text{ kg} = 0.00571 \text{ lb} \cdot \text{s}^2/\text{in}$ |

| Derived quantities | | | |
|---|---|---|---|
| Pressure | Pascal (Pa) Newtons/square meter $(\text{N/m}^2)$ | Pounds/square inch (psi or $\text{lb/in}^2$) | 1 psi = 6895 Pa |
| Energy | Joule (J) Newton-meters (N·m) | Inch-pounds (in·lb) | 1 in·lb = 0.113 J |

*Note*: In regards to mass and force, students (and some engineers) invariably have difficulty with the units of force and mass, probably because common usage is apt to be ambiguous ("a pint's a pound the world around"). Also, it appears that people have better intuition relative to force. If you use the following rules you should have no trouble:

USCS units: Use only pounds, inches, and seconds. For mass, simply divide the weight $W$ (in pounds, on the earth) by 386; i.e., $M = W/386$, where 386 is the acceleration of gravity (on earth) in $\text{in/s}^2$. For mass density, divide the weight density $(\text{lb/in}^3)$ by 386. Then $F$ (lb) will equal $M$ times acceleration $(\text{in/s}^2)$.

SI units: Use newtons, meters, seconds, and kilograms. The weight (force) of one kilogram (on earth) is 9.8 N, where 9.8 is the acceleration due to earth's gravity $(\text{m/s}^2)$. Obviously the mass of a body (kg) is the weight (N) divided by 9.8. Unfortunately, many people discuss the weight of an object in terms of kilograms.

**Table 1.2 Conversion factors and constants**

| | |
|---|---|
| Force | 1 lb = 4.448 N |
| | 1 dyne (dyn) = $2.248 \times 10^{-6}$ lb |
| Length | 1 m = 39.37 in |
| | 1 angstrom (Å) = $10^{-10}$ m = $3.937 \times 10^{-9}$ in |
| | 1 micrometer ($\mu$m) = $10^{-6}$ m = $39.37 \times 10^{-6}$ in |
| Pressure | 1 psi = 6895 Pa |
| | 1 kilogram per square millimeter ($kg/mm^2$) = 1420 psi |
| | 1 atmosphere (atm) = 14.7 psi |
| | 1 torr = 1 millimeter of mercury (mmHg) = 0.01936 psi |
| Energy | 1 J = 1 N·m = 1 watt-second (W·s) = 8.85 in·lb |
| | 1 calorie (cal) = 37.1 in·lb |
| | 1 British thermal unit (Btu) = 9330 in·lb |
| | 1 kilowatthour (kWh) = $3.185 \times 10^7$ in·lb |
| Power | 1 kilowatt (kW) = $8.85 \times 10^3$ in·lb/s = 1.34 horsepower (hp) |
| | 1 hp = 6600 in·lb/s |
| Gravitational constant = $7.06 \times 10^{-4}$ in$^4$/lb·s$^4$ | |

of our U.S. machines, tools, instruments, gages, facilities, designs, and property data are in USCS.

It will be many years before the USCS units are obsolete. You will be faced with a world of inconsistent units; hence, in this book, we will not be consistent. We *strongly* urge that, before starting any problem having inconsistent units, you change all units to either lb·in·s or N·m·s and then proceed; life will be much simpler!

## 1.6 ECONOMICS

In Sec. 1.1 we mentioned the fact that the design of any new device must involve considerations of cost as well as of the more technical aspects. Generally, the viability of any product depends at least as much upon cost as upon function.

For consumer products (things we as individuals might buy), the sales volume is large enough to warrant extensive product engineering and manufacturing automation (mass production). This results in low unit cost where the actual material cost is a significant portion of the total.

Capital equipment, on the other hand, is frequently produced in low volume that cannot justify extensive automation. Hence, the actual material cost may be a very small portion of the total. However, because of the low volume, the cost of functional design engineering may be relatively high.

For any machine part or structural element there is generally a wide variety of actual designs that would be functionally satisfactory—various sizes, shapes, materials, etc. The final selection frequently involves rather arbitrary tradeoffs, or compromises, between, for example, low cost and long life. The analytic methods and tools developed in this book help us to optimize any individual design and can be useful in comparing competing designs.

## 1.7 COMPUTER APPLICATIONS

At the end of each chapter in this book there is a section concerning relevant uses of computers, in particular, the use of small, microprocessor-based personal computers. In some chapters we will present simple application programs that can be readily understood as examples; in other chapters, we will present rather extensive "structural analysis" programs that can be easily used, but that require extensive study to be understood. Each chapter will have a few problems that are computer-oriented.

At the present time, the language most used in personal computers is Beginner's All-Purpose Symbolic Instruction Code (BASIC). Different computers feature somewhat different versions of BASIC, but they all have a great deal in common. All of the programs in this book are written in "NorthStar BASIC" (by North Star Computers, Inc.).

It is not the purpose of this text to teach computer programming; however, for those who have not used BASIC, we will present a very brief introduction to the language. We will not consider disk access because none of the programs in this book requires it.

A BASIC program consists of a number of program lines, each line being numbered (typically, 5, 10, 15, ...). Each line contains one or more program "statements" which instruct the computer what to do. When you RUN the program, the computer will proceed to execute the statements in order of the line number (except when the program statement instructs the computer to do otherwise).

In NorthStar BASIC, only *capital* letters are used; the variables permitted are the following:

Numeric variables: (numbers) denoted by any letter (A, B, X, etc.) or any letter followed by a single digit (A1, Z9, etc.)
String variables: any group of alphanumeric characters enclosed in quotes ("This is a string 1234"), denoted by any letter or letter-digit followed by a dollar sign (A$, B2$, etc.)

Many computer instructions look just like mathematical equations, but they have a somewhat different meaning. For instance, consider a line that might read:

$$\text{LET } X = Y + 2 \qquad \text{(or more simply: } X = Y + 2)$$

This means: Determine the value of the right-hand side ($Y+2$), and assign that value to the variable on the left (X). Thus, a seemingly nonsensical instruction such as $X = X + 1$ is perfectly satisfactory.

To *print* the value of a variable (X0) on the screen, the instruction would be PRINT X0 [or in NorthStar shorthand, ! X0 (preceded naturally by a line number)].

To "document" a program with REMarks that will not be executed, start the line with REM. The rest of that line will be printed when you LIST the program, but will not affect the running of the program—in other words, REM remarks help us remember what we have done.

For the right-hand side of instructions, math symbols have their usual meaning: $+$, $-$, $/$, $(\ )$, $<$, $>$, $=$. $*$ means multiply, and $\hat{\ }$ means exponentiate, for example, $X*X = X\hat{\ }2$. Thus, we might have

$$X = 3.1416*R\hat{\ }2*(SIN(A)+COS(A))\hat{\ }.5$$

The most common "functions" that will be used in our programs are shown below. The computer will evaluate the function, and use the value in the instruction.

| | |
|---|---|
| SIN(A0) | (A0 must be in radians) |
| COS(X) | (X must be in radians) |
| ATN(Y) | (gives arctan in radians) |
| EXP(Z) | (raises e to the power Z) |
| LOG(W) | (evaluates the natural log of W (ln W)) |
| INT(S) | (provides the integer part of S) |
| ABS(S) | (the absolute value of S) |

BASIC permits us to override the normal line-number sequence of operation through a GOTO instruction:

GOTO 120      (jump to line 120)

It also allows us to go to a subroutine, and return at the end of that routine (GOSUB—RETURN):

GOSUB 120      (following the routine starting at 120)
RETURN

In many of our programs, we will arrive at a solution iteratively. BASIC provides a FOR—NEXT loop instruction that is ideal. A typical sequence might be:

20 FOR I=1 TO 10
25 ——

50 NEXT I      (or more simply, NEXT)

This instruction sets I=1, executes the steps between 20 and 50, then sets I=2, executes the steps again, etc. Other variations are:

FOR J=5 TO 50 STEP 5     or     FOR J=1 TO 0 STEP −.2

BASIC provides a conditional statement that allows us to branch or make decisions. The instruction is IF—THEN; that is, IF (this statement is true) THEN (execute this statement). If the first statement is false, proceed to the next line. Typical instructions could be:

IF X+7>Y−6 THEN X=Y

IF X=0 THEN GOTO 120    (or more simply, IF X=0 THEN 120)

An ELSE clause can be added telling what to do if the first statement is false:

IF X=0 THEN 120 ELSE Y=SIN(X)

The second statement in an IF instruction can be another IF statement:

IF X=0 THEN IF Y=>SIN(X) THEN IF Z*Z<W THEN 120

Frequently we find it useful to save space by putting several instructions on a single line. In NorthStar BASIC, this is accomplished with a backslash \ between the instructions; that is,

For I=1 TO 10  \   PRINT I  \  NEXT

In addition to storing numbers as variables, we can store them in vector or matrix form in "arrays." Before an array can be used, we must DIMension it so that BASIC can set aside the appropriate memory space. Arrays can be of any order and size, limited only by core memory. Arrays are dimensioned as follows:

DIM T(25)      [an array of 25 numbers, T(1), T(2), etc.]
DIM T1(3,3)    [a 3×3 array, T1(1,1), T1(1,3), etc.]

Finally, in order to write "interactive" programs that request information from you when you use them, there are INPUT instructions:

INPUT "X=",X

The computer will first print X=, then will assign the number you type to the variable X.

The above is a limited portion of the available instructions; however, just these few should help in understanding the programs throughout this text, and should help in writing the programs associated with some of the computer problems.

To see if you understand the above, determine (by hand) the result from the following (useless) program:

```
10 REM This program is a test
20 Y=0
30 FOR I=1 TO 4
40 FOR J=1 TO 4
50 X=I*J
60 IF X<=10 THEN Y=Y+X
70 NEXT J
80 NEXT I
90 !Y          \REM Print the final value of Y
```

The result should be 60.

Whenever we write a computer program, we must "verify" it in some way. That is, we must determine whether the program is actually doing what we intended, or if it has a "bug" and is outputting erroneous results. Depending on the program, verification can be easy, or very difficult.

There are three verification methods which are useful for the type of programs we are apt to develop in this text:

1. Use the program to solve a problem, or range of problems, where the answer is known.
2. Work an example "by hand" and compare with program output.
3. Write the program in such a way that by altering variables it models a simpler problem where the answer is known.

## 1.8 PROBLEMS

**1.1** When a point moves at constant speed $V$, around a circular path of radius $R$, the point accelerates radially toward the center of the circle. This (centripetal) acceleration has magnitude:

$$a = \frac{V^2}{R} = R\omega^2$$

where $\omega$ is the rotational speed in radians per second (rad/s).

An automobile goes around a curve of 2000-foot (ft) radius at 100 miles per hour (mph). Are the dynamic d'Alembert forces significant relative to the weight forces?

**1.2** A steel "hoop," 10 in in diameter, 1 in wide, and 0.2 in thick, spins at 100 revolutions per minute (rpm). Describe the "loading" on the hoop due to the radial acceleration (Prob. 1.1). Is the load on any element due to acceleration significant relative to the weight of the element?

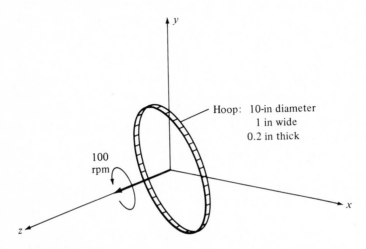

y

Hoop: 10-in diameter
1 in wide
0.2 in thick

100
rpm

x

z

**Figure P1.2**

**1.3** Determine the cost per pound, and the cost per cubic inch (in$^3$) of:
  (*a*) Mild steel
  (*b*) Stainless steel
  (*c*) Aluminum
  (*d*) Fiberglass
  (*e*) Concrete
  (*f*) Clear pine

**1.4** Estimate the cost per pound of the following products:
  (*a*) Automobile
  (*b*) Lawnmower
  (*c*) Can opener
  (*d*) House
  (*e*) Airplane
  (*f*) Boat
  (*g*) Machine tool
  (*h*) Education
  (*i*) Steak
  (*j*) Shoes
  (*k*) Microscope

**1.5** Two equal and opposite forces *F* act on a body as shown in the illustration below. Determine the total moment *M*, due to the two forces, about *A*, *B*, and *C*.

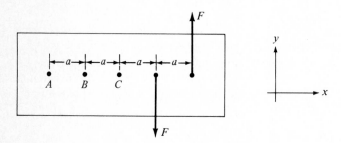

**Figure P1.5**

**1.6** In an attempt to remove a "frozen" (stuck) bolt, a force *F* is applied to a wrench as shown (Fig. P1.6, p. 17). Show the forces and moments acting *on* the portion of the bolt below point *A*. Show the *x*, *y*, and *z* components.

**1.7** A support, made of pipe welded together, is imbedded in concrete ("built-in") at the origin *O*. The pipe weighs 10 lb/ft, and the structure supports the two loads shown. Calculate and show carefully the resultant force and moment acting *on* the ground at *O*. (See Fig. P1.7, p. 17.)

**1.8** When sailboats had wooden masts, they were invariably tapered, being smaller at the top. They were very graceful. Most modern masts are made of aluminum and are not tapered. Why not?

**1.9** The "force" between a car tire and the road is obviously distributed over some fairly large area. For some purposes, we can assume that the total force acts at a point, while for others we would have to consider the actual distribution.

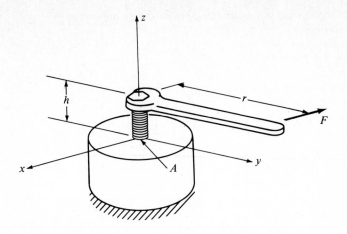

**Figure P1.6**

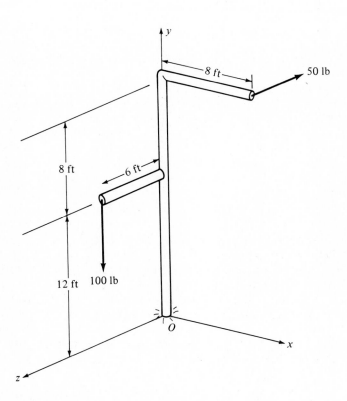

**Figure P1.7**

For which of the following could you make the point load assumption? Why?
(a) Design of front springs
(b) Design of front axle
(c) Design of steering mechanism
(d) Design of brakes
(e) Analysis of friction losses
(f) Estimation of maximum acceleration
(g) Consideration of "ride quality"

**1.10** On the rear wheel of a bicycle, how is the torque transmitted from the hub to the rim?

**1.11** What keeps an automobile tire from sliding around on its rim?

**1.12** Describe how an automobile "jack" works.

**1.13** An electric motor delivers 1 kW at 1800 rpm. How large is the output torque?

**1.14** A 10-hp 3600-rpm motor drives a 6:1 speed reduction unit that is about 85 percent efficient. What will be the output speed and the maximum output torque?

**1.15** A sophisticated "biomechanics force platform" continuously measures three force components $(F_x, F_y, F_z)$ and three moment components $(M_x, M_y, M_z)$ as a person walks across the platform (with only one foot at a time on the platform). At the instant shown in the illustration, the readings are:

$$F_x = +6\,\text{lb} \qquad M_x = -655\,\text{in}\cdot\text{lb}$$
$$F_y = -8\,\text{lb} \qquad M_y = +123\,\text{in}\cdot\text{lb}$$
$$F_z = -160\,\text{lb} \qquad M_z = -20\,\text{in}\cdot\text{lb}$$

Determine the $xy$ location of point $A$ where the "force" between foot and platform can be considered to act.

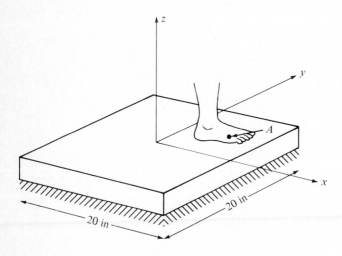

**Figure P1.15**

**1.16** What would be the outcome from running the following?

```
10   FOR A=0 TO 90 STEP 10
20   A0=A/57.3
30   X=COS(A0)  \  Y=SIN(A0)  \  T=Y/X
40   ! A, X, T
50   NEXT
```

**1.17** Determine the output from the following program. Then rewrite the program in simpler form. Find S.

```
5    S=0  \  I=0
10   I=I+1
15   S=S+I*I
20   IF I>10 THEN STOP
25   GOTO 10
```

# TWO

## EQUILIBRIUM AND FREE-BODY ISOLATION

### 2.1 OVERVIEW

In any physical engineering analysis, we first must clearly identify *what* we are going to analyze. This may sound obvious and trite, but it is not. For instance, we could say that we wanted to analyze the forces acting on a car engine. The question is, just *exactly* what constitutes the engine? Does it include the transmission? Or the air-conditioner compressor? When we work any problem, the results will depend on precisely what we mean when we say "engine."

Each entity that we may wish to analyze will be connected, somehow, to the "rest of the world." We must *isolate* the item of interest from the rest of the world by wrapping a fictitious surface around it. This surface is generally called a "control surface" which isolates a "control volume."

All things and interactions within the volume are considered "internal," while all things and interactions outside the surface are "external." The effects due to the rest of the world, via forces which act *across* the control surface, will be considered "external" effects. Thus, in our engine, the forces between connecting rod and crankshaft would be internal forces (occurring in equal and opposite pairs), while the force on an engine mount, supporting the engine, would be an external force.

In every problem we address, we will isolate each item of interest by *carefully* drawing a free-body diagram (FBD) of the item in question and showing *all* the *external* forces and moments that cross the control surface. We will then apply Newton's laws to the isolated body and will be able to determine relationships between those external forces and moments. In order to study any *internal* forces, we will have to isolate a portion of the inner machine such that the forces in question become *external* forces for that FBD.

Early in this text we will isolate whole systems or elements: a beam, a truss bridge, a wrench, etc. Later, we will want to determine what is happening at various locations *within* an element, i.e., at various points along the length of a beam. We will have to isolate a *part* of the beam such that forces that are internal to the beam become external to the FBD.

When we have effects that vary continuously within an element, we will generally isolate a *differential* element of length *dx*, and study the forces acting on it. Finally, in order to understand the effects of loading on the material itself, we will have to isolate and study a differential element of volume *dx dy dz*.

So as we proceed, we will be isolating smaller and smaller free bodies, but we will apply the same laws to all of them: the laws associated with equilibrium.

## 2.2 EQUILIBRIUM

As noted in Chap. 1, this text deals primarily with bodies, machines, and systems that (for all practical purposes) are not accelerating (the introduction of d'Alembert's forces will permit solution of some dynamic problems).

*Newton's second law states that if a particle is not accelerating, then there must be zero net force acting on it.*

Either there are no forces at all, or the forces must sum (vectorially) to zero. This state of zero acceleration, zero force, we call "equilibrium." The particle is in a state of balance with respect to its environment. In similar fashion, a structure or machine that is (for all practical purposes) not accelerating will also be considered as being in equilibrium. A little thought will probably convince you that if a "rigid" body is in equilibrium, then each and every part of that body (if isolated separately), is also in equilibrium.

Figure 2.1*a* shows a particle with forces $F_1$, $F_2$, and $F_3$ acting on it. *If the particle is in equilibrium (zero acceleration) then the vector sum of all forces must be zero.*

In this instance we have, without thinking about it, isolated the particle and have shown all the external forces that cross the boundary. Thus, for a particle:

$$\sum \mathbf{F} = 0 \quad \text{(if it is in equilibrium)} \quad (2.1)$$

Real, three-dimensional bodies can be considered as assemblies of particles. Consider the simplest such assembly made of three particles as shown in Fig. 2.2. We have clearly defined (isolated) the assembly by drawing a control surface around the three particles (the dashed line). Forces which act within the surface are "internal" forces and always occur in equal and opposite pairs

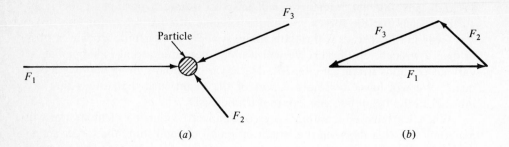

**Figure 2.1** (*a*) A particle is in equilibrium when there is no *net* force acting on it; (*b*) the vector sum of all forces is zero.

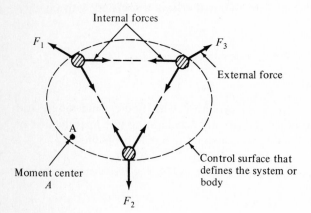

**Figure 2.2** An assembly of particles in equilibrium.

(Newton's third law). The forces which cross the boundary ($F_1$, $F_2$, $F_3$) are the external forces which represent the effect of the rest of the world on our system. Simple logic tells us the following:

Because each particle is in equilibrium, the sum of all forces on it must be zero. Therefore the sum of all forces in Fig. 2.2 must be zero.
All internal forces occur in equal, opposite pairs and must, therefore, sum to zero.
Thus all the *external* forces must also sum to zero.

Thus, Eq. (2.1) holds not only for a particle in equilibrium, but for any assembly of particles in equilibrium.

If now we reconsider the forces acting on the particle in Fig. 2.1, we see that all forces pass through a single point such that there are no moments about any point.

Looking at the assembly of particles in Fig. 2.2, and considering some arbitrary point $A$, we see that many of the forces produce moments about $A$.

Follow this logic:

The moment about $A$ due to the net force acting on any particle is zero
because the net force on that particle must be zero.

The moment about $A$ due to each pair of equal, opposite, internal forces is
zero because they cancel.

Thus the net moment about $A$ due to all *external* forces must be zero.

Clearly, the point $A$ can be anywhere, either on or off the body.

Finally then, when we isolate a body in equilibrium, and show all the
external forces (and moments) that cross the control surface:

1. The vector sum of all external forces must be zero.
2. The vector sum of all moments (about any point or axis), due to external
   effects, must be zero.

$$\sum \mathbf{F} = 0 \qquad \sum \mathbf{M} = 0 \qquad \text{(for a body in equilibrium)} \qquad (2.2)$$

In most problems, there is an obvious $xyz$ coordinate system that is
"natural" for the problem geometry. If we let $F_x$, $F_y$, and $F_z$ be the *components*
of the external forces in the coordinate directions, and $M_x$, $M_y$, and $M_z$ be the
moments due to the external forces about the coordinate axes, then the two
vector equations (2.2) can be replaced with six scalar equations:

$$\sum F_x = 0 \qquad \sum F_y = 0 \qquad \sum F_z = 0$$
$$\sum M_x = 0 \qquad \sum M_y = 0 \qquad \sum M_z = 0 \qquad (2.3)$$

For a general three-dimensional system, the six equations of equilibrium will
allow us to solve for, at most, six unknowns.

Very often, the essence of a problem can be described in two dimensions—
for instance, if the system is symmetrical about the $xy$ plane and all forces lie in
the $xy$ plane. This clearly does not mean that the problem *is* two-dimensional
(we have very few of those), but that for certain purposes it can be modeled in
two dimensions. For this case, involving *coplanar* forces in the $xy$ plane, we
have only three useful equations from the six above:

$$\sum F_x = 0 \qquad \sum F_y = 0 \qquad \sum M_z = 0 \qquad \text{(if all forces lie in the } xy \text{ plane)} \quad (2.4)$$

## 2.3 SUPPORT CONSTRAINTS AND APPLIED FORCES

The effect of the "outside world" on an isolated part or system of parts is via
the forces that cross the control surface. Some of the forces are specified as
known applied forces, while others, which come from supports, are unknown.

Although a support can provide a force (or moment), we actually specify a "displacement constraint"; that is, we limit motion in some direction or rotation about some axis. If, at a point, we specify a displacement constraint, we cannot also specify the force associated with that constraint.

Understanding of the interaction of constraints and applied forces is fundamental to the study of mechanics. In order to put our discussion in context, consider the following example:

**Example 2.1** Consider a 10-ft-long beam, "fixed" at the left-hand end in a rigid support while the right-hand end rests on a block. We apply a load of 1000 lb as shown. We can ask, "What are the support forces and moments at $A$ and $C$?"

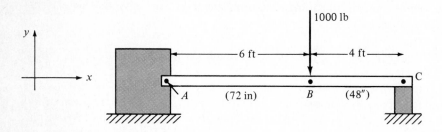

The supports at $A$ and $C$, the attachments to the rest of the world, really define *displacement boundary conditions* for the beam. Looking at the problem it would seem reasonable to assume that the outside world imposes the following conditions:

At $A$: zero vertical displacement and zero rotation about $z$ axis
At $C$: zero vertical displacement
At $B$: externally applied load of $F_y = -1000$ lb

This set of boundary constraints and applied loads are our "independent variables," or those which we can specify independently.

At any point $x$ along the beam we could specify applied loading along any axis $(x, y, z)$ *or* displacements along those axes. We could specify applied moments *or* rotations about any axis—but not both.

For instance, at $A$ we have specified zero vertical displacement. We cannot also specify the vertical force at $A$—that is a "dependent variable" to be solved for. We can specify either the rotation about the $z$ axis (zero in this case) or the applied $M_z$—but not both. At $B$ we can specify the applied vertical load, but must calculate the vertical displacement.

Instead of supporting the right-hand end on a block, we could have used a spring. In this case we would specify neither the displacement nor the force in the $y$ direction; we would specify the "function," or the relationship between the force and the displacement as determined by the size of the spring.

We could generalize the above considerations a bit more: In essence every point in a body can have (at most) six degrees of freedom; that is, it takes six quantities to completely describe the location and aspect of the point ($x$, $y$, $z$ displacements and rotations about the $x$, $y$, and $z$ axes). In the direction of each degree of freedom (displacement or rotation), we can specify the motion, *or* the force (moment), *or* the relation between them.

Later, when we consider computerized methods for structural analysis, we will define the support conditions by stating which degrees of freedom are constrained not to move, and which are free to move.

Returning to the example, having discussed the limitations imposed on supports and forces, let us *isolate* the beam and show all external forces and moments (see the illustration).

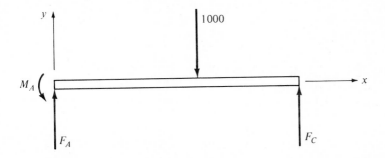

Clearly we have three unknowns, $F_A$, $F_C$, and $M_A$. However, because there are no forces in the $x$ direction, we have only two nontrivial equilibrium equations:

$$\sum F_y = 0 \qquad \sum M_z = 0$$

Thus we have three unknowns but only two equations! If you recall your mathematics, you will remember that this is not a good situation—no solution!

This is an example of a problem that *cannot* be solved using only the equations of equilibrium, and is called "statically indeterminate." The problem can be solved (and we will do so in later chapters), but not by equilibrium alone.

In the "real world," many systems are not determinate. In this case, if we could reduce the unknowns to two, we could solve the problem (since it would then be statically determinate). Looking at the problem physically we see that:

$M_A$ exists because we specified zero rotation at $A$.
$F_A$ exists because we specified zero vertical displacement at $A$.
$F_A$ exists because we specified zero vertical displacement at $C$.

If we were to eliminate any one of those displacement (rotation) constraints, the corresponding force (moment) would be eliminated, and the problem would become determinate.

Let us replace the left-hand support with a simple block support as at $C$:

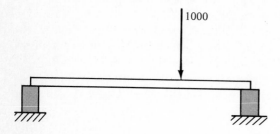

The isolated beam now is:

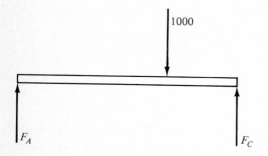

Now we have only two unknowns and solution is easy:

Setting $\quad \sum M_A = 0$

$$F_C(120) - 1000(72) = 0$$
$$F_C = 600 \text{ lb}$$

Setting $\quad \sum F_y = 0$

$$F_A + F_C - 1000 = 0$$
$$F_A = 400 \text{ lb}$$

If we now look again at the original problem, from a purely intuitive point of view, we can determine a priori that it is indeterminate. There are more displacement constraints than are necessary to keep the beam from falling down. For instance, we could completely remove the support at $C$ and the beam would not collapse—sag more, yes; collapse, no. The beam has "redundant" supports (more than the minimum required). Whenever this is the case, the situation is indeterminate.

Very often we desire redundancy so that if one part of a system fails the entire system need not fail. So indeterminacy is a fact of life.

Later we will develop additional equations, involving beam deflections, that will permit us to solve this type of indeterminate problem.       ***

From Example 2.1 we saw that it is most important to determine properly the support displacement constraints and the externally applied forces. In order to streamline and organize our future efforts, it is worthwhile here to agree on a common symbolism to represent the most common types of supports and constraints.

### 2.3.1 Contact Support

This involves physical contact between an isolated body (as a chair leg) and the external world (as the floor) (see Fig. 2.3).

In general, we can have a "normal" force $N$ (normal to the interface contact area) and a friction force $F$. If the surfaces are sliding, or if slipping is

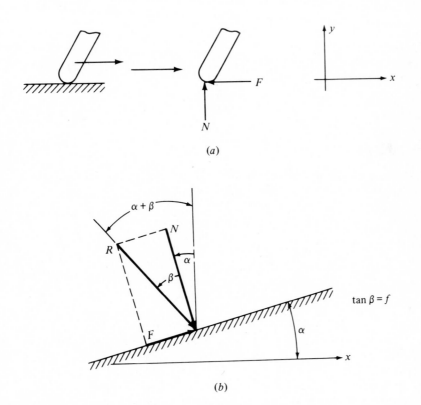

(a)

(b)

**Figure 2.3** Contact support (with friction $F$). (a) General case, (b) sliding up (down) an incline.

imminent, then the friction force equals the coefficient of friction times the normal force ($F = fN$), and acts in a direction to resist the motion.

If it is not known that the surfaces are slipping, then the friction force can take on a variety of values and is determined by other aspects of the problem:

$$\text{Friction force } F \leq \pm fN \tag{2.5}$$

That is, the friction force can range from $-fN$ to $+fN$. The force system is frequently indeterminate such that the exact magnitude and direction of the friction force depends on the past history of the system (just how it arrived at the current position). Ambiguity due to friction is present in almost every mechanical device. It can be quite troublesome in mechanical instruments.

For many cases we will *assume* that $f = 0$ and thus remove the ambiguity (at least from our analysis). We can also obtain limiting solutions by assuming first that $F = -fN$ then $F = +fN$ so that we see the extreme effects of friction.

When sliding is along an incline (Fig. 2.3$b$), as in a screw thread or wedging action, it is often preferable to consider the "resultant" contact force $R$, inclined at an angle $\beta$ to the normal force as shown. The angle $\beta$, called the "friction angle," is the angle whose tangent equals the coefficient of friction.

A more detailed discussion of "friction" will be found in the appendix.

### 2.3.2 Pin Joint

In this type of support, the isolated body is secured to the external world through a "pin" or hinge, and is free to rotate about one axis but constrained in other directions, with no friction at the pin. As shown in Fig. 2.4, rotation is about the $z$ axis.

### 2.3.3 Ball and Socket

This support (Fig. 2.5) permits free rotation (no friction) about all three axes but does not permit any displacement. It can only provide $F_x$, $F_y$, and $F_z$.

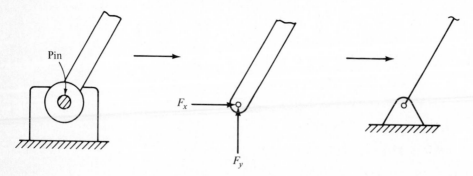

**Figure 2.4.** Pin-joint support with zero friction. Joint can supply $F_x$, $F_y$, $F_z$, $M_x$, and $M_y$. $M_z = 0$. For two dimensions, only $F_x$ and $F_y$ are available.

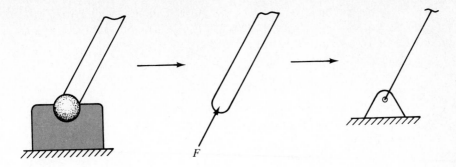

**Figure 2.5** Ball and socket joint.

### 2.3.4 Roller Support

The joint in this kind of support is free to move in one direction and free to rotate about one axis. For the two-dimensional case shown in Fig. 2.6, it can provide only $F_y$. Although drawn as a roller, it can provide either $+F_y$ or $-F_y$.

### 2.3.5 "Built-In" or Fixed End

This connection (Fig. 2.7) to the real world is completely rigid. It cannot be displaced or rotated. In general it can provide three components of force and three components of moment.

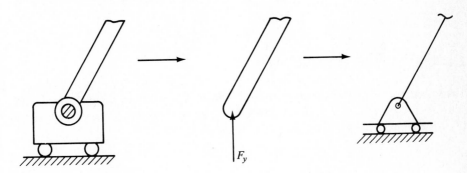

**Figure 2.6** Roller support.

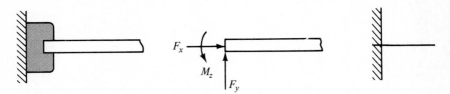

**Figure 2.7** "Built-in" support.

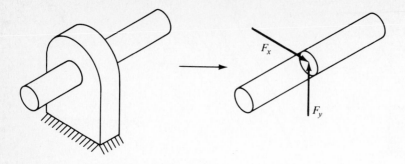

**Figure 2.8** Bearing support (radial).

## 2.3.6 Bearings

A rotating, or "radial" bearing (Fig. 2.8) constrains a shaft against radial motion—i.e., it supplies the necessary radial forces. A "thrust" bearing restricts axial motion providing the necessary axial force. A "linear" bearing permits axial sliding and frequently rotation about the axis.

In all bearing cases, we assume zero friction unless otherwise specified.

## 2.3.7 Springs

"Spring" supports, either linear or rotary, provide a support force (or moment) that is some known function (generally linear) of the spring deformation (Fig. 2.9).

## 2.4 SPECIAL CONFIGURATIONS

There are three special loading situations that occur often enough to warrant special consideration: (*a*) a body loaded by only two forces; (*b*) a body loaded by three forces; (*c*) a sliding belt.

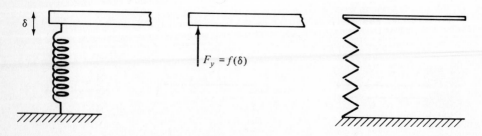

**Figure 2.9** Linear spring support $F_y = f(\delta) = k\delta$.

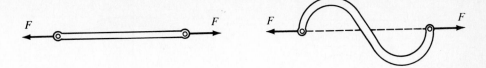

**Figure 2.10** Two-force members.

### 2.4.1 Two-Force Members

Figure 2.10 shows two examples of "weightless" members having forces applied at only two points. For force equilibrium, the forces must be equal and opposite. For moment equilibrium, the forces must be collinear, acting along the line between the two points of application. Whenever you address a structural problem, you should look for two-force members. Knowing the direction of a force is very useful, being equivalent to an equation of equilibrium.

### 2.4.2 Three-Force Members

If a body in equilibrium has only three forces acting on it, as shown in Fig. 2.11, then the three forces *must* lie in a plane (coplanar), and the "lines of action" of the three forces *must* pass through a common point $O$. If the three forces are parallel, the point $O$ is at infinity.

### 2.4.3 Sliding Belts

A common mechanical combination is a belt wrapped around a pulley as shown in Fig. 2.12. If there is no friction, then the belt tension $T_1$ must equal

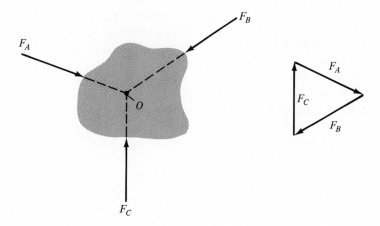

**Figure 2.11** A three-force member.

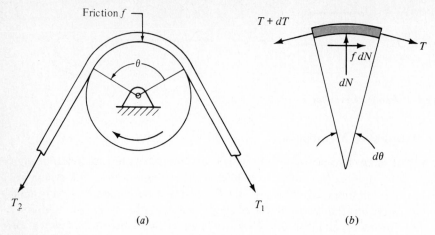

**Figure 2.12** Belt friction: (*a*) Belt and pulley; (*b*) differential element.

$T_2$. If the pulley is on frictionless bearings (Fig. 2.8), again $T_1 = T_2$. However, if there is friction between the belt and the pulley, and if the pulley is restrained, we can have different tensions $T_1$ and $T_2$.

If the belt is slipping, or if it is about to slip, we can calculate a relationship between $T_1$ and $T_2$; otherwise the relationship is ambiguous.

Consider the belt-pulley combination shown where the pulley is rotating clockwise while the belt is held stationary by tension $T_2$ and $T_1$. Because of friction, it is clear that $T_2 > T_1$. So the tension must "vary" from $T = T_1$ to $T = T_2$ as we consider different positions along the belt as defined by the angle $\theta$.

Whenever loading varies from point to point in a body, it is generally useful to isolate a very small section of the body, to write our equations relative to that small section, and to use mathematics to assemble many sections into the whole; this is one of the many reasons for studying calculus. As we proceed through this text, there will be numerous cases where the loading varies and we isolate a *differential element*.

Figure 2.12*b* shows a differential element of the belt, i.e., the element determined by the differential angle $d\theta$. Isolating the element we see the tension $T$ on the right, and $T + dT$ on the left. Note that these are *not* horizontal forces, but slant downwards by $d\theta/2$ on each side. The pulley exerts a normal force $dN$ (why do we write this force as $dN$ not $N$?). If sliding is occurring (or about to occur), then the pulley will also exert a known friction force $f\,dN$. We can now write the equations of equilibrium:

For $\Sigma F_x = 0$:
$$T \cos \frac{d\theta}{2} + f\,dN - (T + dT) \cos \frac{d\theta}{2} = 0$$

For $\Sigma F_y = 0$:
$$dN - T \sin \frac{d\theta}{2} - (T + dT) \sin \frac{d\theta}{2} = 0$$

For small angles ($d\theta$):  $\sin d\theta = d\theta$    $\cos d\theta = 1$

Thus
$$dN - T\frac{d\theta}{2} - T\frac{d\theta}{2} - dT\frac{d\theta}{2} = 0$$

but  $dT\,d\theta \ll T\,d\theta$   and   $dN = T\,d\theta$

Eliminating $dN$ from our equations we have:

$$\frac{dT}{T} = f\,d\theta$$

Integrating $T$ from $T_1$ to $T_2$ as $\theta$ goes from 0 to $\theta$, we finally obtain:

$$\frac{T_2}{T_1} = e^{f\theta} \tag{2.6}$$

Thus the belt tension varies exponentially with both angle of wrap and coefficient of friction. This effect is not only fundamental to the action of belts, but is important to the mechanics of cloth and the strength of knots.

**Example 2.2** A 100-kg steel cylinder, 200 mm in diameter, is placed in a 90° metal "V-block" as shown in the first accompanying illustration. A moment $M_0$ is applied to the cylinder causing it to rotate clockwise. If the coefficient of friction is 0.3 (a reasonable value), how large must $M_0$ be? Start by isolating the

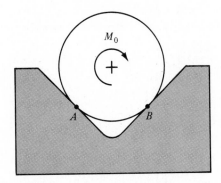

cylinder showing all the externally applied forces. The weight force, assumed to act at the center of gravity of the cylinder is $100(9.8) = 980$ N. There are two normal forces $N_A$ and $N_B$, two friction forces $F_A$ and $F_B$, and the applied moment $M_0$. When slipping occurs, the friction forces equal the normal forces times the friction coefficient. Writing the equilibrium equations, we have:

For $\Sigma F_x = 0$:    $(N_A + 0.3N_A - N_B + 0.3N_B)\sin 45° = 0$

For $\Sigma F_y = 0$:    $(N_A - 0.3N_A + N_B + 0.3N_B)\sin 45° - 980 = 0$

For $\Sigma M_0 = 0$:        $0.1(0.3N_A + 0.3N_B) - M_0 = 0$

where the radius $= 100$ mm $= 0.1$ m. Solving the three equations gives $M_0 = 38.14$ N·m.

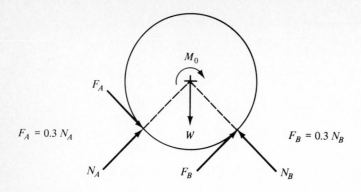

If we solve for $N_A$ and $N_B$, we find $N_A = 444.8$ N and $N_B = 826.7$ N. Many students, when faced with this problem, assume that $N_A = N_B$ because of symmetry. Convince yourself that because of $M_0$, the system is *not* symmetrical.                                                                    ***

**Example 2.3** In order to support a large (500-lb) weight 6 ft from a wall, the structure shown was built. The 6-ft beam *ABC* is pinned to the wall at *A*, and supported by strut *BD* which is pinned at both ends. Determine the forces acting *on* the beam at *A* and *B*. Clearly, if the entire structure is in equilibrium, then each and every part of the structure must also be in equilibrium. Thus we could isolate the entire structure, or strut *BD*, or beam *ABC*, or any segment of any part.

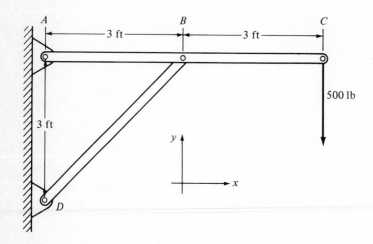

We must make a choice: Either we can start isolating various free bodies and writing equilibrium equations hoping to arrive at a solvable set, or we can sit back and logically determine which free bodies are apt to be most useful; obviously the latter method is preferable. Let us go through that process for this example.

We saw earlier that the equations of equilibrium only give information about forces which are *external* to the free body—those which cross the control surface around our isolated part or system. In this example, the force at $B$, between beam and strut, is *internal* if we isolate the entire structure. In order to calculate the forces at $A$ and $B$, they must be external, so we should probably isolate the beam $ABC$.

If we unthinkingly isolate $ABC$, we have pin joints at $A$ and $B$, each of which can support forces in the $x$ and $y$ directions ($F_x$ and $F_y$). We thus have four unknowns but only three equations of equilibrium.

However, if we realize that strut $BD$ is a two-force member, then the direction of the force at $B$ is known, and only its magnitude is unknown; i.e., we now have only three unknowns. Our *intelligent* isolation then will be:

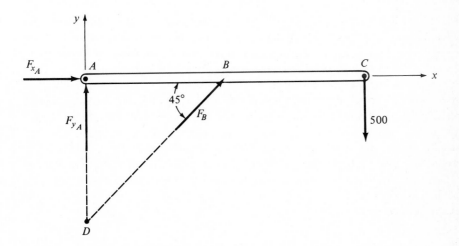

Given the above free-body diagram, we can apply the equations of equilibrium in numerous ways—some requiring less work (hence less chance for error) than others. One general guide is:

> *If the line of action of two or more unknown forces passes through a point, then writing moment equilibrium about that point may well minimize your effort. The intersecting unknowns will not enter the equation and you frequently can avoid simultaneous equations.*

Referring to the free-body diagram, two unknowns ($F_{x_A}$ and $F_{y_A}$) pass through point $A$, two unknowns pass through $B$ ($F_{x_A}$, $F_B$), and two pass through $D$ ($F_{y_A}$, $F_B$). We can write moment equilibrium equations about all three points. We can write the solutions by inspection:

For $\Sigma M_B = 0$: $\qquad\qquad\qquad\qquad F_{y_A} = -500$

For $\Sigma M_D = 0$: $\qquad\qquad\qquad\qquad F_{x_A} = -1000$

For $\Sigma M_A = 0$: $\qquad\qquad\qquad\qquad F_B = \dfrac{1000}{\sin 45} = 1414$

In this example, the unknown forces were shown as acting in the $+x$ and $+y$ directions. The force components at $A$ were found to be negative, indicating that they actually act in the $-x$ and $-y$ directions.

You have a choice of drawing forces in the $+$ directions and letting the equations take care of the signs, or of trying to draw the forces in the correct directions in the free-body diagram and letting minus signs indicate wrong guesses. For simple problems, the latter method helps provide insight and understanding. In complex problems, our intuition may be quite useless, and it only makes sense to draw all forces in the $+$ direction. $\qquad$ ***

The preceding examples were "two-dimensional" in that they could be modeled as lying in the $xy$ plane. But many structures must be treated in three dimensions with the attendant complexity in both sketching the system and solving the equilibrium relations. It is very often advantageous to model a three-dimensional problem as three two-dimensional problems. If three orthogonal views are drawn (looking down the $x$, $y$, and $z$ axes), and the appropriate forces and moments applied, then we have three separate free-body diagrams.

When we look down the $z$ axis, we see force components lying in the $xy$ plane and moments about the $z$ axis. When we look down the $x$ axis, we see force components lying in the $yz$ plane and moments about the $x$ axis, etc.

Given three two-dimensional problems, we have $3(3) = 9$ equations of equilibrium. Clearly, there can be only six "independent" equations, so three must be redundant.


## 2.5 RESULTANTS

We all know that under *some* conditions, two or more forces (moments) can be combined into a single "resultant" force for analytic purposes. Figure 2.13 shows three cases where two forces $F_1$ and $F_2$ are replaced by their resultant $R$. The resultant force must create the same effects as the original forces.

Because forces (and moments) are vector quantities, the resultant will be the vector sum of the individual forces. If the individual forces act at the same point (as in Fig. 2.13$a$), then the resultant acts at the same point. In Fig. 2.13$b$, the forces act at different points, and intersect at some remote point. Because the individual forces create zero moment about that point, the resultant must pass through the remote point as shown.

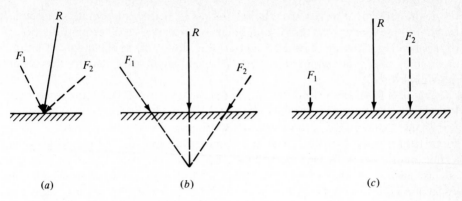

*(a)* *(b)* *(c)*

**Figure 2.13** $R$ is the resultant of $F_1$ and $F_2$.

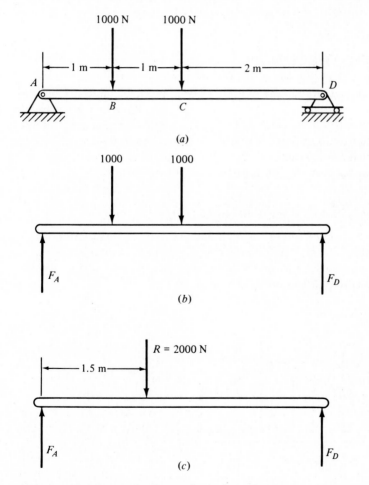

*(a)*

*(b)*

*(c)*

**Figure 2.14** Resultant of loading can be used to calculate *external* forces $F_A$ and $F_D$.

If the individual forces are parallel, as in Fig. 2.13c, the remote point is at infinity. However, we can still use moment equivalence to determine the point of application. Taking moments about the point of application of either individual force, the moment due to $R$ must equal that due to the other individual force.

Figure 2.14 shows a simply supported beam with two 1000-N forces applied. For what purposes can we replace the two forces with the single 2000-N force?

Figure 2.14b and c shows free-body diagrams with two forces and one force, respectively. You can easily show that either loading will give rise to the same support forces $F_A$ and $F_D$. It should be intuitively clear, however, that the effects *within* the beam will be different for the two loadings—i.e., the deflection and internal beam conditions will be different. We can conclude:

> For calculation of effects **external** to an isolated body, we can replace forces with their resultants. For calculation of **internal** effects, we cannot.

## 2.6 DISTRIBUTED FORCES

As we have previously noted, many forces must be considered to be distributed: over a length, over an area, or within a volume. When we want to determine the *external* forces acting on a body, we can replace *any* force distribution by its resultant. Clearly the resultant force must produce the same overall forces and moments as does the original distributed force. The calculation of resultant magnitude and position is very straightforward.

### 2.6.1 Loading Distributed along a Straight Length

Loading which is distributed along a length has the units of force/length (lb/in or N/m). We will call this $q(x)$, indicating that the loading/length $q$ is frequently a function of $x$. Let us find the magnitude and location of the resultant $R$ of the loading diagram shown in Fig. 2.15.

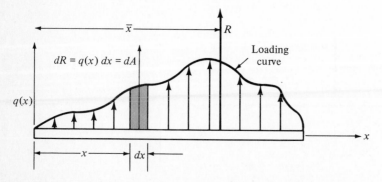

**Figure 2.15** The resultant $R$ of the distributed force equals the area of the diagram and acts at its centroidal position $\bar{x}$.

The magnitude of $R$ is simply the sum of all the differential loads $dR$. But $dR = q(x)\,dx$. From the figure, $q(x)\,dx$ is the area of the loading diagram associated with $dx$. Therefore:

*The magnitude of R equals the area under q(x).*

The moment due to $R$ about any point must equal the sum of the moments due to the differential loads $dR$. Taking moments about the origin:

$$R\bar{x} = \int x\,dR = \int x\,q(x)\,dx = \int x\,dA$$

But $R = A$. Thus

$$\bar{x} = \frac{1}{A}\int x\,dA = \text{centroidal position}$$

*The resultant R acts at the centroid of q(x).*

The resultant of a load distributed over a length equals the area of the loading diagram and acts through the centroid of that area.

## 2.6.2 Load Distributed over a Plane Area

The unit of area-distributed loading is force/area ($\text{lb/in}^2 = \text{psi}$; $\text{N/m}^2 = \text{Pa}$) and, if compressive, is called "pressure." Figure 2.16 shows a "linear" pressure distribution diagram.

In a manner similar to that described in Sec. 2.6.1, it is easy to show that:

*The magnitude of R equals the volume of the pressure distribution diagram and R passes through the centroid of that volume.*

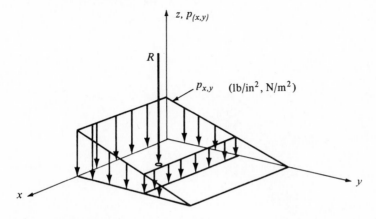

**Figure 2.16** The resultant $R$ of a pressure distribution equals the volume of the diagram and passes through the centroid.

### 2.6.3 Unidirectional Loading Distributed within a Volume

As we all know, the resultant weight of a body of uniform density is simply the weight density (lb/in³ or N/m³) times the volume, and acts at the centroid (center of gravity). For cases of variable density, we must use integral techniques as described in Secs. 2.6.1 and 2.6.2.

## 2.7 FLUID FORCES

Fluids (liquids or gases) acting on the surface of a body can exert both normal forces (generally compressive pressure) and tangential forces (generally viscous "shear" forces).

   If the fluid is stationary relative to the body, only normal forces exist. The in-depth study of such forces is known as "hydrostatics." When fluids move relative to the body, both normal and shear forces can coexist, and the general area is called "hydrodynamics." For this text, we will consider only three simple cases.

### 2.7.1 Liquid Pressure due to Gravity

The pressure in an open (nonpressurized) volume of stationary liquid varies from point to point and is due to the weight of the material above the point in question. This "hydrostatic pressure" due to the gravity is often called the "gravity head," and is equal in all directions at a point.

   If we ignore the atmospheric pressure at the surface of the liquid, the pressure $p$ at depth $d$ is simply $p = \gamma d$, where $\gamma$ is the weight density. Thus as we go deeper down in the liquid, the pressure increases linearly with depth.

   **Example 2.4** A water tank is 6 ft high and 10 ft square as shown in the accompanying illustration. Determine the magnitude and location of the resul-

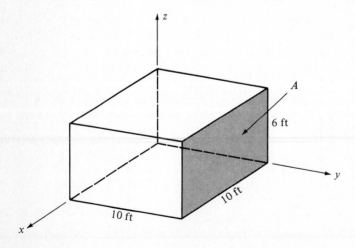

tant pressure force acting on one side of the tank when it is full of water of density 62.4 lb/ft³. Because stationary fluids can only exert force perpendicular to a surface, the resultant $R$ will be a horizontal force. The pressure increases linearly with depth from the surface (in addition to atmospheric pressure) from 0 to $6(62.4) = 374.4$ lb/ft² [or to $72(0.036) = 2.6$ psi]. The pressure diagram is a triangular volume as shown. The resultant, acting at the centroid 2 ft above the base, equals the volume (of the pressure diagram, not the tank!):

$$R = 0.5(374.4)(10)(6) = 11,232 \text{ lb}$$

$$R = 0.5(2.6)(120)(72) = 11,232 \text{ lb}$$

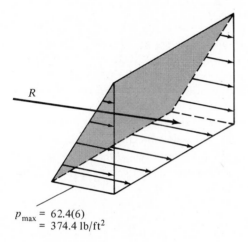

$R$

$p_{\max} = 62.4(6)$
$= 374.4 \text{ lb/ft}^2$

***

## 2.7.2 Fluid Drag

A special type of "pressure" loading is the "drag" force on a body submerged in a moving stream. Detailed analyses of drag forces are given in books on fluid dynamics; here we will just present a very approximate relationship.

The drag force/unit frontal area of the body can be thought of as an equivalent pressure acting on the body in a direction opposite to the flow. This drag pressure $p$ is found empirically (experimentally) to vary primarily with fluid mass density $\rho$, fluid velocity $V$, and body shape. As a first approximation, these are related as follows:

$$p = \tfrac{1}{2}\rho V^2 C_D \tag{2.7}$$

where $C_D$ is an empirical "drag coefficient."

For any "real" problem, you would look up $C_D$ for the appropriate shape, etc. For the purposes of this book (and not a bad approximation), let $C_D = 1$.

### 2.7.3 Buoyancy Forces

A familiar type of fluid force is the "buoyancy" force that acts on any body immersed in a fluid. The force arises from the fact that the fluid pressure is greater on the lower parts of a body than on the upper parts because the depth of the fluid is greater.

We know how pressure varies with depth, so, for any body, we should be able to integrate the differential pressure forces over the area to determine the resultant force.

Fortunately, Archimedes teaches us that this is not necessary because the resultant is an upward force, precisely equal to the weight of the fluid displaced by the body and acting through the center of gravity of the displaced volume.


## 2.8 COMPUTER EXAMPLES

In this chapter we saw that resultant forces, due to distributed loads, act through the centroids of the load distribution diagrams. The process of finding the centroid, if we do not know where it is, requires the evaluation of one or more integrals as seen in Sec. 2.6.

Whenever we are faced with integration of a nonsimple function, it may be much easier to use numerical methods to determine an approximate integral. Here we demonstrate how a very simple program method can be used to determine complex integrals. We will demonstrate on a simple problem, having a known answer, so that we can evaluate the error associated with the numerical solution.

*Problem*: Determine the centroid of a solid hemisphere.

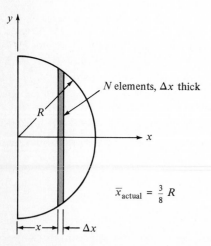

**Figure 2.17** The hemisphere is divided into $N$ elements.

*Solution*: Figure 2.17 shows a section of a hemisphere of radius $R$. We section the volume into $N$ elements of thickness $\Delta x$ where $\Delta x = R/N$. We must evaluate the integral (sum) of $x$ times the element volume, and divide by the integral (sum) of the element volumes to determine the centroid.

Program Listing 2.1 shows the program for computing the centroid using various values of $N$. The error is also computed. Also shown, in Program Output 2.1, is the printout from the program. We see that the accuracy is very good, even for a small number of elements.

The method used here, dividing the body into $N$ elements, can be used for *any* shape we wish to define.

```
10 REM           CENTROID OF HEMISPHERE
20 REM
30 R=1                     \REM  Radius of hemisphere
40 !"   N     X-bar    Error" \!
50 FOR N=10 TO 100 STEP 10 \REM  N = number of elements
60 V=0 \ M=0               \REM  Initialize for next N
70 FOR I=1 TO N
80 X=(I-.5)*R/N            \REM  Position x
90 R1=SQRT(R^2-X^2)        \REM  Radius of element
100 V1=3.14159*R1^2*R/N    \REM  Volume of element
110 V=V+V1                 \REM  Sum of element volumes
120 M1=X*V1                \REM  x times delta-V
130 M=M+M1                 \REM  Sum of M1's
140 NEXT
150 X1=M/V                 \REM  x-bar
160 E=(X1-.375)/.375       \REM  Error, .375 is exact
170 !%5I,N,%10F5,X1,E
180 NEXT
```

**Program Listing 2.1**

| N | X-bar | Error |
|---|-------|-------|
| 10 | .37640 | .00375 |
| 20 | .37535 | .00094 |
| 30 | .37516 | .00042 |
| 40 | .37509 | .00023 |
| 50 | .37506 | .00015 |
| 60 | .37504 | .00010 |
| 70 | .37503 | .00008 |
| 80 | .37502 | .00006 |
| 90 | .37502 | .00005 |
| 100 | .37501 | .00004 |

**Program Output 2.1**

## 2.9 PROBLEMS

**2.1** In developing the equation of equilibrium, we arrived at six scalar equations for the general three-dimensional case, and three equations for the two-dimensional, coplanar case. There are other sets of equations that are equivalent if correctly used. For the two-dimensional case, show that a (static) body is in equilibrium if:

(*a*) For any two points $A$ and $B$, when $\Sigma\, M_A = 0$ and $\Sigma\, M_B = 0$, and there is zero net force in the direction $AB$.

(*b*) The sum of the moments is zero about any three points $A$, $B$, and $C$ not lying in a straight line.

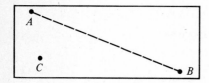

Figure P2.1

**2.2** Show that, for purposes of equilibrium, a force $F$ can be replaced with an appropriate force and moment.

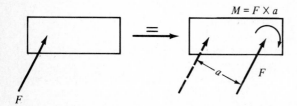

Figure P2.2

**2.3** The frame (or truss) is made of members "pinned" together as shown in the accompanying illustration. Determine the support forces and the forces in each member.

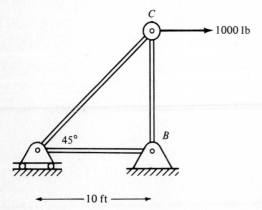

Figure P2.3

**2.4** Determine the force in each of the seven members shown in the accompanying illustration.

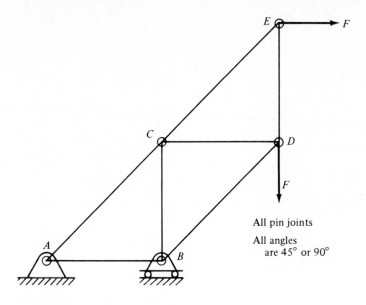

All pin joints

All angles are 45° or 90°

**Figure P2.4**

**2.5** A device for feeding steel bars into a machine tool consists of a 90° "vee" inclined at 30°, as shown in the illustration below. Bars are added at the upper right and roll down to the next bar. Bar No. 1 is the next bar to be fed into the machine. Consider the case of three 100-lb bars:

(*a*) Assuming negligible friction, determine the support forces at *A*, *B*, *C*, and *D*.

(*b*) Bar No. 1 is fed into the machine by sliding it in the *z* direction (perpendicular to the paper). If the friction coefficient is 0.3, estimate the maximum force $F_z$ that might be required to slide bar No. 1.

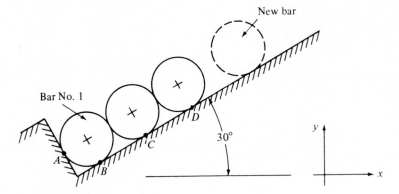

New bar

Bar No. 1

30°

**Figure P2.5**

**2.6** A cylinder "weighing" 100 kg is rolled against a wall as shown in the illustration below.

    (*a*) How large must $M_0$ be to cause the cylinder to rotate?

    (*b*) How large must the friction coefficient be for the cylinder to "climb" the wall?

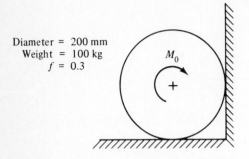

Diameter = 200 mm
Weight  = 100 kg
     *f* = 0.3

$M_0$

**Figure P2.6**

**2.7** Three identical cylinders are "stacked" as shown. How large must the friction coefficient be if they do not fall down?

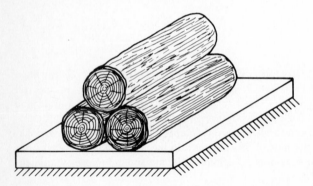

**Figure P2.7**

**2.8** A hydraulic "log-splitter" uses a wedge of angle $2\alpha$ driven by a force $F$ from a hydraulic jack. The friction coefficient between wood and wedge is expected to be about 0.5. What is the maximum value of $\alpha$ such that the wedge will not "pop out" if $F$ is removed? Under this condition, what portion of the energy input to the wedge-wood system goes into frictional energy?

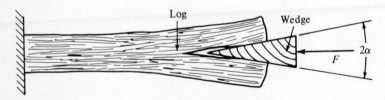

Log

Wedge

$2\alpha$

$F$

**Figure P2.8**

**2.9** To raise weights (jacking), screw threads having a square cross section are often used as shown below. If the *mean* radius of the screw thread is $r$, the "pitch" (axial movement per revolution) is $p$, and the coefficient of friction is $f$, determine the moment $M_0$ required to raise the weight $W$. (*Hint:* Model the problem as a weight being pushed up an inclined plane by a horizontal force.)

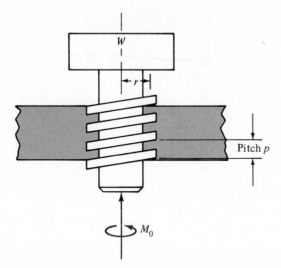

**Figure P2.9**

**2.10** For the screw jack of Prob. 2.9, it is most desirable that the friction be large enough, or the pitch small enough, so that the weight will not go back down when $M_0$ is removed.

(*a*) What is the minimum value of $f$ for this "self-locking" action?

(*b*) What is the maximum efficiency of a self-locking jack? ("Efficiency" is defined as power out/power in.)

**2.11** For most real jacks, the load (see Prob. 2.9) cannot rotate with the jack screw. Accordingly, there must be additional friction losses. Determine an approximate relationship for the value of $M_0$ required to raise $W$. Use the notation of Prob. 2.9.

**2.12** Screw fastenings (as contrasted to screw jacks) have nonsquare thread sections. There are a number of standard thread types, the most common being the 60° thread shown in the accompanying illustration. The peaks and valleys are either flattened (as shown) or rounded.

Threads are defined in terms of the nominal diameter $d$, and the pitch (screw advance per revolution) $p$. An abbreviation indicates the thread type, and a "quality code" may be added.

Thus a "one-inch-eight" is a 1-in-diameter screw having eight threads per inch, and might be labeled 1″-8 UNC-2 ("UNC" meaning Unified Coarse thread standard, and the "2" referring to quality).

Consider a 1″-8 screw having a friction coefficient of 0.3 between nut and bolt. Estimate the torque $M_0$ required to obtain a bolt tension of 5000 lb when the helix angle is assumed to be small:

(*a*) With zero friction under the bolt head

(*b*) With $f = 0.3$ under the bolt head

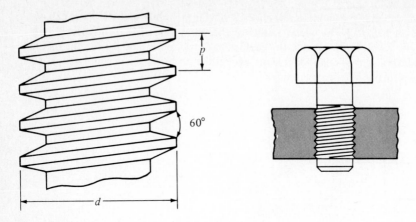

**Figure P2.12**

*Note*: Consider the solution in terms of three angles:

$$\alpha = \tan^{-1}\frac{p}{2\pi r} \qquad \text{(the helix angle)}$$

$$\beta = \tan^{-1}f \qquad \text{(the friction angle)}$$

$$\gamma = 30° \qquad \text{(the outward slope due to thread geometry)}$$

The exact solution is complex, but for small values of $\alpha$, the major effect of $\gamma$ is to increase the normal force in proportion to $1/\cos\gamma$.

Show that $M_0 \cong rT \tan(\alpha + \beta')$ where $\beta' = \tan^{-1}(f/\cos\gamma)$.

**2.13** A standard 13-mm wrench is used to tighten a bolt as shown below. Determine the contact forces between the wrench and the bolt head.

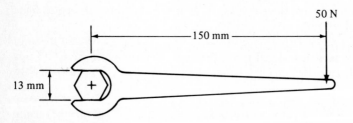

**Figure P2.13**

**2.14** A conventional pipe wrench, or Stilson wrench, is shown in Fig. P2.14, p. 49. If, for analysis, we ignore the screw/nut adjustment mechanism, the wrench can be idealized as shown. For the dimensions given, determine all the forces acting on the $1\frac{1}{2}$-in pipe when $P = 100$ lb.

Must the jaws be serrated as shown, or will friction probably suffice?

**2.15** An idealized pair of "vise-grip" pliers is shown (look at a real pair). For a grip force of 150 N, calculate the forces on the object in the jaws. (See Fig. P2.15, p. 49.)

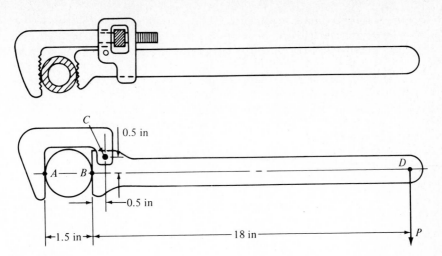

Figure P2.14

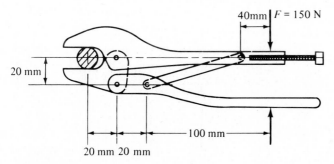

Figure P2.15

**2.16** A "belt-type" pipe wrench is shown. The belt is fastened to the wrench at $B$, wraps about 270° around the pipe, is gripped by the wrench-pipe force at $A$, then goes out a slot in the handle. How large must the pipe-belt-wrench friction coefficient be?

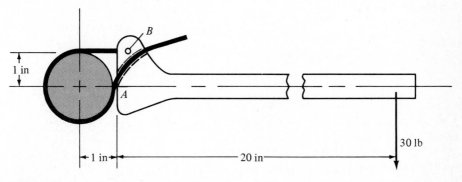

Figure P2.16

**2.17** The inner workings of a worm-gear speed reducer are shown in the accompanying illustration. The worm, characterized by a mean radius $r$ and pitch $p$, is supported by two radial bearings and a thrust bearing. An *input* moment $M_i$ drives the worm screw at $N_i$ rpm.

The gear must have a corresponding pitch, and is characterized by the number of teeth $n$. The contact surface between worm and gear is not only angled because of the screw pitch (helix angle), but is also angled outward by $\gamma$ due to tooth geometry.

In terms of $r$, $p$, $n$, $\gamma$, $M_i$, $N_i$, and the friction coefficient $f$, determine:
(*a*) The output speed $N_o$
(*b*) Output torque $M_o$
(*c*) Efficiency

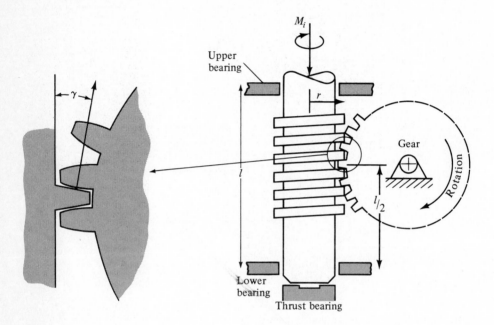

**Figure P2.17**

**2.18** For the worm-gear arrangement of Prob. 2.17, determine the forces acting *on* the worm *by* its bearings when:

$$r = 0.5 \text{ in} \qquad n = 30 \text{ teeth} \qquad \gamma = 20$$

$$p = 0.2 \text{ in} \qquad N_i = 100 \text{ rpm}$$

$$f = 0.1 \qquad M_i = 500 \text{ in} \cdot \text{lb}$$

Vertical distance between bearings: $l = 4$ in

**2.19** A gearbox, with workings similar to those described in Probs. 2.17 and 2.18, is bolted to a nominally flat surface with bolts at $A$ and $B$. The box has an input torque of 500 in·lb, a 30:1 speed ratio, and an efficiency of 39 percent. What can you determine about the support forces at $A$ and $B$? (See Fig. P2.19, p. 51.)

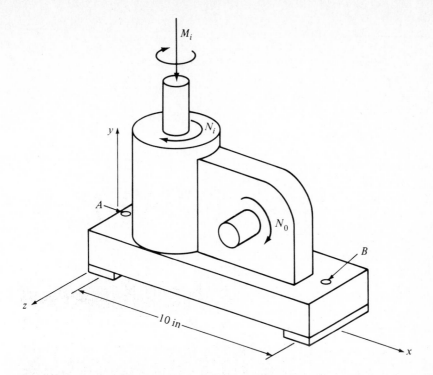

**Figure P2.19**

**2.20** An 1800-rpm motor delivers 0.25 hp. If the tension $T_A$ is 5 lb, how large is $T_B$? How large must $f$ be?

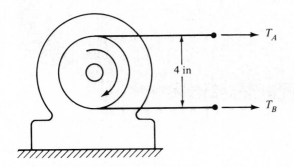

**Figure P2.20**

**2.21** Motor weight can be used to maintain belt tension by mounting the motor on a pivot as shown in Fig. P2.21 (p. 52). When the motor puts out rated power, what will the belt tensions be? How large must $f$ be?

**2.22** A four-wheel drive RV (recreational vehicle) weighs 3600 lb, and has the pertinent dimensions shown in Fig. P2.22 (p. 52). The road/tire friction coefficient is 0.8.

1-kW motor
25-kg weight
3600 rpm

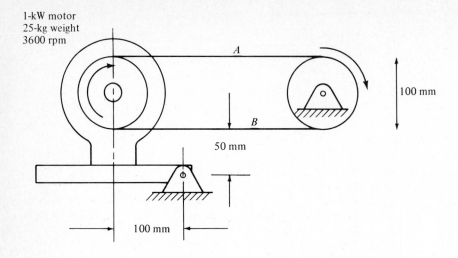

**Figure P2.21**

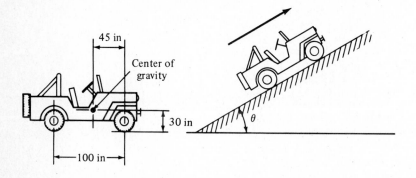

**Figure P2.22**

(*a*) Using only rear-wheel drive, what is the maximum hill angle it can steadily climb?

(*b*) What maximum hill angle with all four wheels?

(*c*) To climb the angle with all four wheels, at 5 mph, assuming 85 percent overall efficiency, how much engine power is required?

**2.23** In building high-speed roadways, it is desirable to "bank" the turns for "best" results at rated speed $V$ and radius of curvature $\rho$. Determine the proper bank angle.

**2.24** A beam, twice as high at $A$ as at $B$ weighs 1000 lb. Find the support forces at $A$ and $B$ (uniform width and density). (See Fig. P2.24, p. 53.)

**2.25** A flagpole, 100 ft high, 10 in in diameter at the base and 5 in in diameter at the top, must withstand wind gusts of 150 mph. Determine the support force and moment required at the base. (Air density = 0.000044 lb/in$^3$.)

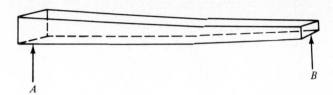

**Figure P2.24**

**2.26** Poured concrete house foundations are made between wooden "forms" that are typically 2 ft wide (perpendicular to the page in the drawing below) and 8 ft high. The forms are set up on the previously poured "footings" (to distribute the total house weight over a considerable area of soil). Then concrete, at about 150 lb/ft³, is poured.

To keep the forms from moving apart, tie wires are inserted across each form at 2-ft vertical spacing as shown at *A, B, C, D,* and *E* in the illustration. Estimate the tension in the five tie wires.

*Note*: This problem is not statically determinate; you cannot yet solve it with precision. Find how the concrete applies force to the form, and then estimate the forces in the wires. What are your simplifying assumptions?

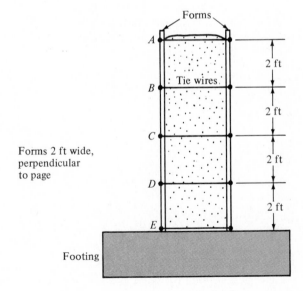

**Figure P2.26**

**2.27** A type of toilet bowl valve found in many American toilets is shown (somewhat idealized) in Fig. P2.27 (p. 54). The float cylinder, of height *H* and cross-sectional area *A*, weighs *W*. The pipe-to-float-seal area is *a*. Determine the range of float heights *H* that would be appropriate, and also determine the force *F* required for flushing; the illustration shows the float lifted off the sealing area.

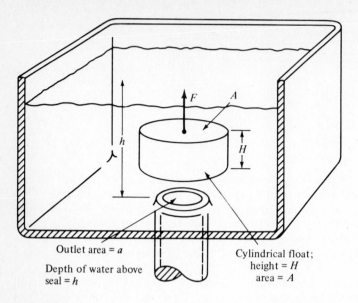

Outlet area = $a$

Depth of water above
seal = $h$

Cylindrical float;
height = $H$
area = $A$

**Figure P2.27**

2.28 through 2.32: Consider the somewhat idealized bicycle shown in Fig. P2.28. The rider
can apply forces on the seat ($W$), the pedal ($P$), and on the handlebars ($F$). All are assumed
to be vertical and static. The front fork is supported by a radial/thrust bearing at $A$, and a
radial bearing at $B$. The large sprocket has 52 teeth, while the small one has 21. The chain
pitch is $\frac{1}{2}$ in (two links/inch).

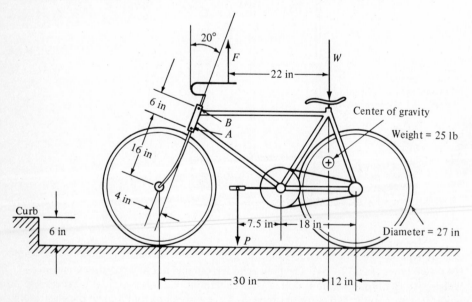

**Figure P.2.28**

**2.28** A 150-lb rider merely sits on the seat ($W = 150$).

(*a*) What are the road/tire contact forces?

(*b*) What are the bearing forces at $A$ and $B$?

**2.29** When "really working," the rider applies 200 lb to the pedal ($P$) and 50 lb to the handlebar ($F$).

(*a*) What are the road/tire contact forces?

(*b*) How large must $f$ be to avoid slipping?

(*c*) What is the chain tension?

**2.30** The air pressure in the tires is 80 psi, and $f = 0.7$ between the tires and the road. The 150-lb rider just sits on the seat. How large a torque, about the $AB$ axis, would be required to turn the wheel for steering?

**2.31** The large sprocket shaft is supported in two bearings at $C$ and $D$, 3 in apart. When a 200-lb load is applied (as in Prob. 2.29), what are the forces acting on the shaft by the bearings?

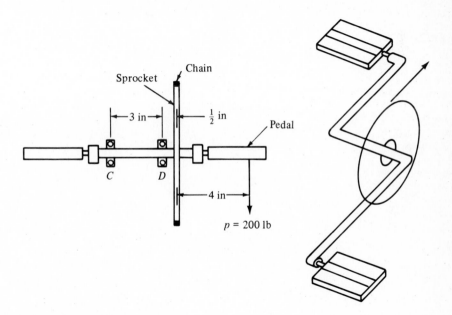

**Figure P2.31**

**2.32** There are various types of bicycle chains; a possible example is shown in Fig. P2.32 (p. 56). For a chain tension of 500 lb, estimate and show the forces acting on a chain pin.

2.33 through 2.34: Write a computer program (and if a computer is available, run the program) to solve the following problems and to determine error as a function of number of elements.

**2.33** Find the centroid of a triangle.

**2.34** Determine the weight of a bar of steel, 40 in long, and tapering linearly from 1 in to 2 in in diameter.

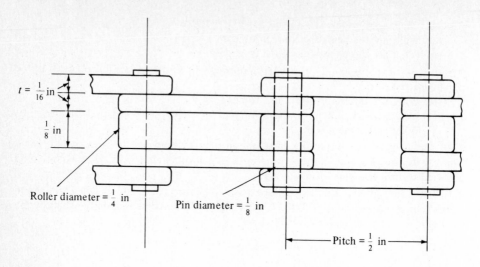

**Figure P2.32**

# THREE

## AXIALLY LOADED MEMBERS AND STRUCTURES

## 3.1 OVERVIEW

In Chap. 2 we saw how Newton's laws could be applied to determine, or at least to estimate, the forces acting on statically determinate bodies. From our experience we know that when we apply a force to some bodies (e.g., a spring, rubber band, or diving board), they "deflect" or "deform."

In fact, *all* bodies deflect and deform whenever forces are applied; sometimes you may not see this motion, but it is there. In this chapter we will be concerned with *axially* loaded members and their deformations, particularly their *elastic* deformations.

The British physicist Robert Hooke (1635–1703) is credited with the first law of "elasticity" for solid bodies (Hooke's law). In essence a deformation is *elastic* if the deformation disappears when the forces that caused the deformation are removed. Few, if any, engineering materials undergo truly elastic deformation; if you look closely enough, you will find some residual deformation. For practical purposes, however, we consider *all* solids to experience some range of elastic deformation. The range of elastic deformation extends to the point where either (*a*) further deformation is permanent and not recoverable when the forces are removed, or (*b*) further deformation causes fracture (it breaks!).

In order to be able to design structural members made from various materials, it is an obvious practice to "characterize" all materials in terms of a set of numerical properties or characteristics. These properties can only be obtained through physical tests on samples of a particular material. We must then assume that the same values of the properties will hold when we build a structural component from a piece of the "same" material.

Unfortunately, no two pieces of material are truly alike, and any material specified will have a range of property values. As might be expected, reduction of property variation requires greater attention to quality control and hence greater material cost. In the United States, material quality control is quite good. It is the writer's experience that uncertainty due to property variation is generally far smaller than uncertainty regarding the loads that the structure will experience.

The overall process of mechanical design and analysis involves both measurement and prediction:

Conduct simple, standardized tests in a laboratory to determine various material properties.

Develop models and theories that will allow us to predict full-scale behavior based on the materials test data.

This entire text revolves around the above two steps. In this chapter we will introduce the two most common materials tests: the "tensile test" and the "hardness test." We will see, in this and later chapters, that tensile test data carry sufficient information for the design of many purely static devices. When oscillatory loading is involved, or when the environmental conditions are harsh (high-temperature, corrosive atmosphere), other tests are required.

Here we will first discuss the tensile test and the tensile properties in some detail, and then use those properties to predict the behavior of structural members. Next, we will study the behavior of structures made from an assembly of axially loaded structural members.

We will study structures that are both determinate and statically indeterminate, and we will introduce the concepts of "stiffness" and "compliance" as they relate to members and structures. We will see that for any elastic structure, at any point on the structure, there exists a very special set of orthogonal axes called the "principal axes." A load applied along one of the principal axes will produce displacement along that axis only.

Finally, we will introduce the hardness test, which serves as a nondestructive substitute for the tensile test, and is commonly used as a production quality control measure.

## 3.2 THE TENSILE TEST

As the name implies, the tensile test involves loading a test specimen in tension (pulling) and observing the resulting deformation. In modern tensile-test machines, either the loading can be preprogrammed (stepwise increasing, steadily increasing, oscillatory), or the motion of the ends of the specimen can be programmed.

Clearly, if we program the load (independent variable), we must measure the deformation (dependent variable) and vice versa. Figure 3.1 shows speci-

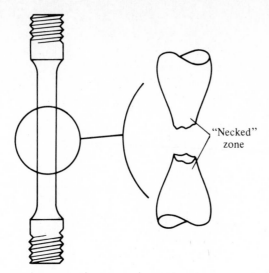

"Necked" zone

**Figure 3.1** Tensile-test specimens.

mens of mild steel before and after testing. Looking at the specimens we can see:

Specimens are designed with care to ensure that the central "gage section" is not affected by the gripping process (threads in this case).
The mild steel specimen "necked-down" considerably prior to fracture; i.e., the deformation was not uniform along the specimen, but localized near the fracture.

Figure 3.2 shows sketches of tensile specimens during the test: (*a*) before localized deformation has occurred, and (*b*) after necking. We can define the pertinent quantities as follows:

$$F = \text{axial load (lb, N)}$$

$$l = \text{gage length between points 1 and 2 (in, m)}$$

$$d = \textit{minimum} \text{ diameter of specimen (in, m)}$$

$$A = \text{actual specimen area (in}^2, \text{m}^2)$$

$$A_0 = \text{initial area}$$

$$l_0 = \text{initial gage length}$$

$$l = l_0 + \Delta l$$

Figure 3.3*a* shows a plot of tensile-test data for a typical steel specimen. For this test, the length was continuously increased and the required force

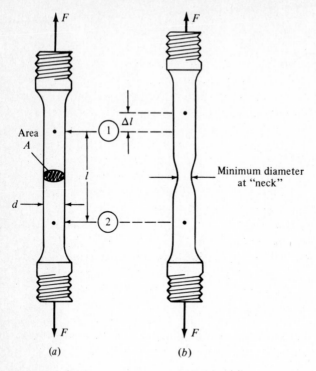

**Figure 3.2** Tensile specimen geometry during test: (*a*) prior to "necking"; (*b*) with "necking."

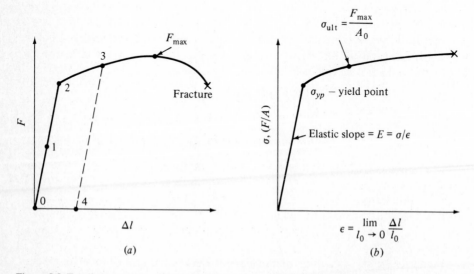

**Figure 3.3** Tensile-test curves: (*a*) load-displacement curve; (*b*) stress-strain curve.

measured. If we consider the force-displacement data, as represented by Fig. 3.3a, we find the following:

1. An apparently straight portion 0-1-2; the deformation at 2 might equal 0.1 to 1 percent of the initial gage length $l_0$, depending on the steel.
2. If we loaded the specimen to point 1 and then released the load, it would appear to return to point 0 (*elastic* behavior). It would also appear to do so if loaded to point 2.
3. If the specimen were loaded beyond the steep, elastic part of the curve to 3, and the load were then released, we would observe a permanent deformation (point 4).
4. The load release line from 3 to 4 is essentially parallel to the elastic line 0-1-2. In fact, the deformation recovery from 3 to 4 is *elastic* recovery.
5. Upon reloading from 4, we retrace the line from 4 to 3 (approximately) and then continue along the original curve.
6. The load goes through a maximum at $F_{max}$; at the same time, necking starts.
7. Continued stretching leads eventually to fracture.

The force $F$ and the change in length $\Delta l$ in Fig. 3.3a obviously depend upon the size of the specimen as well as the material itself. To establish *properties* which are independent of size, we define and plot two new variables—variables that we will use continuously in this text and that you *must* learn and understand:

*Stress*: The applied force divided by the area over which it acts. Stress has the units of force per unit area ($lb/in^2$, psi; $N/m^2$, Pa). In this text, $\sigma$ will represent tensile stress (+) and compressive stress (−).
*Strain*: The change in length divided by the original length of the specimen. Strain, being length divided by length, has no units. We will use $\epsilon$ for tensile strain (+) and compressive strain (−).

For the tensile test:

$$\text{Tensile stress} = \sigma = \frac{F}{A}$$

$$\text{Tensile strain} = \epsilon = \frac{\Delta l}{l_0}$$

(3.1)

Note that the definitions for stress and strain just given are for the very special (although frequently encountered) conditions associated with tensile or compressive loadings only. This is generally called "uniaxial loading." We will see later that there are "shear stress" and "shear strain," where the force and motion are parallel to the area $A$ rather than "normal" to it. Thus, the set of definitions used here, although correct, refer to axial loading only. We shall consider the general "state of stress" in later chapters.

Equations (3.1) involve important implications: By saying that the tensile stress $\sigma = F/A$, we imply that $F$ is uniformly distributed over $A$; and by saying that the tensile strain $\epsilon = \Delta l/l_0$, we imply that $\Delta l$ is uniformly distributed along $l_0$.

Prior to necking, the stress and strain are distributed quite uniformly across and along the specimen. After necking, however, they are not, and the "worst" conditions clearly are at the point of minimum diameter. Hence, following the onset of necking, we modify Eqs. (3.1) as follows:

$$\sigma = \frac{F}{A_{\min}}$$

$$\epsilon = \lim \frac{\Delta l}{l_0} \qquad \text{as } l_0 \to 0 \tag{3.2}$$

In this text, we deal primarily with behavior prior to necking where Eqs. (3.1) and (3.2) are essentially identical. [See Prob. 3.27 for clarification and use of Eqs. (3.2).]

In Fig. 3.3b we have plotted the tensile-test data on stress-strain coordinates. We can now define the "important" properties determined from the tensile test.

### 3.2.1 Elastic Region

The *elastic* region extends from the origin to the "yield point." The stress at which tensile yield occurs will be called the "yield stress" and represented by $\sigma_{yp}$. The curve between the origin and yield is never completely straight; some materials exhibiting more linear behavior than others. We characterize the elastic region in terms of the best straight-line representation. The slope of this line $E$ is the "elastic modulus" (sometimes called "Young's modulus")

$$\frac{\sigma}{\epsilon} = E \qquad \text{(in the } elastic \text{ region for axial loading)} \tag{3.3}$$

where the units for $E$ are $lb/in^2$ or psi (USCS) and $N/m^2$, Pa (SI).

### 3.2.2 Plastic Region

The *plastic* region, which extends from the yield point through fracture, generally slopes upward because of "strain hardening," a mechanism whereby materials (primarily metals) actually become stronger as they are plastically deformed.

We saw that the tensile force $F$ exhibited a maximum at the onset of necking; the stress at the maximum load point is called the "ultimate stress." For convenience this stress is computed using the original cross-sectional area rather than the actual area (a very good approximation):

$$\sigma_{\text{ult}} = \frac{F_{\max}}{A_0} \qquad \text{(the ultimate stress)} \tag{3.4}$$

### 3.2.3 Brittle/Ductile

As you well know, some materials, like the copper in a copper wire, or the steel in a paperclip, can be bent a great deal before breaking. Other materials, like chalk or window glass, break very quickly if we try to bend them.

We will call the former "ductile," and the latter "brittle"—words that you are probably used to. In essence, a brittle material is one that fractures before it yields; i.e., it fractures while still elastic.

### 3.2.4 Energy of Deformation

When a specimen is stretched, the force $F$ obviously puts energy into the specimen. We leave it to you to show (Prob. 3.28) that the area under the $\sigma$-$\epsilon$ curve represents the energy per unit volume required to attain any given strain (see Fig. 3.4):

$$\text{Energy/volume} = \int \sigma \, d\epsilon \qquad (\text{psi, N/m}^2) \qquad (3.5)$$

The "elastic energy," which is recoverable, is stored by virtue of the atoms being pulled away from their normal, equilibrium positions. The "plastic energy," on the other hand, is *not* recoverable, and appears almost instantaneously as heat in the specimen.

We can see from Fig. 3.4 that in a ductile material the energy to fracture is generally very large, while in brittle materials, being restricted to small elastic deformations, the energy to fracture is relatively small.

### 3.2.5 Transverse Deformation

We all know that when we stretch an elastic band it becomes thinner. So it is with most materials. In the tensile test, when we apply a strain in the *axial*

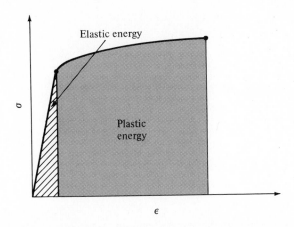

**Figure 3.4** The area under the $\sigma$-$\epsilon$ curve equals energy/unit volume.

direction, we observe a strain (of opposite sign) in the *transverse* (radial) direction. The ratio of the two strain magnitudes is called "Poisson's ratio" (after Simeon Poisson, 1781–1840, yet another Frenchman). We will denote Poisson's ratio by $\nu$. Thus we have

Axial strain:
$$\epsilon_l = \frac{\Delta l}{l_0}$$

Transverse strain:
$$\epsilon_d = \frac{\Delta d}{d_0} \tag{3.6}$$

Poisson's ratio:
$$\nu = -\frac{\epsilon_d}{\epsilon_l}$$

In the elastic region, $\nu$ is typically 0.2 to 0.3 for common engineering materials.

### 3.2.6 Reverse Loading

Tests can, of course, be conducted in compression as well as in tension, and we can plot a compressive stress-strain curve. Let us simply give two broad generalities regarding compression as illustrated in Fig. 3.5:

1. Ductile materials exhibit essentially the same behavior in compression $(-\sigma)$ as in tension $(+\sigma)$. The two yield stresses are the same. However, in compression, tremendous deformation can occur prior to fracture.
2. Brittle materials generally exhibit larger fracture stresses in compression than in tension (frequently by a factor of 3 or more), and larger fracture strains in compression.

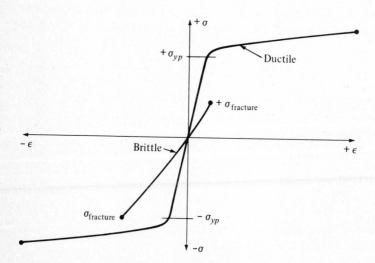

**Figure 3.5** Typical ductile and brittle stress-strain curves showing compressive behavior.

We should note here that the fascinating work of the late Percy Bridgman, Harvard's Nobel prize winner, shows the effect of "hydrostatic" pressure in delaying fracture. Even normally brittle materials like glass can behave like a ductile material when under very high hydrostatic pressure. (P. W. Bridgeman, *Studies in Large Plastic Flow and Fracture*, McGraw-Hill, New York, 1952.)

### 3.2.7 Isotropy

Whenever we characterize a material in terms of the modulus $E$, Poisson's ratio $\nu$, yield stress $\sigma_{yp}$, etc., we imply that these quantities hold in all directions in that piece of material; i.e., we imply that the material is "isotropic." As a first approximation, many engineering materials are isotropic. However, there are always some directional effects due to the processes of manufacture, rolling into long sheets, drawing into long rods, etc.

Composite materials, such as fiber-reinforced plastic (FRP) or steel-reinforced concrete, exhibit *highly* anisotropic behavior. As we will see later, the anisotropic behavior of the composite can frequently be predicted from the relatively isotropic behavior of its constituent materials. Many *living* materials have evolved to a highly anisotropic structure which tends to optimize the overall behavior (bone, wood, etc.).

### 3.2.8 Summary

Because of the numerous quantities discussed in this section, let us list the important properties. For ductile, isotropic materials, we have elastic modulus $E$, Poisson's ratio $\nu$, yield stress $\sigma_{yp}$, and ultimate stress $\sigma_{ult}$. For brittle, isotropic materials, we have elastic modulus $E$, Poisson's ratio $\nu$, tensile fracture stress $+\sigma_f$, and compressive fracture stress $-\sigma_f$.

## 3.3 MATERIAL PROPERTY DATA

Material properties, for a variety of materials, are listed in Appendix 4. These values, taken from a variety of sources, are valid near room temperature. You should be warned that when the temperature of a material exceeds about 40 percent of its *absolute* melting temperature, its properties may change drastically. At these temperatures the thermal motion of the atoms is sufficient to affect the actual deformation mechanisms.

It is quite obvious that at 100 percent of its absolute melting temperature, the material, being liquid, does not correspond very well to our tensile test! Again, properties will vary from specimen to specimen, with temperature, etc. The values listed in Appendix 4 are typical.

Table 3.1 gives an abbreviated set of property data that will be sufficient for your work in this chapter.

**Table 3.1**

| Material (nominal) | Elastic modulus $E$, $10^6$ psi† | Poisson's ratio $\nu$ | Stress limits, 1000 psi† | | Weight density $\gamma$, lb/in$^3$ |
|---|---|---|---|---|---|
| | | | $\sigma_{yp}{}^*$ | $\sigma_{ult}{}^*$ | |
| Structural steel | 30 | 0.27 | 38 | 50 | 0.284 |
| Alloy steel | 29 | 0.3 | 95 | 120 | 0.284 |
| Cast iron | 10 | 0.25 | +25f | −90f | 0.26 |
| Aluminum alloy | 10 | 0.32 | 45 | 70 | 0.10 |
| Brass | 15 | 0.3 | 60 | 74 | 0.31 |
| Nylon | 0.4 | 0.4 | +8f | . . . | 0.04 |
| Wood (fir) | 2 | . . . | +8f | . . . | 0.02 |
| Concrete | 3 | 0.15 | . . . | −5f | 0.09 |
| Glass (silica) | 10 | 0.2 | +10f | . . . | 0.08 |

*Note*: See Appendix 4 for additional material data.

*For brittle materials, stress limit values are fracture stresses: tensile, +xxf, and compressive, −xxf.

† For SI units:

$$E \text{ (in GPa)} = 6.895 \times E \text{ (in } 10^6 \text{ psi)}$$

$$\sigma \text{ (in MPa)} = 6.895 \times \sigma \text{ (in 1000 psi)}$$

## 3.4 AXIALLY LOADED MEMBERS

Given the tensile-test data for any material, we can now predict the elastic behavior of any "axially loaded" member of the same material. By axially loaded we imply a member that is relatively long, nominally straight, and loaded in tension or compression along its length (axis). In Fig. 3.6 a rod or strut of length $L$ and cross-sectional area $A$ is subject to an axial force $F$. Here, we will assume that any nonuniform loading near the ends can be ignored, so that the entire member can be thought of as being equivalent to a tensile specimen. By direct comparison we have:

$$\sigma = \frac{F}{A} \qquad \epsilon = \frac{\delta}{L} \qquad E = \frac{\sigma}{\epsilon} = \frac{F/A}{\delta/L} \qquad (3.7)$$

Thus
$$\delta = \frac{FL}{AE} \qquad \text{if } \frac{F}{A} < \sigma_{yp}$$

If we reverse the force $F$ $(-F)$ and compress the rod, we will observe a compression deformation $(-\delta)$, a shortening of the rod. Because of Poisson's ratio, the rod will grow in diameter in proportion to the change in length. As we all know, if we push too hard on a long rod, it will "buckle." In a later chapter we will discuss determination of the "critical buckling load" for axially loaded members ($F_{crit}$). For now we will simply add buckling as a constraint on

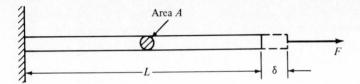

**Figure 3.6** Extension of an elastic member.

the maximum load that can be applied and still remain elastic. In compression:

$$\delta = \frac{FL}{AE} \quad \text{if } \left|\frac{F}{A}\right| < \sigma_{yp} \text{ and if } |F| < F_{crit} \tag{3.8}$$

**Example 3.1** A "truss" is a structure made entirely from axially loaded members, pinned together at their ends. The members do not bend, but can extend or compress.

The simple "truss" shown in the accompanying illustration is made from two tubular steel members having cross-sectional areas of $2\,\text{in}^2$ each. We will assume that the end fittings are properly made "pin joints." Assume that the rod $BC$ will not buckle.

$$E = 30 \times 10^6 \text{ psi} \qquad \sigma_{yp} = 60,000 \text{ psi}$$

Find the motion of point $B$.

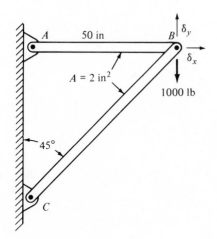

The overall solution strategy is quite straightforward:

1. Find the forces in members $AB$ and $BC$.
2. Determine the "deformations" of $AB$ and $BC$.
3. Find the "motion" or "displacement" of $B$.

Note that we have carefully distinguished between "deformation," and "motion" or "displacement." "Deformation" refers to the change in size or shape of a part when loaded. "Displacement" or "motion" refers to the final location (in a fixed coordinate system) of a specific point on the member.

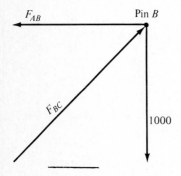

Isolating pin $B$ we find:

$$F_{AB} = +1000 \text{ lb} \quad \text{(tension)}$$

$$F_{BC} = -1414 \text{ lb} \quad \text{(compression)}$$

Now using Eq. (3.7) or (3.8) we compute the deformations in each rod:

$$\delta_{AB} = \frac{1000(50)}{2(30 \times 10^6)} = +8.33 \times 10^{-4} \text{ in} \quad \text{(extension)}$$

$$\delta_{BC} = \frac{-1414(70.71)}{2(30 \times 10^6)}$$

$$:= -16.66 \times 10^{-4} \text{ in} \quad \text{(compression)}$$

Rod $AB$ is now *slightly* longer than it was (0.00167 percent) while $BC$ is slightly shorter. Point $B$ must move such that the ends of $AB$ and $BC$ "fit together." Rod $AB$ (by itself) is free to swing in an arc about point $A$, while $BC$ can rotate about $C$. This effect (greatly exaggerated) is shown in the illustration (*a*, p. 69).

Because we are (generally) dealing with *very small* deformations, we can approximate the circular arcs with straight lines (an excellent approximation). Looking only in the vicinity of $B$, we can draw the approximate displacement diagram (*b*, p. 69) and analyze the appropriate triangles to obtain:

$$\delta_x = \delta_{AB} = 8.33 \times 10^{-4} \quad \text{(to the right)}$$

$$\delta_y = \delta_{AB} + 1.414\delta_{BC} = 31.9 \times 10^{-4} \quad \text{(down)}$$

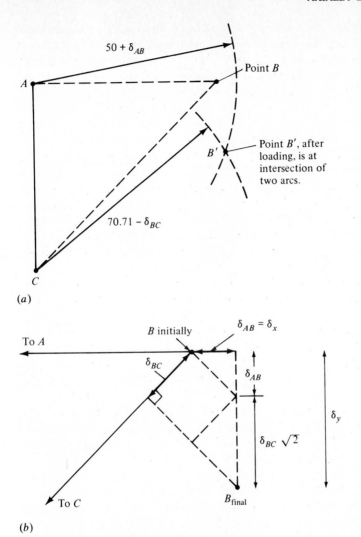

(a)

(b)

$***$

We should note that solution of Example 3.1 required three separate, independent steps:

1. Use of equilibrium to determine forces
2. Use of force-deformation relationships to determine the "deformations"
3. Use of the conditions of "geometric fit" (pieces must fit together) to determine the "displacements"

It is also important to note that in Example 3.1 we have used two approximations that frequently evoke question and concern, especially among students who are trying to understand the subject matter thoroughly. Those approximations were:

We assumed implicitly that the deformation of the truss geometry was sufficiently small that the force calculations (based on undeformed truss angles of 45° and 90°) were correct.

We assumed, again because the deformations were small, that motion along an arc could be replaced by straight-line motion.

These are clearly assumptions, not the reality. Throughout this text (unless otherwise noted) we will make these assumptions. If you are dubious, and if you have some spare time, you can put your calculator to work and re-solve the example in an iterative fashion, i.e., recalculate the forces with point $B$ displaced by $\delta_x$ and $\delta_y$, etc. You will find that the answers we obtained are in error by about 0.002 percent!

In Sec. 3.5, where we deal with "springs" that certainly *could* have large deformations, we will still make the small-deformation approximations.

In a computer example at the end of this chapter we will determine an "exact" solution and see how large the errors are.

In Chap. 4 we will develop computerized "structural-analysis" programs to solve for the forces and deflections in two- and three-dimensional trusses. The programs, which can handle quite complex problems, incorporate the small-deflection assumptions discussed above.

## 3.5 SPRINGS

We have all seen various types of springs in various situations: large "coil springs" and "leaf springs" in the suspensions of automobiles, trucks, and trains; small coil springs in bicycle seats and ball-point pens; springs in clocks, springs in switches, etc.

Springs are used in a great many mechanical devices either to store energy (to be used later), or to provide "compliance" between two parts of a system (a car suspension spring permits the wheel to "comply" with the road imperfections while the vehicle moves in a straighter line). While springs serve many diverse purposes, they are all essentially elastic elements. When properly used, the spring material is never loaded to a stress approaching its yield point.

The behavior of a spring is given in terms of its force-deflection characteristic, just as in a tensile test. Although springs can have characteristics with many different shapes, we will consider only those with linear behavior—i.e., where the force is proportional to the deflection as shown in Fig. 3.7. It is clear that:

$$F = k\delta \tag{3.9}$$

where $k$ = the "spring constant" (lb/in, N/m).

If we think about it a bit, we realize that all structural members of any device act like springs, the only differences being that some are stiffer than others, and some are capable of larger elastic deformation than others.

In Fig. 3.7 we have defined the spring constant $k$ as being force per unit

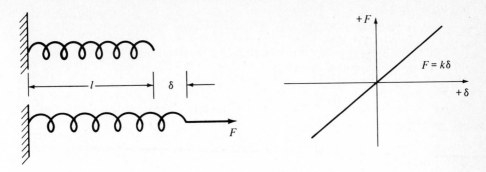

**Figure 3.7** The deflection of a linear spring is proportional to the force acting on it.

deflection; $k$ represents the spring *stiffness*. There are occasions when it is more convenient to talk in terms of deflection/force ($1/k$). This quantity, the inverse of the spring constant or spring stiffness $k$, is called "compliance," $C$:

$$\text{Compliance } C = \frac{1}{k} = \frac{\delta}{F} \quad \text{(in/lb, m/N)} \tag{3.10}$$

So far, we have associated $k$, the spring constant, with the stiffness of a single structural or machine element. Frequently, structural elements and machine parts are connected together in such a way that they undergo a sort of mutual deformation (such as point $B$ in Example 3.1). In such cases it becomes useful to be able to describe and work with the "stiffness" of the *total structure*, i.e., of the assembled structure.

In Chap. 4 we will see how to combine the stiffness characteristics for individual elements into a "stiffness matrix" describing the entire multielement structure. In this chapter we will use the simple spring equation (3.9) and will reserve the matrix formulation for Chap. 4 and later.

**Example 3.2** A rigid (relative to the springs) beam is supported on two springs having $k = 20$ lb/in. A 90-lb load is placed as shown in the illustration below. Assuming that the beam does not move sideways, how far down does point $B$ move?

From equilibrium of the isolated beam shown in the illustration we

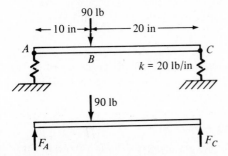

calculate $F_A$ and $F_C$:

$$F_A = 60 \text{ lb} \qquad F_C = 30 \text{ lb}$$

Knowing the force-deformation relationship for a spring [Eq. (3.9)], we can solve for displacements $\delta_A$ and $\delta_C$:

$$\delta_A = \frac{60}{20} = 3 \text{ in} \qquad \delta_C = \frac{30}{20} = 1.5 \text{ in}$$

Finally, knowing that the beam is rigid, points $A$, $B$, and $C$ must lie on a straight line. We can calculate $\delta_B$ from the displacement geometry sketch shown below.

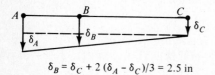

$$\delta_B = \delta_C + 2(\delta_A - \delta_C)/3 = 2.5 \text{ in}$$

We could say that the "effective" spring constant at $B$ (force/deflection) was $90/2.5 = 36$ lb/in.
***

Example 3.2 was statically determinate; we could first solve for the forces, then for the spring deformations, and finally for the desired deflection. This is the simplest type of deflection problem. Consider now an example that is not determinate.

**Example 3.3** The problem shown below looks only slightly different from that given in Example 3.2. We now have three equally spaced springs at $A$, $B$, and $C$, and the load is placed at point $D$. We now ask, "What is the deflection at $D$?"

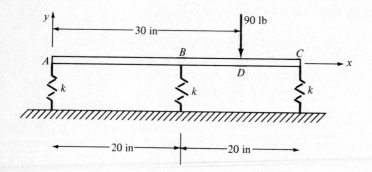

Isolating the beam as shown in the accompanying illustration on top of p. 73, and *assuming* that all springs are compressed, we can write the equilibrium equations:

For $\Sigma F_y = 0$: $\qquad\qquad\qquad\qquad F_A + F_B + F_C = 90$

For $\Sigma M_C = 0$: $\qquad\qquad\qquad\qquad 40F_A + 20F_B = 10(90)$

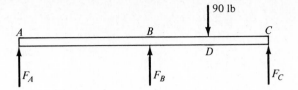

We have three unknowns with only two equations. If we were to write other equilibrium equations, such as $\Sigma M_A = 0$, they would not be independent and would be of no help. Because we cannot solve for the forces directly, we must write the deformation and deflection equation in terms of the unknown forces.

The force-deformation relationships give:

$$\delta_A = \frac{F_A}{k} \qquad \text{(compressive)}$$

$$\delta_B = \frac{F_B}{k} \qquad \text{(compressive)}$$

$$\delta_C = \frac{F_C}{k} \qquad \text{(compressive)}$$

We now have five equations, but we also have six unknowns ($F_A$, $F_B$, $F_C$, $\delta_A$, $\delta_B$, $\delta_C$); still we cannot solve the problem.

The final relationships we can use are those that prescribe one deflection in terms of another in order that *ABDC* lie on a straight line: the conditions of *geometric fit* (see the illustration below). Point *A* has moved to *A'*, *B* to *B'*, etc. There are many ways to write the conditions of geometric fit, some easier to use than others. For instance, because *B* is half-way from *A* to *C*,

$$\delta_B = \frac{\delta_A + \delta_C}{2}$$

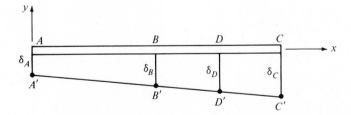

We now have sufficient equations to solve for the forces in each spring and all the pertinent deflections. Sometimes solution of numerous simultaneous equations can be a time-consuming challenge!

For a problem of this type, writing the equilibrium equations in terms of the unknown *displacements*, rather than the unknown *forces*, can frequently expedite matters. Writing the equilibrium and geometric fit equations in an organized fashion, we have:

$$\delta_A + \delta_B + \delta_C = \frac{90}{k} \qquad \left(\sum F_y = 0\right) \tag{1}$$

$$2\delta_A + \delta_B + 0 = \frac{45}{k} \qquad \left(\sum M_C = 0\right) \tag{2}$$

$$\delta_A - 2\delta_B + \delta_C = 0 \qquad \text{(geometric fit)} \tag{3}$$

These can be solved in many standard ways (in Chap. 4 we will introduce the "gaussian elimination" method that is readily programmable on a computer). In this case, we are lucky; if we subtract Eq. (1) from Eq. (3), we can solve for $\delta_B$ directly. Equation (2) then gives $\delta_A$, and either Eq. (1) or (3) gives $\delta_C$:

$$\delta_A = 0.375 \qquad \delta_B = 1.5 \qquad \delta_C = 2.625 \qquad \text{(in)}$$

Finally, we find $\delta_D = 2.062$ in.                                   \*\*\*

In Example 3.3, we saw a typical indeterminate problem. It was necessary to use three distinct steps to develop three sets of relationships, and then organize a method for solving them. If I may quote *The Introduction to the Mechanics of Solids*, edited by S. H. Crandall and N. C. Dahl (McGraw-Hill, New York, 1959):

> We dignify these three steps with an equation number because they are fundamental to all work in the mechanics of deformable bodies . . .
>
>    1. Study of forces and equilibrium requirements.
>    2. Study of deformation and conditions of geometric fit.          (3.11)
>    3. Application of force-deformation relations.

## 3.6 SUPERPOSITION

The solution of many elastic-deformation problems can be simplified by dividing the problem into two or more simpler problems, solving those, and then reassembling the total problem solution.

When we have more than one load applied to a structure, one method for dividing the problem is to consider the effect of each load independently and then add the results. This process, called "superposition," is perfectly valid as long as the deformations are all *linear* elastic and as long as they are *small*. If we apply $F_1, F_2, \ldots, F_n$ loads to a linear elastic structure, it does not matter whether we apply them all at once, in the sequence $F_1, F_2, \ldots, F_n$, or in the reverse sequence $F_n, \ldots, F_1$. The contribution to the deflection at any point, due to a given load, is independent of the loading sequence.

If, on the other hand, the system is not linear elastic, then the deformation can depend on the sequence, and superposition is not valid. Because we deal primarily with linear elastic behavior in this book, superposition can be used in many instances.

There is no "magic method" to determine how or when to use super-position in any given problem. We must use our reasoning powers to determine if there is a way to divide the problem to make it more tractable. Consider a case in point:

**Example 3.4** Let us return to Example 3.3 and see how superposition could help us. In this case, instead of applying multiple loads, we are applying a single force of 90 lb at point $D$. It is applied to a rigid (nondeformable) beam $ABC$. We can replace the 90-lb load with any statically equivalent set of loads without altering the problem (the inverse of replacing a set of loads with their resultant).

In particular, we can replace the single load with a 90-lb load at the central point $B$, together with a 900 ($10 \times 90$) in·lb moment also acting at point $B$ as shown in the accompanying illustration.

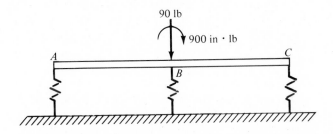

We can now solve the problem, almost by inspection:

Because of symmetry, the 90-lb load will push the beam straight downward without any rotation. The resulting force in each spring will be $90/3 = 30$ lb, and the entire beam moves down $30/20 = 1.5$ in.

Because of the moment alone, the beam rotates about $B$, such that $A$ moves up a distance $\delta_m$ while $C$ moves down $\delta_m$. Moment equilibrium about $B$ gives:

For $\Sigma M_B = 0$: $\qquad 2(20)(k\delta_m) = 900 \qquad \delta_m = 1.125$

Summing the deflections due to the force and the moment we have:

$$\delta_A = 1.5 - 1.125 = 0.375$$

$$\delta_B = 1.5 + 0 = 1.5$$

$$\delta_C = 1.5 + 1.125 = 2.625 \qquad\qquad ***$$

The method of superposition, which we have demonstrated for simple springs, works for any linear elastic structure. Frequently the method will provide no help, but it is always worthwhile to see if we can reduce a complex problem to a series of simpler ones. (The same comment holds for many aspects of life.)

**Example 3.5** As another spring example, consider two springs, of stiffness $k_1$ and $k_2$, connected end to end as shown in the accompanying illustration. In the first case (a), the load $F$ is applied at the top and we wish to know the deflection at the top. In (b), the top and bottom of the assembly are fixed, a load $F$ is applied at the junction between the two springs, and again we want to find the deflection where $F$ is applied.

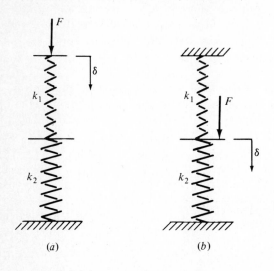

(a)  (b)

Case (a) is statically determinate; both springs carry the same load $F$, and the total deflection is simply the sum of the individual deflections:

$$\delta = \delta_1 + \delta_2 = \frac{F}{k_1} + \frac{F}{k_2} = F\left(\frac{1}{k_1} + \frac{1}{k_2}\right)$$

The assembly has an effective spring constant $k_e$, where

$$k_e = \frac{1}{1/k_1 + 1/k_2}$$

Case (b) is indeterminate and we must use the three steps of Eq. (3.11).

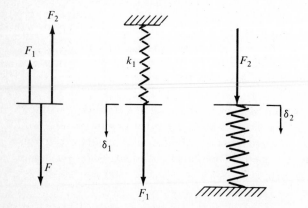

As shown in the illustration, equilibrium gives:

$$F = F_1 + F_2$$

Force/deformation gives:

$$F_1 = k_1\delta_1 \qquad F_2 = k_2\delta_2$$

Geometric fit gives:

$$\delta_1 = \delta_2 = \delta$$

Thus

$$\delta = \frac{F}{k_1 + k_2}$$

and

$$k_e = k_1 + k_2 \qquad\qquad ***$$

## 3.7 COMPLIANCE IN ARBITRARY DIRECTIONS (MOHR'S CIRCLE)

We know that if we pull on a coil spring it will extend in the direction of the force. However, when we dealt with the truss of Example 3.1, we saw that a vertical force applied at point $B$ did *not* produce a straight downward motion, but also produced motion to the right. Sometimes when we apply a force to an elastic system or structure, the deflection at the force will be in the direction of the force, and sometimes it will not.

In general, we will want to apply forces to a structure in *any* direction, and calculate the resulting displacements.

As we proceed through this text, we will find numerous occasions when we are able to compute some quantity relative to one set of $xyz$ axes, but are really interested in that quantity relative to some other set of axes $(x'y'z')$ that are rotated relative to the first set. Let us start here, where we are dealing with simple, easy-to-imagine, spring systems, and see how the structure compliance varies in different directions. In later chapters, we will be able to apply the same analysis, by analogy, to other, less intuitive, quantities.

If we look back at the simple spring shown in Fig. 3.7, it makes no sense to ask about stiffness along any direction other than the horizontal; it has zero stiffness (infinite compliance) in the vertical direction. Any angularity of the applied force $F$ would cause a *large* vertical displacement until the force and the spring were again aligned. We need at least two springs if we are to discuss stiffness and compliance in various directions.

Figure 3.8a shows two springs 1 and 2 having spring constants $k_1$ and $k_2$ oriented along orthogonal axes that we will call the I-II axes as distinguished from the $xy$ axes which have been rotated an angle $+\alpha$ about the $z$ axis.

A force $F_x$ is applied along the $x$ axis. We want to determine the deflections in the $x$ and $y$ directions due to $F_x$. [*Note*: In the usual cartesian coordinate system convention, we use the right-hand rule to determine positive rotational direction. In this case, the thumb of the right hand points in the $+z$ direction, and the fingers indicate positive rotation (counterclockwise).]

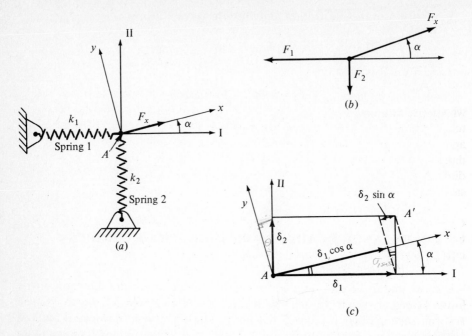

**Figure 3.8** Derivation of deflections along the $xy$ axes: (a) problem definition; (b) force equilibrium; (c) displacement geometry.

The solution is quite straightforward. We can calculate the force components in the I-II directions, then determine the spring deformations, and finally determine the displacement of point $A$.

Figure 3.8b shows the forces acting at point $A$ where $F_1$ and $F_2$ are the tensile forces in springs 1 and 2, respectively. Equilibrium of point $A$ gives:

$$F_1 = F_x \cos \alpha \qquad F_2 = F_x \sin \alpha$$

Knowing $k_1$ and $k_2$, the spring deformations are:

$$\delta_1 = \frac{F_1}{k_1} = \frac{F_x \cos \alpha}{k_1}$$

$$\delta_2 = \frac{F_2}{k_2} = \frac{F_x \sin \alpha}{k_2}$$

Figure 3.8c shows the geometric relationships when point $A$ moves to point $A'$. The motions in the $x$ and $y$ directions are:

$$\delta_x = \delta_1 \cos \alpha + \delta_2 \sin \alpha$$

$$\delta_y = \delta_2 \cos \alpha - \delta_1 \sin \alpha$$

Combining the above equations, we have:

$$\delta_x = \frac{F_x \cos \alpha \cos \alpha}{k_1} + \frac{F_x \sin \alpha \sin \alpha}{k_2}$$

$$\delta_y = \frac{F_x \sin \alpha \cos \alpha}{k_2} - \frac{F_x \sin \alpha \cos \alpha}{k_1}$$

For the special case of $k_1 = k_2$ we find that $\delta_x = F_x/k$ while $\delta_y = 0$. Thus, when we have equal spring constants in two orthogonal directions, the "system" spring constant is the same in every direction. No matter which way we apply a force, we will get the same magnitude deflection, and it will be in the direction of the force. It is obvious that the stiffness should be the same in the I direction and the II direction, but it is not at all obvious that it should be the same in *all* directions.

The above case ($k_1 = k_2$) is a very special case and not at all general. Consider now the case of unequal spring constants, and forces applied along both the $x$ and the $y$ directions ($F_x$ and $F_y$). With two applied forces, equilibrium now gives:

$$F_1 = F_x \cos \alpha - F_y \sin \alpha \qquad F_2 = F_x \sin \alpha + F_y \cos \alpha$$

Combining $F_1$ and $F_2$ with the force-deformation relations and the geometric fit relations, we obtain:

$$\delta_x = \frac{F_x \cos \alpha \cos \alpha}{k_1} - \frac{F_y \sin \alpha \cos \alpha}{k_1} + \frac{F_x \sin \alpha \sin \alpha}{k_2} + \frac{F_y \sin \alpha \cos \alpha}{k_2}$$

$$\delta_y = \frac{F_x \sin \alpha \cos \alpha}{k_2} + \frac{F_y \cos \alpha \cos \alpha}{k_2} - \frac{F_x \sin \alpha \cos \alpha}{k_1} + \frac{F_y \sin \alpha \sin \alpha}{k_1}$$

Each of the above equations has two terms defining the effect of $F_x$ on the deflection, and two terms defining the effect of $F_y$. Combining terms, we have:

$$\delta_x = F_x \left( \frac{\cos \alpha \cos \alpha}{k_1} + \frac{\sin \alpha \sin \alpha}{k_2} \right) - F_y \left( \frac{\sin \alpha \cos \alpha}{k_1} - \frac{\sin \alpha \cos \alpha}{k_2} \right)$$

$$\delta_y = F_x \left( \frac{\sin \alpha \cos \alpha}{k_2} - \frac{\sin \alpha \cos \alpha}{k_1} \right) + F_y \left( \frac{\cos \alpha \cos \alpha}{k_2} + \frac{\sin \alpha \sin \alpha}{k_1} \right)$$

In order to develop a more concise method for calculating the deflections in the $x$ and $y$ directions, we will rewrite the above equations in terms of "compliance" (inverse of $k$) $C_{ij}$, where $C_{ij}$ relates the motion in the $i$ direction to a force in the $j$ direction—that is, $C_{ij} = \delta_i/F_j$:

$$\delta_x = C_{xx}F_x + C_{xy}F_y$$

$$\delta_y = C_{yx}F_x + C_{yy}F_y$$

where $C_{xx} = \delta_x/F_x$, $C_{yy} = \delta_y/F_y$, $C_{xy} = \delta_x/F_y$, $C_{yx} = \delta_y/F_x$, $C_{11} = \delta_1/F_1 = 1/k_1$, and $C_{22} = \delta_2/F_2 = 1/k_2$.

The displacement equations could be written in matrix form:

$$\{\delta\} = [C]\{F\}$$

where $[C]$ is the compliance matrix. The terms of the matrix are:

$$C_{xx} = \quad C_{11}\cos\alpha\cos\alpha + C_{22}\sin\alpha\sin\alpha$$

$$C_{yy} = \quad C_{11}\sin\alpha\sin\alpha + C_{22}\cos\alpha\cos\alpha$$

$$C_{xy} = -C_{11}\sin\alpha\cos\alpha + C_{22}\sin\alpha\cos\alpha$$

$$C_{yx} = \quad C_{xy}$$

Because $C_{xy} = C_{yx}$, we need only three quantities to describe the compliance in any direction. The compliance equations become even simpler when we introduce the following trigonometric identities:

$$\cos\alpha\cos\alpha = 0.5(1 + \cos 2\alpha)$$

$$\sin\alpha\sin\alpha = 0.5(1 - \cos 2\alpha)$$

$$\sin\alpha\cos\alpha = 0.5\sin 2\alpha$$

We now can put the compliance relations into final form:

$$C_{xx} = 0.5(C_{11} + C_{22}) + 0.5(C_{11} - C_{22})\cos 2\alpha$$

$$C_{yy} = 0.5(C_{11} + C_{22}) - 0.5(C_{11} - C_{22})\cos 2\alpha \qquad (3.12)$$

$$C_{xy} = \qquad\qquad - 0.5(C_{11} - C_{22})\sin 2\alpha$$

Equations (3.12) are a set of "transformation" equations that occur repeatedly in applied mechanics. Formally, they are the transformation equations for a second-order "tensor." A very useful characteristic of these equations is that if a plot is made with $C_{ii}$ and $C_{jj}$ plotted horizontally, and $C_{ij}$ plotted vertically, the result is a circle centered on the horizontal axis. The graphical representation is attributed to Otto Mohr (1835–1918) and is called "Mohr's circle."

Figure 3.9 shows the Mohr's circle for compliance, where the "normal compliance" ($C_{11}$, $C_{22}$, $C_{xx}$, $C_{yy}$, $C_{ii}$) is plotted horizontally, and the "cross compliance" ($C_{xy}$, $C_{yx}$, $C_{ij}$) is plotted vertically.

When dealing with transformation problems that can be represented by the Mohr's circle, we gain considerable insight by studying the geometrical relationships on the circle rather than by using the equivalent equations (3.12). Hence, in many problems throughout the rest of this book, we will ask you to draw the pertinent Mohr's circle, and to determine critical quantities from the circle geometry.

There are a number of points to be made regarding Mohr's circle that will aid in the construction and interpretation of them:

1. The center of the circle is *always* on the abscissa (horizontal axis).
2. For Mohr's compliance circles, the entire circle will lie to the right of the origin (but not for other types of Mohr's circles).
3. Because of the "double-angle" relationship, we must rotate twice as far on the circle as in the real coordinate system. For instance, looking back at Fig.

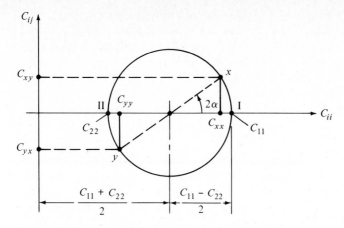

**Figure 3.9** Mohr's circle for elastic compliance.

3.8, to go from the $x$ axis to the I axis we rotate clockwise through an angle $\alpha$. On the Mohr's circle of Fig. 3.9, to go from $x$ (representing the $x$ axis) to I (representing the I axis), we rotate clockwise (the *same* direction) by $2\alpha$ (twice as far).

4. Each point on the circle (I, II, $x$, $y$) represents the normal compliance $C_{ii}$ and the cross compliance $C_{ij}$ along that axis.

5. Any pair of orthogonal axes (I-II, or $xy$), being 90° apart, will plot 180° apart, at opposite ends of a diameter, on the Mohr's circle.

6. Now comes an aspect of Mohr circles that is, at first, confusing for some students: For any set of $xy$ axes, being at opposite ends of a diameter, one axis ($x$ or $y$) must be above the abscissa and the other ($y$ or $x$) must lie below. However, because $C_{xy} = C_{yx}$, the signs of both $C_{xy}$ and $C_{yx}$ must be the *same*. That is, we *cannot* assign + to points above the axis and − to points below; either both are +, or both are −. So we must use a sign convention that is in agreement with Eqs. (3.12) as follows: (*a*) Points to the right of the origin represent (+) normal compliance. (*b*) For cross compliance (assuming a standard $xyz$ set of coordinate axes, and that we are rotating about the $z$ axis): If $C_{xy} = C_{yx}$ is positive, plot $x$ *down*; if $C_{xy} = C_{yx}$ is negative, plot $x$ *up*. As shown in Fig. 3.9, $C_{xy} = C_{yx}$ and is negative.

7. The I-II axes are those for which the cross compliance is zero, and therefore are very special axes. We will call them the "Principal Axes," and will, in general, plot I to the right of II.

*Note*: We make the following statement, but make it without proof:

At any point, in any elastic system, there exists a set of three orthogonal principal axes (I, II, III). Along these axes the cross compliance is zero; i.e., a force along any principal axis will produce displacement *only* along that axis.

The equations we developed [Eqs. (3.12)], and the Mohr's circle representation, hold *only* for rotation about one of the principal axes.

Early in this section we considered the case of two identical springs 1 and 2. We saw then that the total stiffness was the same in every direction. In terms of the Mohr's circle, the case of equal stiffness (and therefore equal compliance) will plot as a *point* (a circle of zero radius). For this case, the compliance is the same in all directions and there is never any cross compliance.

**Example 3.6** For the spring system shown in the accompanying illustration, determine the displacement of point $A$ along the $x'$ and $y'$ directions when the force $F$ of 1000 N is applied.

$$C_{11} = C_{xx} = \frac{1}{k_1} = 10^{-4}$$

$$C_{22} = C_{yy} = \frac{1}{k_2} = 0.5 \times 10^{-4}$$

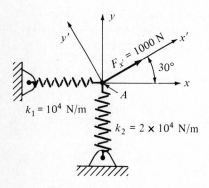

Plot the Mohr's circle as shown below, passing through $C_{11}$ and $C_{22}$. To determine the position of the $x'$ axis, rotate counterclockwise $2 \times 30°$ from the I

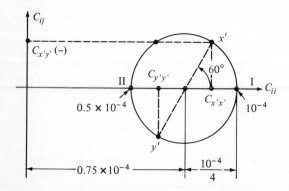

axis. From the geometry of the circle:

$$C_{x'x'} = 0.75 \times 10^{-4} + 0.25 \times 10^{-4} \cos 60 = 0.875 \times 10^{-4}$$

$$|C_{x'y'}| = 0.25 \times 10^{-4} \sin 60 = 0.217 \times 10^{-4}$$

$x'$ is plotted *up*; therefore $C_{x'y'}$ is negative.

Finally, now that we know the compliance and cross compliance, we can compute the deflections due to the 1000-N force:

$$\delta_{x'} = C_{x'x'} F_{x'} = 0.0875 \text{ m}$$

$$\delta_{y'} = C_{y'x'} F_{x'} = -0.0217 \text{ m} \qquad\qquad ***$$

**Example 3.7** For another Mohr's circle example, one which is a bit different, let us consider again the structure described in Example 3.1, which is reproduced below. The problem is to find the magnitude and direction of the principal compliances.

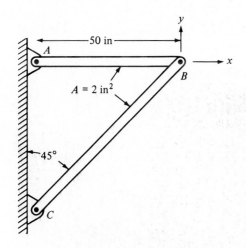

In Example 3.1 we found that when $F_y = -1000$ lb, the deflections were:

$$\delta_x = 0.000833 \text{ in} \qquad \delta_y = -0.00319 \text{ in}$$

The compliances are therefore:

$$C_{yy} = \frac{\delta_y}{F_y} = \frac{-0.00319}{-1000} = +31.9 \times 10^{-7} \text{ in/lb}$$

$$C_{xy} = \frac{\delta_x}{F_y} = \frac{0.000833}{-1000} = -8.33 \times 10^{-7} \text{ in/lb}$$

In order to plot the circle, we also must know $C_{xx}$; i.e., we must calculate the motion in the $x$ direction due to $F_x$. Whenever $F_x$ is applied, only rod $AB$ is loaded so that:

$$\delta_x = \frac{F_x L}{AE} \qquad C_{xx} = \frac{L}{AE} = 8.33 \times 10^{-7}$$

We can now plot the $x$ and $y$ points and draw the circle (see the illustration below). Because $C_{xy}$ is negative, we will plot $x$ upward.

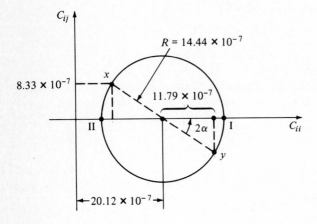

From the geometry of the circle just illustrated,

$$\tan 2\alpha = \frac{8.33}{11.79} = 0.707$$

$$2\alpha = 35.24$$

$$\alpha = 17.62$$

The center of the circle is at

$$\frac{C_{xx} + C_{yy}}{2} = 20.12 \times 10^{-7}$$

The circle radius is $14.44 \times 10^{-7}$; thus the principal compliances are:

$$C_{11} = (20.12 + 14.44) \times 10^{-7} = 34.56 \times 10^{-7}$$

$$C_{22} = (20.12 - 14.44) \times 10^{-7} = 5.68 \times 10^{-7}$$

The principal directions are shown in the final illustration accompanying this example.

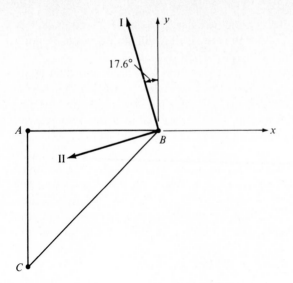

*** 

## 3.8 CYLINDERS AND HOOPS

One of the structural geometries that we often encounter in practice is a circular hoop-type element loaded radially outward as shown in Fig. 3.10. Here we see an internal pressure loading $p$ (force/area) acting on the entire inner surface of the ring (this could arise if we forced the ring over a solid shaft of radius slightly greater than the initial, internal ring radius).

In addition to the situation above, we could have an outward radial *body force* due to acceleration if the ring were rotating about its axis. This would be typical in a flywheel, a grinding wheel, or any rotating cylinder. We could equally well have inward radial loading due to external pressure.

For the geometry shown in Fig. 3.10, where there are no axial loads, the

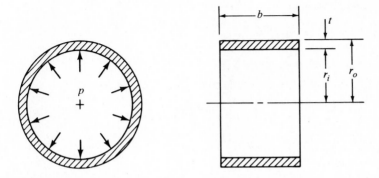

**Figure 3.10** Hoop-type element uniformly loaded with internal pressure $p$ (psi, Pa).

stresses that arise are often called "hoop stresses." Clearly, the hoop of Fig. 3.10 will tend to grow in size such that the entire ring will undergo tensile hoop stresses, $\sigma$.

To calculate the hoop stress $\sigma$, we must isolate a free body and then apply the equilibrium relations. If we isolate the entire hoop, we do not see $\sigma$ because the tangential hoop stresses are all internal. We *must* (in our imagination) cut the hoop in such a way that the forces due to $\sigma$ become *external* forces. There are various sections of the hoop that we could isolate; some are easier to analyze than others.

Perhaps the simplest situation is if we assume that the pressure $p$ is caused by a fluid inside the hoop. Then we isolate one-half of the hoop *including* the fluid within a semicircular portion as shown in Fig. 3.11. Now, when we show the external forces, we see the pressure $p$ pushing upward on the fluid, the hoop tensions $T$, and any radial forces $F_r$ acting on the cut surfaces of the hoop.

Writing the equilibrium relation for the vertical direction, we have:

For $\Sigma F_y = 0$: $\qquad\qquad\qquad 2T = p(2r_i b)$

where $r_i$ is the inner radius and $b$ is the hoop width. The hoop stress $\sigma$, which we wish to calculate, is simply the tension $T$ divided by the hoop cross-sectional area $bt$; thus:

$$\sigma = \frac{T}{bt} = \frac{pr_i b}{bt} = \frac{pr_i}{t} \qquad\qquad (3.13)$$

where $t$ is the wall thickness.

In writing Eq. (3.13), we assumed implicitly that the force $T$ was uniformly distributed over the area $bt$. This is only true when $t \ll r_i$. Hence, Eq. (3.13) holds only for "thin-walled" hoops or tubes where $t/r_i$ is less than perhaps 0.1 or 0.2. When $t \ll r_i$, there is little difference between the inner radius $r_i$, the

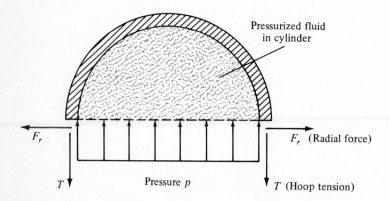

**Figure 3.11** Half of the hoop, with the fluid, is isolated.

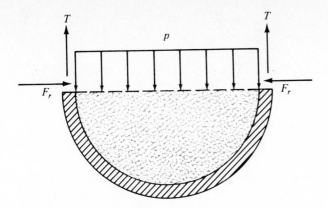

**Figure 3.12** Isolation of lower half of hoop proves $F_r = 0$.

outer radius $r_o$, or the mean radius $r$. Therefore, for thin-walled tubes we can write a very useful, approximate, relation:

$$\sigma \cong \frac{pr}{t} \tag{3.14}$$

where $r$ is any radius.

If now we consider equilibrium in the horizontal direction, we see that $F_t = F_r$; but that does not do us much good. Here we can use a very powerful set of logic based on symmetry.

The complete hoop in Fig. 3.10 is clearly symmetrical about a diameter; so whatever happens to the top half (Fig. 3.11) must also happen to the bottom half. Figure 3.12 shows the lower half with the loading applied; clearly $F_r$ on the bottom part must be the equal and opposite reaction to the $F_r$ acting on the top half. The *only* possibility for $F_r$ to be symmetrical top and bottom, and also equal and opposite top and bottom, is $F_r = 0$! Thus, because of symmetry, there can be no radial forces acting on the cross section.

When the cylinders or hoops become "thick," the stresses become larger at the inner radius $r_i$ than at $r_o$. The complete equations for "thick-walled cylinders" are given in Appendix 3.

**Example 3.8** The steel ring shown in the accompanying illustration on p. 88 is subject to an internal pressure of 500 psi. How large is the hoop stress $\sigma$, and how much does the radius grow because of pressure (assuming elastic behavior)?

From Eq. (3.14), the hoop stress is:

$$\sigma = \frac{pr}{t} = \frac{500(5)}{0.1} = 25,000 \text{ psi}$$

The tangential strain $\epsilon$ is $\sigma/E$. This is the change in circumference divided by

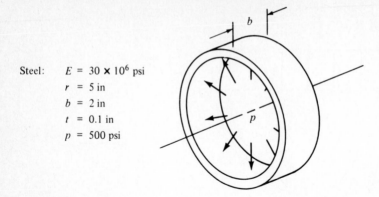

Steel:  $E = 30 \times 10^6$ psi
  $r = 5$ in
  $b = 2$ in
  $t = 0.1$ in
  $p = 500$ psi

the original circumference, and is also equal to the change in radius divided by the original radius. The change in radius $\delta_r$, then, is:

$$\delta_r = \frac{r\sigma}{E} = \frac{5(25,000)}{30 \times 10^6} = 0.00417 \text{ in} \qquad \qquad ***$$

It is important when studying an example such as the one above, where there is a numerical answer, to give thought to the size of the result. In this way we can develop a bit of judgment and predictive ability.

## 3.9 GENERAL EXAMPLES

In order to bring together some of the various aspects of Chaps. 1, 2, and 3, let us look at a few general examples. As we go through them, we will point out the three steps of Eq. (3.11).

**Example 3.9** The support column shown on p. 89 is made from steel tubing which is filled with concrete. When loaded with a force of 100,000 lb, how much does the column compress (assuming elastic compression)?

*Equilibrium requirements*: Let the compressive force on the concrete be $F_c$, and that on the steel tube be $F_s$; if we isolate the top plate, we have:

$$F_c + F_s = 100,000$$

*Deformation and geometric fit*: The shortening of the concrete $\delta_c$, must be equal to that of the steel $\delta_s$:

$$\delta_c = \delta_s = \delta$$

*Force-deformation relations*:

Concrete:
$$\delta_c = \frac{F_c L_c}{A_c E_c}$$

Steel tube:
$$\delta_s = \frac{F_s L_s}{A_s E_s}$$

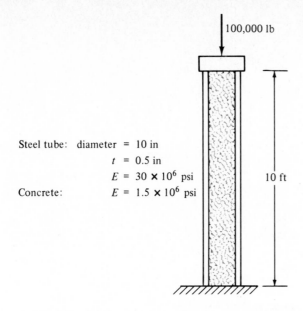

Steel tube: diameter = 10 in

$t$ = 0.5 in

$E$ = 30 × 10$^6$ psi

Concrete: $E$ = 1.5 × 10$^6$ psi

100,000 lb

10 ft

Now we must solve the four simultaneous equations. This example is typical of a class of problems where two or more elements undergo equal deformations, and their individual loads add together. By equating the deformations we obtain a relation among the forces:

$$\frac{F_c}{F_s} = \frac{A_c E_c}{A_s E_s} = 0.213$$

Combining this ratio with the equilibrium equations gives:

$$F_c = 17,570 \qquad F_s = 82,430$$

Finally, solving for either deformation, we have:

$$\delta_c = \delta_s = 0.022 \text{ in}$$

This is an example of a *composite* structure, where two (or more) materials are combined into a single entity. The three steps of Eq. (3.11) allow us to determine the properties of the composite from the properties of the constituent materials. ✱✱✱

**Example 3.10** Three identical springs ($k$) are joined at 120° to each other at point $A$, as shown in the accompanying illustration (top of p. 90). Find the displacement of $A$ when a force $F$ is applied at any angle $\theta$.

First, we should note from symmetry that the system must have the same deflections, along $F$, whenever $F$ is aligned along any of the springs. This means that the system compliance is the same when $\theta = 90°$, 210°, and 330°.

Referring to Mohr's circle, the only way to have three equal normal compliances is if the circle is actually only a point. Therefore the compliance

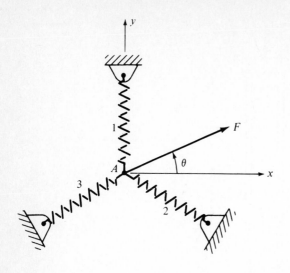

must be the same in all directions; and the displacement must have the same magnitude, and be in the direction of $F$, for all angles $\theta$. Do you see how the Mohr's circle allowed us to draw rather sweeping conclusions from rather meager information?

For simplicity, let us calculate the displacement when $F$ is vertically downward.

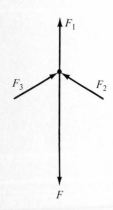

*Equilibrium requirements:*

For $\Sigma F_x = 0$:                    $F_2 = F_3$

For $\Sigma F_y = 0$:                $F_1 + 0.5(F_2 + F_3) = F$

*Deformation and geometric fit:*

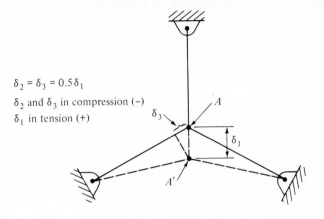

$\delta_2 = \delta_3 = 0.5\delta_1$

$\delta_2$ and $\delta_3$ in compression (−)

$\delta_1$ in tension (+)

*Force-deformation relations:*

$$\delta_1 = \frac{F_1}{k} \qquad \delta_2 = \frac{F_2}{k} \qquad \delta_3 = \frac{F_3}{k}$$

Combining and solving gives:

$$F = k(\delta_1 + \delta_2)$$

$$= k\left(\delta_1 + \frac{\delta_1}{2}\right)$$

$$\delta_1 = \frac{F}{1.5k} \qquad\qquad\qquad ***$$

**Example 3.11** A 10-in-long tapered cylinder, as shown in the accompanying illustration, has a large diameter of 1 in, and a small diameter of 0.5 in. The taper is straight (linear). How much does the cylinder stretch when a load of 1000 lb is applied? Is the deformation equivalent to that for a straight cylinder of 0.75 in diameter? (Steel; $E = 30 \times 10^6$ psi)

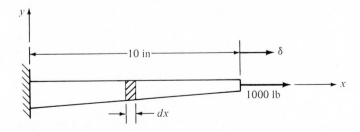

We must note that the stress *varies* from a low value at $x = 0$ to a high value at $x = 10$. Whenever stress varies, we must write our relationships for a *differential* element and use integration to obtain final answers (or use numerical methods).

If we consider an element of length $dx$ at position $x$, we can calculate its contribution ($d\delta$) to the total deformation $\delta$:

$$d\delta = \frac{F\,dx}{AE}$$

But
$$A = \frac{\pi d^2}{4} = \frac{\pi(1 - 0.05x)^2}{4}$$

$$\delta = \frac{4F}{\pi E} \int_0^L \frac{dx}{(1 - 0.05x)^2} = \frac{4F}{\pi E} \left(\frac{-1}{0.05}\right) \int_0^{10} \frac{-0.05\,dx}{(1 - 0.05x)^2}$$

$$= \frac{80F}{\pi E} = 0.00052 \text{ in}$$

For the uniform rod of 0.75 in diameter:

$$\delta = \frac{FL}{AE} = 0.00075 \text{ in}$$

Obviously the two deformations are not equal. What would happen if the area varied linearly from end to end?

\*\*\*

## 3.10 THE HARDNESS TEST

When machine parts and structural members are made from a piece of any given material, we need to know if the material properties of that specific piece are the same as those used in the design calculations. However, the tensile test is a "destructive" test in that the specimen must be destroyed in order to obtain the desired information. We therefore need a "nondestructive" test that will not harm the part, and which is quick and inexpensive.

The "hardness test," which is universally used as a quality control measure, is such a test, and provides information relating to the yield stress of the material. It is often assumed that if the hardness (and thus the yield stress) is within reasonable bounds, then the other properties probably are also.

The hardness test is an "indentation" test: A small, hard, indenter (spherical, or diamond-shaped) is pressed against the material surface to produce a small, permanent indentation. The "hardness' is a measure of the material's resistance to indentation.

There are various commercial hardness-testing machines. In all cases, the applied load and the resulting indentation are determined. In most methods, the load is fixed and the size of the indentation is measured. The actual "hardness number" will vary with the method. For most systems the hardness is simply the applied load divided by the (projected) area of the indentation;

i.e., force/area, or stress. The units used do not fit any standard set; they are in kilograms per square millimeter.

Hardness tests using kilograms per square millimeter are: Brinell, Vickers, and Knoop. When a hardness indentation is made, the material is not free to move sideways as in tension and compression testing. Consequently, the hardness values are greater than the yield stress. As a very good approximation,

$$H \cong 3.1\sigma_{yp} \qquad (3.15)$$

To keep the units straight, you can use the following:

$$H \text{ (in psi)} = 1420H \text{ (in kg/mm}^2)$$

$$H \text{ (in MPa)} = 9.79H \text{ (in kg/mm}^2)$$

The hardness test most used in production is the Rockwell system. Here the numbers are quite arbitrary and can be converted to more fundamental quantities only through empirical relations.

The actual size of the hardness indentation can vary from perhaps 0.03 in across (large) to 0.001 in across (microhardness test).

## 3.11 COMPUTER EXAMPLES

### 3.11.1 Truss Errors

Early in this chapter we noted the errors in the truss calculations due to the assumption of small deformations. Consider here an iterative program which allows us to account for the changes in truss angles. Let us look again at Example 3.1. Shown below are the original shape and the deformed shape under load.

In the program, we calculate the forces in terms of the angles, and then determine the deformed shape. From the deformed shape, we determine new angles and repeat the process. We continue to iterate until the situation stabilizes.

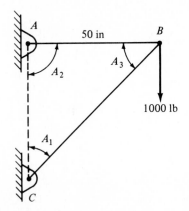
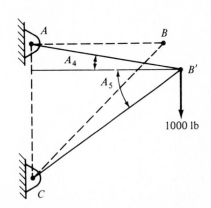

```
10 REM                    EXAMPLE 3.1   (EXACT)
20 REM
30 L=50 \ A=2 \ F=1000
40 FOR I=6 TO 2 STEP -2
50 E=30*10^I \ !\!"E = ",E                    \REM  Vary E, Print E
60 A1=45 \ A2=90 \ A3=45                       \REM  Initial angles
70 C=57.29578                                  \REM  Degrees/rad.
80 !"       Delta-x       Delta-y       Angle-1      Angle-2"
90 %15F6,F*L/(A*E),-(1+2*2^.5)*F*L/(A*E), 45,90,"   (approximate)"
100 L1=L \ L2=L*2^.5 \ L3=L                    \REM  Initial lengths
120 A4=90-A2 \ A5=90-A1                         \REM  Pertinent angles
160 C1=SIN(A4/C)+COS(A4/C)*SIN(A5/C)/COS(A5/C)  \REM  Solve for F1 & F2
170 F1=F/C1                                     \REM  in terms of A4 & A5
180 F2=F1*COS(A4/C)/COS(A5/C)
220 L1=L1*(1+F1/(A*E))                          \REM  New lengths L1 & L2
230 L2=L2*(1-F2/(A*E))
270 S=(L1+L2+L3)/2                              \REM  Triangle geometry
280 R=(S-L1)*(S-L2)*(S-L3)/S \ R=R^.5
290 A9=A2        \REM  (For comparison)
300 A1=2*ATN(R/(S-L1)) \ A1=A1*C                \REM  New angles A1 & A2
310 A2=2*ATN(R/(S-L2)) \ A2=A2*C
350 X=L1*COS((90-A2)/C) -L                      \REM  Deflections x & y
360 Y=-L1*SIN((90-A2)/C)
400 !%15F6,X,Y,A1,A2                            \REM  Print for this iteration
440 IF ABS(A2-A9) > .000001 THEN 100            \REM  Iterate if required
450 NEXT
```

**Program Listing 3.1**

E =  30000000

| Delta-x | Delta-y | Angle-1 | Angle-2 | |
|---|---|---|---|---|
| .000833 | -.003190 | 45.000000 | 90.000000 | (approximate) |
| .000835 | -.003191 | 45.002314 | 89.996344 | |
| .000835 | -.003191 | 45.002314 | 89.996344 | |

E =  300000

| Delta-x | Delta-y | Angle-1 | Angle-2 | |
|---|---|---|---|---|
| .083333 | -.319036 | 45.000000 | 90.000000 | (approximate) |
| .082318 | -.318819 | 45.230387 | 89.635266 | |
| .082458 | -.318404 | 45.230223 | 89.635741 | |
| .082458 | -.318404 | 45.230223 | 89.635741 | |

**Program Output 3.1**

E = 3000

| Delta-x | Delta-y | Angle-1 | Angle-2 | |
|---------|---------|---------|---------|---|
| 8.333333 | -31.903560 | 45.000000 | 90.000000 | (approximate) |
| .135139 | -29.820225 | 68.074847 | 59.255949 | |
| 3.240303 | -27.059464 | 66.689440 | 63.057976 | |
| 2.799645 | -28.401485 | 67.752225 | 61.723724 | |
| 2.952160 | -28.198282 | 67.621826 | 61.963639 | |
| 2.923931 | -28.264887 | 67.672739 | 61.894826 | |
| 2.931864 | -28.252297 | 67.664093 | 61.909001 | |
| 2.930205 | -28.255785 | 67.666689 | 61.905317 | |
| 2.930634 | -28.255052 | 67.666173 | 61.906125 | |
| 2.930542 | -28.255235 | 67.666305 | 61.905930 | |
| 2.930560 | -28.255192 | 67.666276 | 61.905976 | |
| 2.930560 | -28.255198 | 67.666288 | 61.905970 | |
| 2.930560 | -28.255198 | 67.666288 | 61.905970 | |

**Program Output 3.1** (*continued*)

Program Listing 3.1 shows the program for this problem and Program Output 3.1 the output for three different elastic moduli. It should be clear that the errors are indeed small unless the modulus becomes small (as in a rubberlike material).

### 3.11.2 Tapered Rod

An aluminum rod ($E = 10^7$ psi) 20 in long is tapered linearly from a 2 in diameter at one end to a 1 in diameter at the other. The rod is subject to a tensile load of 10,000 lb. Although this problem can be solved analytically, write a program to determine the elastic deformation of the rod.

The appropriate program is given in Program Listing 3.2 and the output in Program Output 3.2. It is often quicker to solve a problem like this numerically than analytically.

```
10 REM            TAPERED ROD PROBLEM
20 REM
30 REM  20" aluminum rod, tapered from R=1" to R=.5"; 10,000 lb load
40 N=100
50 X1=20/N                  \REM  Delta-x
60 FOR I=1 TO N
70 X=X1*(I-.5)              \REM  Position x
80 R=1-.5*X/20             \REM  Radius
90 A=3.14159*R*R           \REM  Area at x
100 D1=10000*X1/(A*10000000)  \REM  Stretch of delta-x (FL/AE)
110 D=D+D1                 \REM  Sum the deflections
120 NEXT
130 !"Deflection = ",D," inches"
```

**Program Listing 3.2**

`Deflection = 1.2732221E-02 inches`

**Program Output 3.2**

## 3.12 PROBLEMS

**3.1** A tensile test was carried out on the low-alloy steel specimen shown below. Prior to "necking," the change in the 2-in gage length was measured; following necking, the minimum diameter was measured. The force-deformation data shown below were recorded. Plot the stress-strain curve. Determine $E$, $\sigma_{yp}$, $\sigma_{ult}$, and $\sigma_{fract}$.

(See Prob. 3.27 for the computation of strains during necking.)

| $F$, 1000 lb | $\Delta L$, in | $D_{min}$, in |
|---|---|---|
| 2.00 | 0.0007 | |
| 4.00 | 0.0015 | |
| 6.00 | 0.0021 | |
| 8.00 | 0.0027 | |
| 10.00 | 0.0035 | |
| 10.13 | 0.0100 | |
| 11.08 | 0.100 | |
| 12.13 | 0.2 | (Necking) |
| 10.38 | $\cdots$ | 0.430 |
| 9.91 | $\cdots$ | 0.400 |
| 9.48 | Fracture | 0.372 |

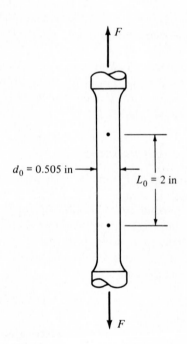

**Figure P3.1**

**3.2** A 40-ft length of steel rod is hung vertically, suspended from the top. How much does it stretch due to its own weight under the following conditions?

(*a*) When the rod section is constant along the length

(*b*) When the rod tapers linearly from diameter $D$ at the top to $D/3$ at the bottom

**3.3** A 500-lb weight is to be lifted from the ocean depths as shown in the accompanying illustration. When the slack is taken out of the cable, there are just 10,020 ft of cable off the winch drum. What length of cable must be wound onto the drum before the weight

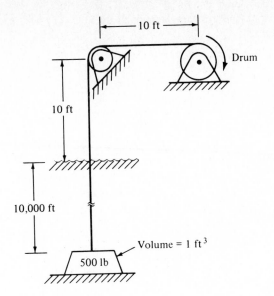

Figure P3.3

leaves the bottom? The cable has a $\frac{1}{4}$ in diameter; a 1-ft length of cable weighs about 0.1 lb, and has a stiffness of 60,000 lb/in.

**3.4** A "belt," made from brass shim stock, goes over a 10-in-diameter pulley and supports a 100-lb weight as shown in the accompanying illustration. Initially, the pulley is free to rotate. Then a moment $M_0$ is applied to the pulley causing it to rotate against the belt friction.

  (*a*) How large is $M_0$?
  (*b*) How far down does the weight move when $M_0$ is applied?

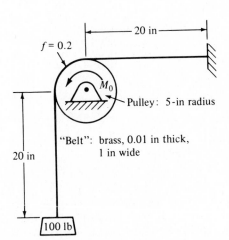

Figure P3.4

**3.5** A frame, made from two members pinned together, is loaded with 500 lb at point $B$. Find the $x$ and $y$ displacements of point $B$.

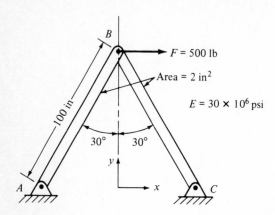

**Figure P3.5**

**3.6** Solve Prob. 3.5 for the case where member $BC$ has been strengthened by doubling its area to $4\,\text{in}^2$.

**3.7** Five tubular steel members are pinned together and support a 2-kN load as shown in the accompanying illustration. For all members, the steel tubing area $= 400\,\text{mm}^2$. When the load is applied, what is the motion of:
 (a) Point $C$
 (b) Point $B$
 (c) Point $D$

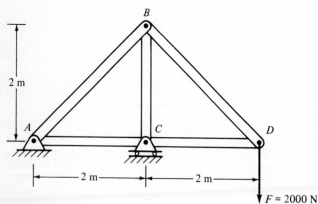

**Figure P3.7**

**3.8** Five wooden members are fabricated into a truss as shown below. Assume that the joints act as pin joints, and that there is no buckling. For all members, $A = 16\,\text{in}^2$, $E = 1.5 \times 10^6\,\text{psi}$. When the 500-lb load is applied, what is the motion of:
 (a) Point $D$
 (b) Point $C$

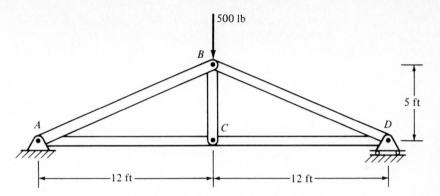

**Figure P3.8**

**3.9** A typical automotive valve assembly is shown in the accompanying illustration. For the spring, $k = 200$ lb/in and its free length $= 3.75$ in. What is the minimum force $F$, acting on the pushrod, that can open the valve?

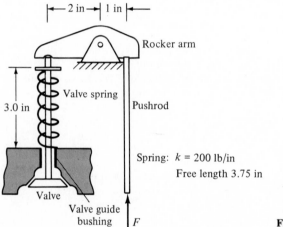

**Figure P3.9**

**3.10** A relatively rigid bar is supported horizontally by two springs (see the illustration below). Where should a force $F$ be placed $(a/L)$ such that the bar moves downward without rotating?

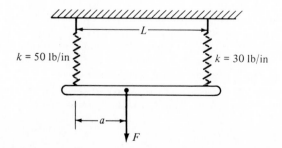

**Figure P3.10**

**3.11** A weighing platform is supported through two levers and a spring as shown below. When a load $F$ or $F'$ is applied, the spring is compressed $\delta$. Determine $\delta$ when $F$ is applied and when $F'$ is applied.

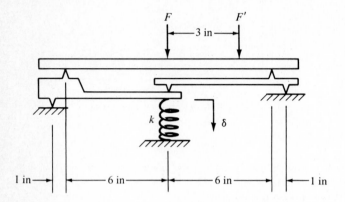

**Figure P3.11**

**3.12** A pontoon bridge (supports floating in the water) can be modeled as a series of bridge sections supported by springs of stiffness $k$. The spring force is due to the water being displaced by the float. In the instance illustrated below, the floats are all 4 ft square and 1 ft high. As a 200-lb man walks across the bridge, how far does point $B$ move downward?

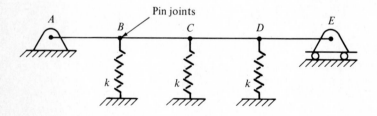

**Figure P3.12**

**3.13** A motor drives the chain sprocket at $B$ in the accompanying illustration at 600 rpm. A chain belt connects the drive sprocket to the driven sprocket ($C$) which is connected to a load (winch). An "idler" sprocket at $A$ can move horizontally compressing the spring $k$, but cannot move vertically. When the motor is stopped, and the chain is slack, point $A$ is 2 in to the left of the line tangent to the two sprockets ($\delta_0 = 2$ in). When the machine runs, the tight chain forces the idler to the right. By measuring $\delta$ we can determine power. When $\delta = 1.5$ in, how much power is being transmitted?

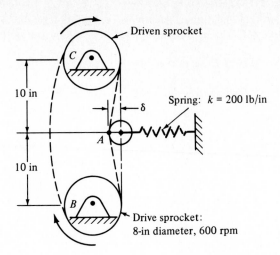

Figure P3.13

**3.14** A "torsional spring" provides a moment, or torque, proportional to the angle of rotation between its ends (as in a clock spring):

$$M_t = k_\phi \phi \qquad k_\phi = \text{in·lb/rad, N·m/rad}$$

Two members of length $L$ are pinned together as shown below and joined by a torsional spring at $B$. Initially, $A$, $B$, and $C$ are collinear; when $F$ is applied, how far does $B$ move?

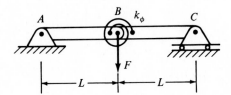

Figure P3.14

**3.15** A "cantilever beam" of length $L$ is modeled as four rigid sections, pinned together and connected with torsional springs as shown below. When $F$ is applied, how far down does it move? (See Prob. 3.14.)

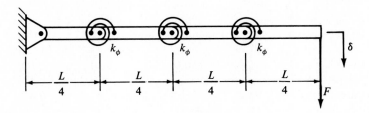

Figure P3.15

**3.16** A rigid member $AB$ is pinned at $A$ and held in a perfectly vertical position by the spring $k$ (when $F = 0$). A downward load $F$ is applied at a small eccentricity ($e$) as shown ($e \ll L$).

(a) Determine the equilibrium position of $B$ (as denoted by $\delta$) in terms of $F$, $k$, $L$, and $e$.

(b) What happens as $F$ is continually increased (even when $e$ is *very* small)?

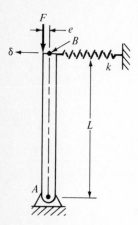

**Figure P3.16**

**3.17** A "toggle" mechanism is made from two 10-in members pinned together. A spring is attached to their midpoints such that when $F = 0$, the members are at 30° as shown in the accompanying illustration. Make a plot of the force (vertical axis) vs. the height of point $B$ as the 30° angle is forced to zero (use large-angle relationships).

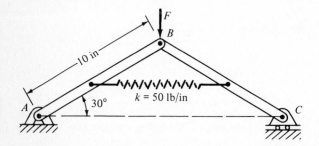

**Figure P3.17**

**3.18** Four 10-in rigid members are pinned together and held apart by the spring shown in Fig. P3.18 (p. 103). When $F$ is applied, how far down does it move?

**3.19** A rigid member is supported by three springs, as shown in Fig. P3.19 (p. 105). When $F = 0$, the member $ABC$ is horizontal. When $F$ is applied, determine the deflections of $A$, $B$, and $C$.

**3.20** Two $\frac{1}{8}$-in-thick steel supports are joined by a bolt inside a sleeve separator, as shown in Fig. P3.20 (p. 103). During assembly, after the nut is "snug," or "handtight," wrenches are used to turn the nut an additional one-quarter turn. See Prob. 2.12 for screw thread definitions (ignore deformation of the $\frac{1}{8}$-in strips).

(a) What are the forces and stresses on the bolt and sleeve?

(b) How far ($\delta$) is the sleeve compressed?

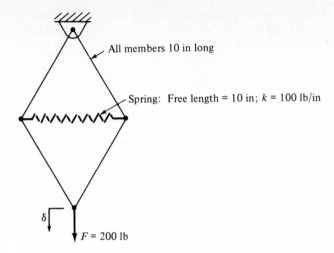

All members 10 in long

Spring: Free length = 10 in; $k$ = 100 lb/in

$\delta$

$F$ = 200 lb

**Figure P3.18**

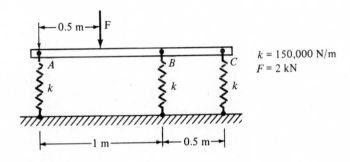

$k$ = 150,000 N/m
$F$ = 2 kN

**Figure P3.19**

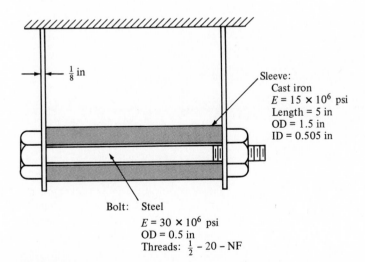

Sleeve:
Cast iron
$E = 15 \times 10^6$ psi
Length = 5 in
OD = 1.5 in
ID = 0.505 in

Bolt: Steel
$E = 30 \times 10^6$ psi
OD = 0.5 in
Threads: $\frac{1}{2}$ – 20 – NF

**Figure P3.20**

**3.21** A composite support is made from a 1-in-diameter steel rod within a 2-in-diameter aluminum tube, as shown below. If initially the two members are just the same length, what portion of the 50,000 lb does each carry?

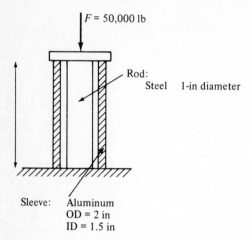

$F$ = 50,000 lb

Rod:
Steel    1-in diameter

Sleeve:    Aluminum
OD = 2 in
ID = 1.5 in

**Figure P3.21**

**3.22** Solve Prob. 3.21 for the case where initially the aluminum tube is found to be 0.004 in shorter than the steel rod.

**3.23** The drive rod on a very large reciprocating compressor is essentially a large bolt and sleeve assembly, as shown in the accompanying illustration. Prior to use, a hydraulic jack device is used to stretch the bolt elastically while the nut is tightened. This provides a "preload" between the parts. If the preload is adjusted to give a *prestress* of 10,000 psi to the bolt (when $F = 0$), how large must $F$ be to reduce the force in the sleeve to zero?

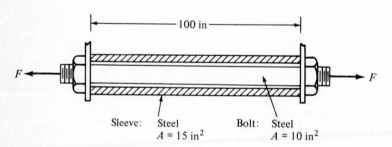

$\longmapsto$——— 100 in ———$\longmapsto$

$F$

$F$

Sleeve:    Steel
$A = 15 \text{ in}^2$

Bolt:    Steel
$A = 10 \text{ in}^2$

**Figure P3.23**

**3.24** Two springs are joined at point $A$, as shown below. Draw the Mohr's circle for the compliance of point $A$. When $\alpha = 30°$, and $F = 50$ lb, how large is the displacement of $A$ in the direction of $F$?

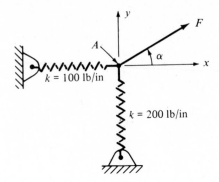

**Figure P3.24**

**3.25** Four members are pinned together as shown below. For all members, $A = 4\,\text{in}^2$, $E = 30 \times 10^6$ psi. At what angle(s) $\alpha$ can we apply $F$ such that point $C$ moves in the direction of $F$?

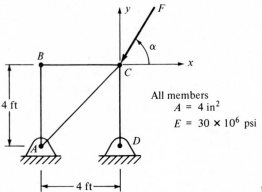

All members
$A = 4\,\text{in}^2$
$E = 30 \times 10^6$ psi

**Figure P3.25**

**3.26** Two members are pinned together as shown below. For both members, the material is steel tubing and $A = 3\,\text{in}^2$. Determine the magnitude and direction of the principal compliances at $B$.

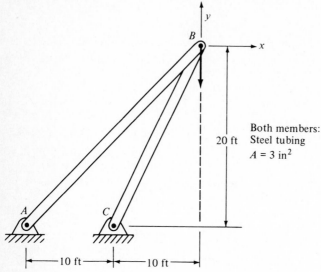

Both members:
20 ft    Steel tubing
        $A = 3\,\text{in}^2$

|← 10 ft →|← 10 ft →|

**Figure P3.26**

**3.27** When a tensile specimen necks, the maximum strain occurs at the minimum diameter. During any strain increment, if we consider an extremely short specimen length $L$ as shown below, the contribution to the total strain $\epsilon$ is

$$d\epsilon = \frac{dL}{L}$$

The total strain is the integral of $dL/L$ as $L$ goes from the initial value $L_0$ to the final value $L_f$; that is,

$$\epsilon = \ln (L_f) - \ln (L_0) = \ln \frac{L_f}{L_0}$$

During plastic deformation, the volume of the small element shown remains essentially constant so that the ratio of lengths $(L)$ can be replaced by the ratio of diameters. Show that

$$\epsilon = 2 \ln \frac{D_0}{D_f}$$

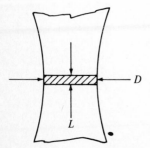

**Figure P3.27**

**3.28** Referring to Fig. 3.4, show that when a member is loaded in tension or compression the energy input, per unit volume of material ($u$), equals the area under the $\sigma$-$\epsilon$ curve. Show that for a linear elastic material:

$$u = \frac{\sigma^2}{2E}$$

**3.29** The Hoover Dam is 726 ft high, and can deliver about 1350 megawatts (MW). The "penstocks," steel tubes that deliver the water to the turbines, are approximately 550 to 590 ft below the water level in Lake Mead. These tubes are 30 ft in diameter with walls about 2 in thick. How large is the tensile hoop stress in the steel?

If you visit Hoover Dam, take the trip to the bottom and walk along those penstocks.

**3.30** Write a computer program to solve Prob. 3.2. If you use 100 elements, how large is the error?

**3.31** Write a program to solve Prob. 3.8 taking into account the changes in angles.

**3.22** Write a program to give the deflection delta ($\delta$) for a range of loads $F$ for Prob. 3.16.

**3.33** Use the computer to solve Prob. 3.17.

# FOUR

## MATRIX STRUCTURAL ANALYSIS (AXIAL LOADING)

### 4.1 OVERVIEW

In Chap. 3, we saw that the solutions of problems in applied mechanics generally involve three sets of relationships:

1. The equations of equilibrium
2. Relations between deformation and geometric fit
3. Force-deformation relationships

If a structure, such as a truss, is statically determinate, we can determine the forces in the various members using only the equilibrium equations. Then we can proceed to find the deformations and thus the deflections. This is relatively straightforward.

However, even with a determinate problem, if the system has many parts, and we want to calculate deflections, the solution can be quite difficult and time-consuming.

But, if the structure is indeterminate, the process of solution can be formidable.

A variety of methods have been developed to treat complex problems in mechanics. Classical "energy methods," which considerably simplify the work, are usually presented in texts such as this. However, even using energy methods, the work involved in solving for deflections in a truss having five or six members is substantial (and the potential for error is high).

During the past two decades, as computers have become ever more powerful and available, a great effort has gone into the development of computer-based methods for solving complex, previously unsolvable, problems in applied mechanics. Many of these methods can be grouped under the general heading of "finite-element methods" (FEM), methods which deal with *finite* elements as opposed to the *infinitesimal* elements treated through calculus. In general, when using finite-element methods, the pertinent equations are solved at a discrete number of points rather than everywhere. Thus the solutions are generally approximate; the greater the number of points, the better the approximation.

Finite-element programs are available to treat not only the analysis of elastic structures and components, but also problems in structural dynamics, heat transfer, fluid flow, and the like.

In Appendix 2, there is an introduction to energy methods as they relate to applied mechanics. Energy methods are fundamental to advanced work in the field, and, indeed, form the basis of many finite-element developments. However, as the availability of computer hardware and appropriate software becomes greater and greater, it is reasonable to expect that the practitioner of engineering mechanics will rely more on computer methods and less on energy methods.

Accordingly, in the body of this text, we will bypass the energy methods and go directly to very simple applications of the finite-element method as applied to trusses and space frames.

Clearly, in an elementary mechanics text we are not going to plumb the depths of finite-element analysis; however, we will introduce you to the *method*, and will develop simple yet powerful computer programs. You will be able to solve problems in minutes, on a personal computer, that would otherwise require days or months.

The subset of finite-element analysis that we will consider could formally be called "matrix structural analyses using the displacement method." Our "elements" will be complete struts or beams, and we will obtain (within the limits of the theory) exact solutions. The large body of finite-element work deals with continua which are divided into small, but finite, elements, and results in approximate solutions.

Although we will be dealing with a particular subset of finite-element analysis, we will be using the finite-element method, and will label our developments "FEM."

In this chapter we will consider structures made entirely from axially loaded, pin-jointed members. These are our familiar "truss" problems. We will be able to handle both determinate and indeterminate problems with equal ease. We will also be able to treat cases where some of the members are not the proper length to fit together easily, and also problems where some of the members are either heated or cooled subsequent to assembly.

We will start here with the simple two-dimensional truss structures, and will present the major steps in the FEM. These will form a framework which will accommodate more complex structural analysis and finite-element problems. We will then extend the truss-type analysis to three dimensions.

Later, after we treat the *bending* of beams, we will extend the FEM to include both axial loading and bending loads in two dimensions. These structures are usually called "frames," as opposed to trusses where the loading is only axial.

Finally, after we have considered *torsion*, we will be able to consider "space frames"—three-dimensional structures involving axial loads, bending, and twisting.

The solution of problems via FEM clearly must use precisely the same relationships that we have been using: equilibrium, deformation and geometric fit, and the force-deformation relations. The power of FEM lies in the organization of the problem statement, and methods of equation solution, that are amenable to computer automation. We know that if we had to work a problem with perhaps 25 unknowns, we could (hopefully) assemble 25 equations and (in due time) solve them. The organization inherent in the FEM allows this to happen automatically.

As we develop the FEM, we will see that there are four general steps:

1. Define the structure.
2. Assemble the appropriate set of equilibrium and deformation equations.
3. Solve the resulting set of simultaneous equations.
4. Manipulate the results to give the desired output.

It is often easiest, when developing a general theory or procedure, to consider it in light of a simple example. Hence, although we are aiming at

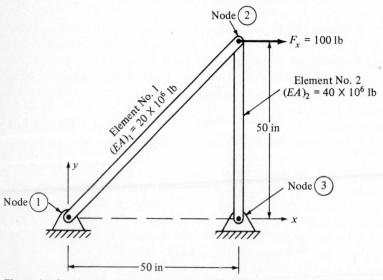

**Figure 4.1** Simple truss for the structural analysis.

eventually being able to handle complex trusses, we will use the simple example shown in Fig. 4.1 for discussion. As we proceed, we will assemble the necessary data in a manner that will readily allow it to be input to a computer.

## 4.2 DEFINING THE STRUCTURE

Any truss is composed of a number of *elements* that we will number $i = 1, 2, 3, \ldots, I$. For FEM purposes, the element number $i$ will identify the individual elements. For each element, we must (obviously) state the product $EA$ which represents the stiffness per unit length of the element. Implicit in the above is the requirement that the element be uniform along its length (non-uniform elements can be treated if replaced with "equivalent" uniform struts). Our example has two elements, No. 1 having $EA = 20 \times 10^6$ lb, and No. 2 having $EA = 40 \times 10^6$ lb.

The elements in a truss are joined to each other, and to the outside world by pin joints at their ends. We will call these connecting points "nodes," and will identify them by numbers $j = 1, 2, \ldots, J$. In diagrams we will circle the node numbers to distinguish them from element numbers.

We can now describe the topography of the truss by giving the coordinates of each node. If we then state which elements lie between which nodes, we have assembled the structure.

We could assemble the data for the example as follows:

Number of *elements* = 2

Number of *nodes* = 3

| Element No. | Located between nodes | | $EA \times 10^{-6}$, lb |
|:---:|:---:|:---:|:---:|
| 1 | 1 | 2 | 20 |
| 2 | 2 | 3 | 40 |

| Node | Coordinates | |
|:---:|:---:|:---:|
| | $x$ | $y$ |
| 1 | 0 | 0 |
| 2 | 50 | 50 |
| 3 | 50 | 0 |

If we had a truss with many elements and nodes, the above lists would simply be longer. These data describe the physical structure itself—its location, size, shape, and stiffness. What are *not* yet represented are the effects of the outside world. We have not told how the structure is supported, or how it is loaded (we will leave the effects of misfit and temperature till later).

In this particular FEM, the effects of the outside world, through support constraints and loads, can be specified *only* at nodes. Thus, at *any* node we can define externally applied loads, *or* externally imposed displacement constraints.

Specification of applied loads is very simple: At each node we input the *x* and *y* components of the externally applied force. If the force is zero or unknown, we input 0.

If the node is supported from the outside, the force will be unknown, and again we will input 0. For this example, the only "independent," externally applied load is 100 lb, in the +*x* direction, at node 2.

In order to define the displacement constraints in a way that the computer can understand, we will use the concept (which you have seen before) of degrees of freedom (dof). You know that, in general, a moving point or mass particle has 3 degrees of freedom, typically in the *x*, *y*, and *z* directions. That is, three numbers will completely define the position of that point. A body at a point will have up to 6 dof: positions along the *x*, *y*, and *z* axes, and rotation about them.

For our FEM analyses, we will define the possible dof at each node, and then state which dof are actually "active," and which are constrained not to move. We will see that this defines the support structure.

For two-dimensional truss problems, a node can move only in the *xy* plane; hence each node has just 2 dof. Either a node *can* move in the *x* direction (active) or it *cannot* (inactive); and likewise for the *y* direction. We will code the dof activity as follows:

Inactive dof: (zero motion, zero freedom) is labeled 0

Active dof: (motion permitted) is labeled 1

Consider our example: At node 1, a fixed support, there can be *no* motion in either the *x* or *y* directions; therefore, both dof are inactive (0). Both dof at node 3 are also inactive (0). However, at node 2 there is no external constraint, so both dof are active (1).

If we recall our earlier discussions regarding support forces and constraints (Chap. 2), we will remember that we could specify *either* a force along a dof *or* a displacement along that dof, not both. Here, if you specify zero displacement along a dof (inactive, 0), then it makes no sense to also specify an external load along that same dof. The force would simply be passed to the foundation and would not affect any of the members. Thus, along inactive dof, specify zero (unknown) force.

For our example, we can now include information on dof activity and external loading in the following table of nodal data.

| Node | Deg. of freedom | | Applied forces | | Coordinates | |
|------|------|------|------|------|------|------|
|      | $x$ | $y$ | $F_x$ | $F_y$ | $x$ | $y$ |
| 1 | 0 | 0 | 0 | 0 | 0 | 0 |
| 2 | 1 | 1 | 100 | 0 | 50 | 50 |
| 3 | 0 | 0 | 0 | 0 | 50 | 0 |

The data in the table, plus the element data already discussed, completely define the "truss problem." In fact, given a set of coded data, you should be able to work backwards and reconstruct the problem! Given a structural analysis program, the only intellectual work involves inputting the node and element data; then the computer will take over and do all the work!

Different types of supports will, of course, have different dof activity. Figure 4.2 shows the activity coding for the three most common nodes.

It might seem that the supports shown in Fig. 4.2 are not sufficiently general to allow us to define some types of problems. Usually, if you put your

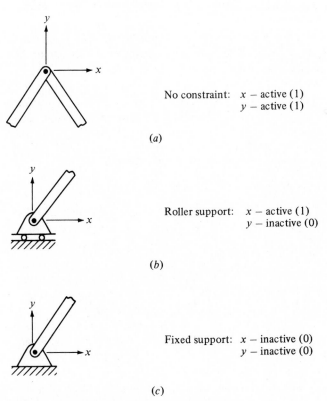

No constraint: $x$ – active (1)
$\phantom{No constraint:}$ $y$ – active (1)

(a)

Roller support: $x$ – active (1)
$\phantom{Roller support:}$ $y$ – inactive (0)

(b)

Fixed support: $x$ – inactive (0)
$\phantom{Fixed support:}$ $y$ – inactive (0)

(c)

**Figure 4.2** "Activity" coding for truss pin joints.

ingenuity to work, you can define, or at least approximate, other support systems. Figure 4.3 shows two such conditions.

Figure 4.3a shows a roller support at some angle to the $xy$ axes. The joint is free to move tangent to the surface, but not normal to it. We can easily approximate that condition by adding an extra element at an appropriate angle as shown. If the element is sufficiently stiff, the joint will only move in a direction perpendicular to the new element, and compression of the element will be negligible.

Figure 4.3b shows the case of imposed displacement ($\delta$) in the $y$ direction and free movement in the $x$ direction. Again we add an extra, stiff element, but now we load it sufficiently to cause the desired deflection $\delta$. If the new element is sufficiently stiff, the added force $F$ will be large compared with other forces in the system, and the support node will move almost as desired.

It has been said that "nature abhors a vacuum"; by the same token, structural analysis programs abhor an active dof where the motion is not limited at all. For instance, consider a vertical member, pinned at the top, and supporting a weight at the bottom. The bottom of the member can, in fact, move in both the $x$ and $y$ directions; hence we would normally consider both dof to be active. However, not only can the node move in the $x$ direction, but the motion is completely unconstrained in that the smallest force will produce a displacement. That is (for small angles), the horizontal stiffness is zero. A zero-stiffness term of this type will produce a singularity in the formalized solution procedure and must be avoided. In this instance we could simply declare the $x$ dof to be inactive; in other cases, an extra, stabilizing element can

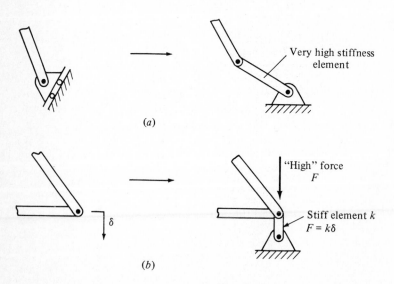

Figure 4.3 Support constraints approximated by addition of relatively stiff elements.

be added. Of course, the programs can be written to identify such singularities.

At this stage we have seen how to define completely a problem in numerical code. We leave it to you, the reader, to consider Example 4.1. Given the data tables, you should be able to draw the diagram and vice versa.

**Example 4.1** The data listed below are precisely as they would be entered on the terminal of a computer:

Number of *elements* = 5

Number of *nodes* = 4

| Node | Deg. of freedom | | Applied forces | | Coordinates | |
|---|---|---|---|---|---|---|
| | $x$ | $y$ | $F_x$ | $F_y$ | $x$ | $y$ |
| 1 | 0 | 0 | 0 | 0 | 0 | 0 |
| 2 | 1 | 1 | 100 | 0 | 50 | 70 |
| 3 | 1 | 1 | 0 | −200 | 100 | 0 |
| 4 | 1 | 0 | 0 | 0 | 50 | 0 |

| Element No. | Located between nodes | | $EA \times 10^{-6}$, lb |
|---|---|---|---|
| 1 | 1 | 2 | 40 |
| 2 | 2 | 3 | 40 |
| 3 | 2 | 4 | 40 |
| 4 | 1 | 4 | 60 |
| 5 | 3 | 4 | 60 |

From the coded data given, you should be able to construct the structure with supports and loading as shown in the accompanying illustration (see page 116).

\*\*\*

At this point we have learned nothing new; we have simply established a coding system which, in essence, allows us to describe a free-body diagram in simple numerical terms. These coded data are sufficient for solution of a problem, and they are all that would be input to a FEM program.

Before we proceed to establish the inner workings of the FEM, we might look again at Example 4.1. This is basically a very simple problem; it is statically determinate, so we could find all the forces readily; we could easily calculate the deformation of each element. However, it would take some mental exercise to calculate the displacement of node 3. How long would it take you?

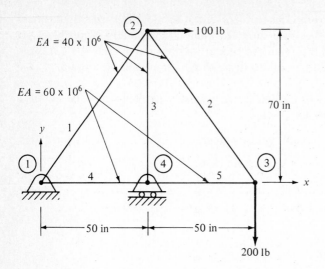

By typing the data shown into a microcomputer containing a FEM program, we can determine all forces, and all displacements in a couple of seconds of computing time! Perhaps this will give you courage to press on!

## 4.3 THE DISPLACEMENT METHOD

Now let us delve into the intellectually challenging aspects of the FEM.

We know that in the solution of any truss problem:

1. Equilibrium must be satisfied at each node.
2. Each node can move only in the directions of active degrees of freedom.
3. The force in each element depends on its stiffness and on the relative motion of its two ends.

A nicely organized method that puts all of the above together, a method ideally suited to computers, is the "displacement method."

Basically, this method involves the use of nodal displacements, rather than forces, as the primary variables in the analysis. We can determine the total force acting on each element in terms of its stiffness and the motions of its end nodes. We can then introduce these element forces into equilibrium relations for the nodes. This provides a set of "equilibrium equations" where the unknowns are nodal displacements rather than element forces. We must then solve that set of equations.

It would seem at the outset that (for the case of a two-dimensional truss) the number of simultaneous equilibrium equations would equal twice the number of nodes $I$ ($\Sigma F_x = 0$, $\Sigma F_y = 0$ at each node). However, only *active* dof can move. The displacement of an inactive dof is zero and need not be solved for. Hence, the number of equilibrium equations equals the number of active dof, which we will number $m = 1, 2, \ldots, M$.

For the example we are considering, there are three nodes with a total of 6 dof. However, only 2 dof are active ($M = 2$), so we need only $M = 2$ equations. Thus, the initial problem is reduced in complexity from solving for a large number of force components (six) to solving a smaller number of displacements (two).

Let us note again that if we specify a displacement constraint, we cannot also specify an external load along that dof; the load will be a dependent variable to be calculated. Also, if we specify an external load, the displacement along that dof is the dependent variable. Accordingly, we can specify only $M$ external loads. In our example, the most we can specify are the two load components at node 2.

We see then that we should be able to write $M$ equilibrium equations in terms of the $M$ unknown nodal displacements and the $M$ externally applied loads. Let us see how we can organize the writing of this set of equations.

> Let: $u_m$ = displacement along the $m$th dof
>
> $F_m$ = externally applied force along the $m$th dof

At any node, the externally applied forces must be carried by the elements which meet at that node. Thus we can say:

At any node: The sum of the forces acting *on* the elements at that node must equal the sum of the externally applied forces at that node (vector sums).

Thus, for each of the $M$ active dof, in the direction of that dof, we can equate the externally applied load (if any) to the loads acting on the elements. The loads acting on each element, in turn, depend on its stiffness and the motions of its ends.

So for our example, we can write an equilibrium equation for each of the ($M = 2$) active dof in the structure:

$$\left. \begin{array}{l} k_{11}u_1 + k_{12}u_2 = F_1 \\[2mm] k_{21}u_1 + k_{22}u_2 = F_2 \end{array} \right\} \quad M \text{ equilibrium equations}$$

The coefficients ($k_{ij}$) can be thought of as "stiffness coefficients" giving the influence of motion along one active dof on the force along that (or another) active dof:

$$k_{ii} = \frac{\text{force acting along dof } i}{\text{unit displacement of dof } i}$$

$$\text{or } k_{ij} = \frac{\partial F_i}{\partial u_j}$$

$$k_{ij} = \frac{\text{force acting along dof } i}{\text{unit displacement of dof } j}$$

The values of the stiffness coefficients obviously depend upon the stiffness of the elements at that node, and how they are geometrically arranged (we will soon see how to determine them).

For the general case, written in matrix notation, the equations of equilibrium are:

$$[k]\{u\} = \{F\} \tag{4.1}$$

The matrix $[k]$, giving the overall force-displacement characteristics of the structure is called a "stiffness matrix," and because it describes the whole assembly, is called the "global stiffness matrix." All diagonal terms must be real and positive; other terms can often be zero.

James Clerk Maxwell (Scottish, 1831–1879) was the first to show that for any linear elastic structure, $k_{ij} = k_{ji}$, i.e., stiffness matrices are symmetrical.

You should realize that once we have established the global stiffness matrix, we have, in effect, established the equilibrium equations.

The general method for establishing $[k]$ is to consider each element of the structure by itself, in turn, and:

1. Develop a "local" stiffness matrix $[k_1]$ relating forces and motions at the two ends of the element.
2. Place appropriate values from the local stiffness matrix $[k_1]$ into the global matrix $[k]$.

### 4.3.1 Local Stiffness Matrix $[k_1]$

Consider the general two-dimensional truss element shown in Fig. 4.4. It has two nodes, 1 and 2. It has four possible dof which are numbered 1, 2, 3, and 4. The element is of length $L$, stiffness $AE$, and is at an angle $\alpha_1$ to the $x$ axis (and at the complementary angle $\alpha_2$ from the $y$ axis). The local stiffness matrix $[k_1]$

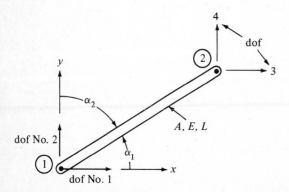

**Fig. 4.4** Geometry for calculating the local stiffness matrix $[k_1]$.

that we would like to establish is:

$$[k_1] = \begin{bmatrix} k_{11} & k_{12} & k_{13} & k_{14} \\ k_{21} & k_{22} & k_{23} & k_{24} \\ k_{31} & k_{32} & k_{33} & k_{34} \\ k_{41} & k_{42} & k_{43} & k_{44} \end{bmatrix}$$

where the subscripts 1 to 4 are the numbers of the dof.

Consider the calculation of two terms, $k_{11}$ and $k_{21}$; these represent the forces along dof 1 and 2 due to a unit displacement along dof 1. Figure 4.5 shows the geometry. A unit displacement along dof 1 produces a compression of the strut where $\delta = 1 \cos \alpha_1 = \cos \alpha_1$. To produce this compression, there must be an axial load $F = AE\delta/L = AE(\cos \alpha_1)/L$. The component of $F$ along dof 1 is $F_1 = F \cos \alpha_1$. Thus:

$$k_{11} = \frac{AE}{L} \cos \alpha_1 \cos \alpha_1$$

The force component along dof 2 gives us:

$$k_{21} = \frac{AE}{L} \cos \alpha_1 \sin \alpha_1$$

$$= \frac{AE}{L} \cos \alpha_1 \cos \alpha_2$$

If we calculate all terms, being careful with sines, signs, and cosines, we have:

| dof | 1 | 2 | 3 | 4 |
|---|---|---|---|---|
| 1 | $C_1C_1$ | $C_1C_2$ | $-C_1C_1$ | $-C_1C_2$ |
| 2 | $C_2C_1$ | $C_2C_2$ | $-C_2C_1$ | $-C_2C_2$ |
| 3 | $-C_1C_1$ | $-C_1C_2$ | $C_1C_1$ | $C_1C_2$ |
| 4 | $-C_2C_1$ | $-C_2C_2$ | $C_2C_1$ | $C_2C_2$ |

$$[k_1] = \frac{AE}{L} \times$$

where $C_1 = \cos \alpha_1$ and $C_2 = \cos \alpha_2 = \sin \alpha_1$.

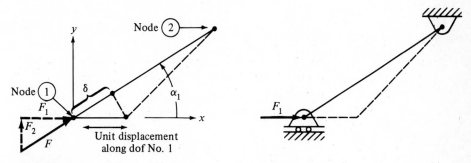

**Figure 4.5** Calculation of $k_{11}$ and $k_{21}$.

We see that all of the above terms are readily calculated from the data that describe the system. For our example, the local stiffness matrices for the two elements are:

$$[k_1]_1 = 282{,}885 \begin{bmatrix} 0.5 & 0.5 & -0.5 & -0.5 \\ 0.5 & 0.5 & -0.5 & -0.5 \\ -0.5 & -0.5 & 0.5 & 0.5 \\ -0.5 & -0.5 & 0.5 & 0.5 \end{bmatrix}$$

$$[k_1]_2 = 800{,}000 \begin{bmatrix} 0 & 0 & 0 & 0 \\ 0 & 1 & 0 & -1 \\ 0 & 0 & 0 & 0 \\ 0 & -1 & 0 & 1 \end{bmatrix}$$

### 4.3.2 Global Stiffness Matrix

We now must assemble the global matrix $[k]$ from the $(I)$ local stiffness matrices $[k_1]_i$ associated with the $(I)$ elements. For our example, $M = 2$; therefore, the global matrix is only $2 \times 2$. Only the *active* dof are included in $[k]$. So we must sum the various local stiffness coefficients that act at each active dof. This is equivalent to summing the forces that act along each active dof.

For the example, both active dof are at global node 2 (Fig. 4.1). The matrix we need is:

$$[k] = \begin{bmatrix} k_{11} & k_{12} \\ k_{21} & k_{22} \end{bmatrix}$$

where 1 and 2 represent the active dof at node 2 (1 = x dof, 2 = y dof).

At this point it is very easy to confuse the local dof numbering and the global dof numbering. Fundamentally, we find which local-element dof corresponds with a global dof, and add the local coefficient to the global stiffness matrix.

For some it is useful to consider the global matrix to be three-dimensional, consisting of $I$, $M \times M$ layers, one for each element. On each layer, we include only the terms from the local matrix that correspond to the active dof. At each of the $M \times M$ locations we then sum the contributions due to the $I$ layers.

First, consider element 1, and find which local dof corresponds with one of the global dof:

Local dof 1 is inactive (ignore)
Local dof 2 is inactive (ignore)
Local dof 3 corresponds to global dof 1
Local dof 4 corresponds to global dof 2

So a local matrix term with subscript 33 is a part of the global matrix term having subscript 11, and must be added to the global 11 term; that is, for element 1, the corresponding matrix terms are:

Local 33 corresponds to global 11
Local 44 corresponds to global 22
Local 34 corresponds to global 12

Thus the four terms within the dashed lines in $[k_1]_1$ are put into $[k]$.

For element 2 it just happens (for this particular example) that the local/global correspondence is exactly the same as for element 1:

Local dof 3 corresponds to global dof 1
Local dof 4 corresponds to global dof 2

The four terms within the dashed lines in $[k_1]_2$ are added to $[k]$. Finally, with the terms added,

$$[k] = \begin{bmatrix} 141,443 & 141,443 \\ 141,443 & 941,443 \end{bmatrix}$$

Putting it all together, we can write the equilibrium equations as:

$$\begin{bmatrix} 141,443 & 141,443 \\ 141,443 & 941,443 \end{bmatrix} \begin{Bmatrix} u_1 \\ u_2 \end{Bmatrix} = \begin{Bmatrix} 100 \\ 0 \end{Bmatrix}$$

In expanded form, this becomes:

$$141,443 u_1 + 141,443 u_2 = 100$$

$$141,443 u_1 + 941,443 u_2 = 0$$

which are readily solved to give:

$$u_1 = 0.000833 \text{ in}$$

$$u_2 = -0.000125 \text{ in}$$

The structural analysis programs that we will use will calculate the $(I)$ local stiffness matrices, put appropriate terms into the global matrix, and then store local stiffness data for future use.

## 4.4 SOLVING THE EQUILIBRIUM EQUATIONS

Having established the global stiffness matrix $[k]$, we can now proceed to solve for the $M$ unknown displacements $u$. In our example, where $M = 2$, the solution was trivial. When $M > 5$, the work, if done without computer help, can be considerable, and by the time $M = 10$, we might well give up! Using computer solutions, however, $M$ can be as large as desired, limited only by computer capability and cost. In some applications, $M$ may exceed 10,000 (not recommended for a small machine)!

For truss-type problems the $M \times M$ stiffness matrix may be almost completely filled with nonzero values. Hence, we simply store $[k]$ in an $M \times M$ array. For finite-element analyses involving solid bodies, where the elements tend to be small polyhedra and $M$ becomes very large, the stiffness matrix is

$$
\begin{bmatrix} x & x & x & x \\ x & x & x & x \\ x & x & x & x \\ x & x & x & x \end{bmatrix}
\rightarrow
\begin{bmatrix} x & x & x & x \\ 0 & x & x & x \\ 0 & 0 & x & x \\ 0 & 0 & 0 & x \end{bmatrix}
$$

**Figure 4.6** Using gaussian elimination, the square global stiffness matrix is made triangular.

often very sparsely filled, with values in a narrow "band" near the diagonal. For such cases only the banded values are stored for efficient use of high-speed memory.

There are various methods for solving a set of $M$ simultaneous equations. The method currently most used in FEM programs is the "gaussian elimination" method (named for Karl Friedrich Gauss, German, 1777–1855). This technique, eminently suited to computer manipulation, involves:

1. Altering the equations to reduce $[k]$ to a triangular matrix as shown in Fig. 4.6
2. Starting at the bottom row (Fig. 4.6), where there is only one coefficient, solve for the $M$th $u$; then go to the next row upward and solve for that displacement, etc.

**Example 4.2** As an example of the gaussian elimination method, consider the arbitrary set of four equations shown below:

$$[k]\{u\} = \{F\}$$

$$
\begin{bmatrix} 2 & 3 & 6 & 2 \\ 3 & 2 & 4 & 1 \\ 6 & 4 & 5 & 0 \\ 2 & 1 & 0 & 3 \end{bmatrix}
\begin{Bmatrix} u_1 \\ u_2 \\ u_3 \\ u_4 \end{Bmatrix}
=
\begin{Bmatrix} 16 \\ 0 \\ 12 \\ 0 \end{Bmatrix}
$$

The value of a set of simultaneous equations is not changed by subtracting a multiple of one row from another row. Therefore, execute the following:

1. Subtract $\frac{3}{2}$ of row 1 from row 2 to make $k_{21} = 0$.
2. Subtract 3 of row 1 from row 3 to make $k_{31} = 0$.
3. Subtract 1 of row 1 from row 4 to make $k_{41} = 0$.

We now have:

$$
\begin{bmatrix} 2 & 3 & 6 & 2 \\ 0 & -2.5 & -5 & -2 \\ 0 & -5 & -13 & -6 \\ 0 & -2 & -6 & 1 \end{bmatrix}
\begin{Bmatrix} u_1 \\ u_2 \\ u_3 \\ u_4 \end{Bmatrix}
=
\begin{Bmatrix} 16 \\ -24 \\ -36 \\ -16 \end{Bmatrix}
$$

Subtract multiples of row 2 from rows 3 and 4 to make $k_{32}$ and $k_{42}$ both zero.

This gives:

$$\begin{bmatrix} 2 & 3 & 6 & 2 \\ 0 & -2.5 & -5 & -2 \\ 0 & 0 & -3 & -2 \\ 0 & 0 & -2 & +2.6 \end{bmatrix} \begin{Bmatrix} u_1 \\ u_2 \\ u_3 \\ u_4 \end{Bmatrix} = \begin{Bmatrix} 16 \\ -24 \\ +12 \\ +3.2 \end{Bmatrix}$$

Finally subtract two-thirds of the third row from the fourth giving the triangular matrix in the equations:

$$\begin{bmatrix} 2 & 3 & 6 & 2 \\ 0 & -2.5 & -5 & -2 \\ 0 & 0 & -3 & -2 \\ 0 & 0 & 0 & 3.933 \end{bmatrix} \begin{Bmatrix} u_1 \\ u_2 \\ u_3 \\ u_4 \end{Bmatrix} = \begin{Bmatrix} 16 \\ -24 \\ +12 \\ -4.8 \end{Bmatrix}$$

From the bottom row we find $\quad u_4 = -1.22$

then from row 3 $\qquad\qquad\qquad u_3 = -3.19$

then from row 2 $\qquad\qquad\qquad u_2 = +16.96$

and from row 1 $\qquad\qquad\qquad u_1 = -6.65$ $\qquad\qquad\qquad$ **\*\*\***

The above process is readily translated into a recursive computer program that rapidly solves for the displacements $u_m$. (See the structural analysis program listings at the end of this chapter.)

If the displacements are all that we require, the problem is solved. Generally, however, we also want to know the forces acting on each element so that we can compute stresses, etc. We discuss force determination next.

## 4.5 DETERMINING THE ELEMENT FORCES

Clearly, if we know the motions of every node, we can compute the change in length of each element, and thus the forces that must be acting on that element.

In order to compute the global stiffness matrix $[k]$, we have already had to calculate the $(I)$ local stiffness matrices $[k_1]_i$. We can write the equilibrium equations for each element as:

$$\{F_1\} = [k_1]\{u_1\}$$

Referring to Fig. 4.4, we can solve for the four forces $(F_1, F_2, F_3, F_4)$ along the four dof for each element. Note that the only nonzero displacements are along the active nodes.

Returning to our example, for element 1:

$$\begin{Bmatrix} F_1 \\ F_2 \\ F_3 \\ F_4 \end{Bmatrix} = 282{,}885 \begin{bmatrix} 0.5 & 0.5 & -0.5 & -0.5 \\ 0.5 & 0.5 & -0.5 & -0.5 \\ -0.5 & -0.5 & 0.5 & 0.5 \\ -0.5 & -0.5 & 0.5 & 0.5 \end{bmatrix} \begin{Bmatrix} 0 \\ 0 \\ 0.000833 \\ -0.000125 \end{Bmatrix}$$

These evaluate to (as shown in the accompanying illustration):

$F_1 = -100$

$F_2 = -100$

$F_3 = +100$

$F_4 = +100$

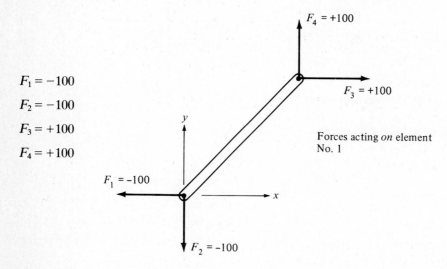

Forces acting *on* element No. 1

It is left to the reader to show that on element 2, the forces along the four dof are:

$$F_1 = 0$$

$$F_2 = 100$$

$$F_3 = 0$$

$$F_4 = -100$$

## 4.6 TEMPERATURE AND DISPLACEMENT INITIAL CONDITIONS

In the foregoing analyses we have made the following implicit assumptions:

1. Initially the structure "fits together" perfectly so that there can be no initial internal loads due to improper fit.
2. The temperatures of all elements remain constant.

For many real problems, these assumptions are not justified, and we would like to include such effects in our analysis. Fortunately, it is very easy to do so;

let us reconsider the basic structural equilibrium equation (4.1), and apply it to any individual element:

$$\{F_1\} = [k_1]\{u_1\} \tag{4.2}$$

This equation states that the forces $\{F_1\}$, acting *on* an element at its nodes, equals the local element stiffness matrix $[k_1]$ times the nodal displacements $\{u_1\}$. Clearly, when the displacements $\{u_1\} = 0$, then $\{F_1\} = 0$. This is in agreement with the assumptions listed above; initially there are no displacements and no forces.

Suppose, however, that a particular element was too long to fit between its assigned nodal positions; we could *make* it fit by applying an initial force $F_0$ just large enough to compress it to the proper length. We would have to add the proper components of any such initial force to Eq. (4.2), thus:

$$\{F_1\} = [k_1]\{u_1\} + \{F_{1_0}\} \tag{4.3}$$

When we assemble the $M$ equilibrium equations for the total structure, the appropriate initial forces (if any) must be added. The equations to be solved, then, are:

$$[k]\{u\} = \{F\} - \{F_0\} \tag{4.4}$$

The initial forces $F_0$ could be due to the elements actually being too long by amounts $\delta_0$, or could be due to thermal expansion by an amount equivalent to $\delta_0$. The magnitude of $F_0$ along any element axis, then, is:

$$F_0 = \frac{AE}{L}(\delta_0 + \alpha L \, \Delta T) \tag{4.5}$$

where $\alpha$ is the coefficient of thermal expansion and $\Delta T$ is the temperature increase.

For the finite-element program, we can introduce $F_0$ by either (*a*) calculating $F_0$ ourselves, and subtracting the proper components of $F_0$ from the externally applied forces, or (*b*) stating $\delta_0$ and $\alpha \, \Delta T$ for each element and letting our friendly computer take care of the necessary manipulation. Which would you prefer? Clearly, the chances for error are much smaller if we choose (*b*). Accordingly, when you run the FEM programs, you will be asked to input both $\delta_0$ and $\alpha \, \Delta T$ for each element. For many cases they are simply zero.

## 4.7 THREE-DIMENSIONAL TRUSS-TYPE STRUCTURES

While the two-dimensional analysis is useful for a great many problems, the world is three-dimensional, and many structures must be studied in a full three-dimensional fashion.

What is the difference between our two-dimensional FEM program and one for three-dimensional structures? Really not very much. The essential changes should be clear to anyone who has followed along to this point:

1. Each node now has 3 dof that can be either active (1) or inactive (0).
2. Each node now has 3 coordinates $(x, y, z)$.
3. We can now apply forces in the three coordinate directions at each node.
4. Each local stiffness matrix $[k_1]$ is now $6 \times 6$ rather than $4 \times 4$.

All else is essentially the same.

The only problem you might encounter in writing a three-dimensional FEM is in establishing the local stiffness matrix $[k_1]$.

Figure 4.7 shows a general three-dimensional element. Its direction is given in terms of the three angles shown $(\alpha_1, \alpha_2, \alpha_3)$. The stiffness matrix coefficients now involve the three direction cosines:

$$C_1 = \cos \alpha_1 = \frac{x}{L}$$

$$C_2 = \cos \alpha_2 = \frac{y}{L}$$

$$C_3 = \cos \alpha_3 = \frac{z}{L}$$

The matrix is:

$$[k_1] = \frac{AE}{L} \begin{bmatrix} C_1C_1 & C_1C_2 & C_1C_3 & -C_1C_1 & -C_1C_2 & -C_1C_3 \\ C_2C_1 & C_2C_2 & C_2C_3 & -C_2C_1 & -C_2C_2 & -C_2C_3 \\ C_3C_1 & C_3C_2 & C_3C_3 & -C_3C_1 & -C_3C_2 & -C_3C_3 \\ -C_1C_1 & -C_1C_2 & -C_1C_3 & C_1C_1 & C_1C_2 & C_1C_3 \\ -C_2C_1 & -C_2C_2 & -C_2C_3 & C_2C_1 & C_2C_2 & C_2C_3 \\ -C_3C_1 & -C_3C_2 & -C_3C_3 & C_3C_1 & C_3C_2 & C_3C_3 \end{bmatrix}$$

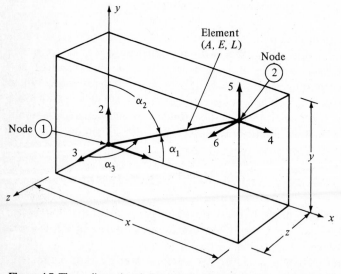

**Figure 4.7** Three-dimensional element showing the 6 dof and defining the three angles $\alpha_1, \alpha_2, \alpha_3$.

Looking at the upper left-hand $3 \times 3$ matrix within the dashed lines of $[k_1]$, we see that:

$$[k_1]_{ij} = \frac{AE}{L} C_i C_j$$

where $i$ and $j = 1$, 2, and 3. The same grouping is repeated in the other three corners with the signs as shown.

We see, then, that it will be very simple for a computer to calculate the local stiffness matrices. The remainder of the problem is the same as before.

## 4.8 SUMMARY

As an introduction to computer-based finite-element methods for structural analysis, we have studied two- and three-dimensional truss-type structures made of *elements* pinned together at *nodes*.

We defined the structure by giving coordinates of all nodes, stating which element lies between which nodes, and giving the stiffness of each element.

The possible degrees of freedom were defined for each node, and we stated which dof were free to move (active, 1), and which were constrained by the structure supports not to move (inactive, 0).

Elements which were too long or short to fit together properly were handled by simply stating how much too long (or short) they were. Temperature changes, of one element relative to the others, after assembly, are part of the initial data.

Having established a coding method for organizing all the input data, we then considered the steps involved in solving a problem:

1. Establishing local stiffness matrices $[k_1]_i$, and then assembling the global stiffness matrix $[k]$
2. Establishing the external loading vector $\{F\}$ and the initial forces due to misfit and temperature $\{F_0\}$
3. By use of the gaussian elimination method, solving for the displacements $\{u\}$ of all active degrees of freedom
4. With knowledge of the nodal displacements $\{u\}$, and the local stiffness matrices $[k_1]_i$, solving for the forces acting on each element $\{F_1\}_i$

Later we will extend this analysis to include both bending and twisting. The method of solution will not change one iota; the number of degrees of freedom will increase, and the stiffness matrices will have more terms, but the method will be precisely the same.

Because this entire chapter is computer-related, we will not have a separate section on computer examples.

## 4.9 STRUCTURAL ANALYSIS PROGRAM LISTINGS

```
20  !"                    STRUCTURAL ANALYSIS"
30  !"               (finite element-method)"            \!
60  !"               2-DIMENSIONAL STRUCTURES"
70  !"                   Axial loading only"\!
90  REM
100 !" ***********************************************************************"\!\!
110 REM
120 REM
130 REM   1.  DEFINE THE STRUCTURE
140 REM       a. Number of elements and nodes
150 REM       b. Input element data
160 REM       c. Input node data
170 REM       d. Input element data
180 REM       e. Determine and number the active degrees of freedom (1,2,--M)
190 REM       f. Dimension arrays
200 REM       g. Establish the applied force (moment) vector {f}
210               GOSUB 780
220 REM
230 REM
240 REM   2.  ESTABLISH THE GLOBAL STIFFNESS MATRIX  (k)
250 REM        For each element,
260 REM       a. Determine element constants and store in array
270 REM       b. Establish local stiffness matrix (k1)
280 REM       c. Put appropriate terms into global matrix (k)
290               GOSUB 1590
300 REM
310 REM
320 REM   3.  ESTABLISH THE INITIAL FORCE VECTOR  {f0}
330 REM        For each element,
340 REM       a. Calculate initial forces due to temperature or misfit
350 REM          and put into vector {f0} and final element force
360 REM          array E1
370 REM       b. Subtract initial forces from the applied forces
380 REM          {f}={f}-{f0}
390               GOSUB 1960
400 REM
410 REM
420 REM   4.  SOLVE FOR DISPLACEMENTS {u} BY GAUSSIAN ELIMINATION
430 REM          (k)*{u}={f}
440 REM       a. Triangularize stiffness matrix {k}
450 REM       b. Calculate displacements {u}
460 REM       c. Put displacements into nodal array N1
470               GOSUB 2230
```

**Program Listing 4.1**

```
480 REM
490 REM
500 REM    5.  DETERMINE FORCES ON EACH ELEMENT  (at both nodes)
510 REM        For each element\
520 REM          a. Determine nodes N1 AND N2
530 REM          b. Establish local displacement vector {u1}
540 REM          c. Establish local stiffness matrix (k1)
550 REM          d. Calculate forces due to displacements {f1}
560 REM                {f1}=(k1)*{u1}
570 REM          e. Add to initial forces in array E1
580                     GOSUB 2610
590 REM
600 REM
610 REM    6.  PRINT OUT RESULTS
620 REM          a. Nodal motion
630 REM          b. Forces on each element at both ends
640                     GOSUB 2950
650 REM
660 REM
670 REM    7.  FURTHER CALCULATIONS
680 REM          a. Other loadings for this structure?
690                     GOTO 3320
700 REM
710 REM
720 REM *****************************************************************************
730 REM
740 REM    1.  DEFINE THE STRUCTURE
750 REM
760 REM          **** a. Number of elements and nodes ****
770 REM
780 !"             Input data to define the problem"\!\!
790 INPUT"                Number of ELEMENTS = ",I9      \REM I=ELEMENTS
800 INPUT"                Number of NODES    = ",J9      \REM J=NODES
810 !\!
820 REM          **** b. Dimension arrays ****
830 REM
840 DIM N(J9,6)     \REM NODAL INPUT DATA
850 DIM N1(J9,2)    \REM NODAL DISPLACEMENTS FOR OUTPUT
860 DIM E(I9,5)     \REM ELEMENT INPUT DATA
870 DIM E1(I9,6)    \REM ELEMENT FORCES FOR OUTPUT
880 DIM K0(I9,6)    \REM ELEMENT CONSTANTS
890 DIM M(6)        \REM GLOBAL DOF NUMBERS FOR LOCAL ELEMENT DOF
900 DIM K1(4,4)     \REM LOCAL ELEMENT STIFFNESS MATRIX
910 DIM F1(4)       \REM LOCAL FORCE VECTOR
920 DIM U1(4)       \REM LOCAL DISPLACEMENT VECTOR
```

**Program Listing 4.1** (*continued*)

```
930 REM
940 REM            #### c. Input nodal data ####
950 REM
960 !"Input data for each NODE:"
970 !"    For degrees of freedom, 'active'=1; 'nonactive'=0"\!\!
980 !"         Deg of free.      Applied forces        Coordinates"
990 !"Node     x     y        Fx         Fy          x        y"\!
1000 FOR J=1 TO J9
1010 !J,\ !TAB(11), \ INPUT1" ", N(J,1) \ !TAB(16), \ INPUT1" ",N(J,2)
1020 !TAB(27), \ INPUT1" ",N(J,3) \ !TAB(39), \ INPUT1" ", N(J,4)
1030 !TAB(51), \ INPUT1" ", N(J,5) \ !TAB(62), \ INPUT1" ", N(J,6)
1040 INPUT"    OK? (CR/N)",Z$
1050 IF Z$="N" THEN 1010
1060 NEXT \ !\!
1070 REM
1080 REM            #### d. Input element data
1090 REM
1100 !"Input data to define ELEMENT location, stiffness, and initial conditions"
1110 !"     Note multipliers for E#A"
1120 !\!
1130 !"Element    Located     E#A        Delta-      Temp."
1140 !"  No.      Between    #(10^-6)     Zero       Strain"
1150 !"           Nodes      (lb,N)      (in,m)     #(10^6)"\!
1160 FOR I=1 TO I9
1170 !"    ",I,TAB(13),\INPUT1"",E(I,1)\!TAB(17),\INPUT1"",E(I,2)
1180 !TAB(22), \ INPUT1"   ",E(I,3)
1190 !TAB(35), \ INPUT1"   ",E(I,4)
1200 !TAB(49), \ INPUT1"   ",E(I,5)
1210 E(I,3)=E(I,3)#10^6 \ E(I,5)=E(I,5)#10^-6
1220 INPUT"   OK?  (CR/N) ",Z$
1230 IF Z$="N" THEN 1170
1240 NEXT \!\!\!
1250 REM
1260 REM            #### e. Determine and number active degrees of freedom ####
1270 REM
1280 FOR J=1 TO J9
1290 FOR K=1 TO 2
1300 IF N(J,K)=0 THEN 1320
1310 M=M+1 \ N(J,K)=M
1320 NEXT
1330 NEXT
1340 !"        There are ",M," degrees of freedom"\!
1350 !"             Wait for Computation" \ !\!
1360 REM
1370 REM            #### f. Dimension arrays ####
```

**Program Listing 4.1** (*continued*)

```
1380 REM
1390 DIM K(M,M)        \REM GLOBAL STIFFNESS MATRIX
1400 DIM F(M)          \REM APPLIED FORCE VECTOR
1410 DIM F0(M)         \REM INITIAL FORCE VECTOR
1420 DIM U(M)          \REM DISPLACEMENT VECTOR
1430 REM
1440 REM          **** g. Establish the applied force vector {f} ****
1450 REM
1460 FOR J=1 TO J9 \ FOR K=1 TO 2
1470 IF N(J,K)=0 THEN 1500
1480 M=N(J,K)
1490 F(M)=N(J,K+2)
1500 NEXT
1510 NEXT
1520 RETURN
1530 REM
1540 REM **********************************************************************
1550 REM
1560 REM    2.  ESTABLISH THE GLOBAL STIFFNESS MATRIX (k)
1570 REM
1580 REM          **** a. Determine and store element constants ****
1590 REM
1600 FOR U=1 TO M \ FOR V=1 TO M \ K(U,V)=0 \ NEXT \ NEXT
1610 REM
1620 FOR I=1 TO I9
1630 N1=E(I,1) \ N2=E(I,2)               \REM N1 IS NODE AT LOCAL ORIGIN
1640 REM
1650 FOR K=1 TO 2 \ M(K)=N(N1,K) \ M(K+2)=N(N2,K) \ NEXT
1660 X=N(N2,5) -N(N1,5)
1670 Y=N(N2,6) -N(N1,6)
1680 L=SQRT(X*X+Y*Y)
1690 C1=X/L
1700 C2=Y/L
1710 S0=E(I,3)/L \ S1=SQRT(S0)
1720 K0(I,1)=S1*C1
1730 K0(I,2)=S1*C2
1740 K0(I,3)=C1
1750 K0(I,4)=C2
1760 K0(I,5)=S0
1770 K0(I,6)=L
1780 REM
1790 REM          **** b. Establish local stiffness matrix (k1) ****
1800 REM
1810 GOSUB 3140               \REM LOCAL STIFFNESS MATRIX SUBROUTINE
1820 REM
```

**Program Listing 4.1** (*continued*)

```
1830 REM          **** c. Put appropriate terms into global matrix (k) ****
1840 REM
1850 FOR U=1 TO 4 \ FOR V=1 TO 4
1860 M1=M(U) \ M2=M(V)
1870 IF M1=0 THEN 1890 \ IF M2=0 THEN 1890
1880 K(M1,M2)=K(M1,M2)+K1(U,V)
1890 NEXT \ NEXT
1900 NEXT
1910 RETURN
1920 REM
1930 REM *************************************************************************
1940 REM
1950 REM    3.  ESTABLISH THE INITIAL FORCE VECTOR {f0}
1960 REM
1970 REM          **** a. Calculate initial forces, put in {f0} and E1 ****
1980 REM
1990 FOR I=1 TO I9
2000 N1=E(I,1) \ N2=E(I,2)              \REM N1 IS NODE AT LOCAL ORIGIN
2010 FOR K=1 TO 2 \ M(K)=N(N1,K) \ M(K+2)=N(N2,K) \ NEXT
2020 F=-K0(I,5)*(E(I,4)+E(I,5)*K0(I,6))
2030 IF F=0 THEN 2100
2040 FOR U=1 TO 2
2050 F0(M(U))=F0(M(U))-F*K0(I,U+2) \ E1(I,U+2)=-F*K0(I,U+2)
2060 NEXT
2070 FOR U=3 TO 4
2080 F0(M(U))=F0(M(U))+F*K0(I,U)   \ E1(I,U+2)=+F*K0(I,U)
2090 NEXT
2100 NEXT
2110 REM
2120 REM          **** b. Subtract initial forces {f}={f}-{f0} ****
2130 REM
2140 FOR K=1 TO M \ F(K)=F(K)-F0(K) \ NEXT
2150 RETURN
2160 REM          .
2170 REM *************************************************************************
2180 REM
2190 REM    4.  SOLVE FOR DISPLACEMENTS (u) BY GAUSSIAN ELIMINATION
2200 REM
2210 REM          **** a. Triangularize global matrix (k) ****
2220 REM
2230 FOR I=1 TO M
2240  FOR J=I+1 TO M
2250   Z=-K(J,I)/K(I,I)
2260    FOR K=I TO M
2270     K(J,K)=K(J,K)+Z*K(I,K)
```

**Program Listing 4.1** (*continued*)

```
2280    NEXT
2290    F(J)=F(J)+Z*F(I)
2300   NEXT
2310 NEXT
2320 REM
2330 REM            **** b. Calculate displacement vector (u) ****
2340 REM
2350 FOR I=M TO 1 STEP-1
2360 Z=0
2370 FOR J=I+1 TO M
2380  Z=Z+U(J)*K(I,J)
2390 NEXT
2400 U(I)=(F(I)-Z)/K(I,I)
2410 NEXT
2420 REM
2430 REM          c. Put displacements in nodal array N1 ****
2440 REM
2450 M1=1
2460 FOR J=1 TO J9 \ FOR K=1 TO 2
2470 IF N(J,K)<>M1 THEN 2500
2480 N1(J,K)=U(M1)
2490 M1=M1+1
2500 NEXT \ NEXT
2510 RETURN
2520 REM
2530 REM *****************************************************************************
2540 REM
2550 REM
2560 REM   5.  DETERMINE FORCES ON EACH ELEMENT
2570 REM
2580 REM
2590 REM          **** a. Determine nodes N1 and N2 ****
2600 REM
2610 FOR I=1 TO I9
2620 N1=E(I,1) \ N2=E(I,2)
2630 E1(I,1)=N1 \ E1(I,2)=N2
2640 REM
2650 REM          **** b. Establish local displacement vector (u1) ****
2660 REM
2670 FOR K=1 TO 2 \ M1=N(N1,K) \ U1(K)=U(M1) \ NEXT
2680 FOR K=1 TO 2 \ M1=N(N2,K) \ U1(K+2)=U(M1) \ NEXT
2690 REM
2700 REM          **** c. Establish local stiffness matrix (k1) ****
2710 REM
2720 GOSUB 3140            \REM LOCAL STIFFNESS MATRIX SUBROUTINE
2730 REM
```

**Program Listing 4.1** (*continued*)

```
2740 REM          **** d. Calculate local forces {f1} due to {u1} ****
2750 REM          **** e. Add {f1} to initial forces in E1 ****
2760 REM
2770 FOR U=1 TO 4
2780 F1(U)=0
2790 FOR V=1 TO 4
2800 F1(U)=F1(U)+K1(U,V)*U1(V)
2810 NEXT
2820 E1(I,U+2)=E1(I,U+2)+F1(U)
2830 NEXT
2840 NEXT
2850 RETURN
2860 REM
2870 REM
2880 REM **********************************************************************************
2890 REM
2900 REM    6.  PRINT OUT THE RESULTS
2910 REM
2920 REM          **** a. Nodal displacements ****
2930 REM          **** b. Forces at both ends of each element ****
2940 REM
2950 !"        Nodal displacements (in,m)"
2960 !"Node          x          y"\!
2970 FOR J=1 TO J9
2980 !J,     " ",%13F7, N1(J,1),N1(J,2)
2990 NEXT
3000 !\!\!
3010 Z3=0
3020 !"                Nodal forces   (lb,N)"
3030 !"El.   Node          Fx          Fy"
3040 !
3050 FOR I=1 TO I9
3060 !%2I,I,%10I, E1(I,1), %13F1, E1(I,3),E1(I,4)
3070 !%2I,I,%10I,  E1(I,2), %13F1,E1(I,5),E1(I,6)
3080 NEXT
3090 !\!\INPUT "OK? (CR/N) ",Z$
3100 !
3110 IF Z$="N" THEN 2950
3120 !\!\!\RETURN
3130 REM
3140 REM **********************************************************************************
3150 REM
3160 REM                    LOCAL STIFFNESS MATRIX ROUTINE
3170 REM
3180 FOR U=1 TO 4 \FOR V=1 TO 4 \ K1(U,V)=0 \ NEXT\NEXT
3190 FOR U=1 TO 2 \ FOR V=1 TO 2
```

**Program Listing 4.1** (*continued*)

```
3200 K0=K0(I,U)*K0(I,V)
3210 K1(U,V)=K0
3220 K1(U,V+2)=-K0
3230 K1(U+2,V)=-K0
3240 K1(U+2,V+2)=K0
3250 NEXT \ NEXT
3260 RETURN
3270 REM
3280 REM *****************************************************************************
3290 REM
3300 REM    7.  FURTHER CALCULATIONS
3310 REM
3320 REM          **** a. Other loadings for this structure? ****
3330 REM
3340 !\ INPUT "Run other loadings for this structure? (CR/N) ",Z$
3350 IF Z$="N" THEN CHAIN "2.01*"
3360 FOR I=1 TO I9 \ FOR K=3 TO 6 \ E1(I,K)=0 \ NEXT \ NEXT
3370 FOR U=1 TO M \ F0(U)=0 \ F(U)=0 \ NEXT
3380 !\!
3390!"                 Applied forces (lb)"
3400!"Node #              Fx          Fy"
3410 FOR J=1 TO J9
3420 IF N(J,1)=0 THEN IF N(J,2)=0 THEN 3500
3430 !"    ",J,
3440!TAB(21), \ INPUT1" ",N(J,3) \ !TAB(32), \ INPUT1" ",N(J,4)
3450 FOR K=1 TO 2 \ M1=N(J,K)
3460 F(M1)=N(J,K+2)
3470 NEXT
3480 INPUT"    OK? (CR/N) ",Z$
3490 IF Z$="N" THEN 3430
3500 NEXT
3510 !\!
3520 !"Element   Delta-Zero   T-Strain *(10^6)"\!
3530 FOR I=1 TO I9
3540 !%7I,I,\ INPUT1"          ",E(I,4)
3550 !TAB(27),
3560 INPUT1"",E(I,5) \ INPUT"  OK?  (CR/N) ",Z$
3570 E(I,5)=E(I,5)*10^-6
3580 IF Z$="N" THEN 3540
3590 NEXT
3600 !\!
3610 GOTO 290
```

**Program Listing 4.1** (*continued*)

```
20 !"                    STRUCTURAL ANALYSIS"\!
50 !"       3-dimensional structures;  axial loading only"\!\!
70 REM
80 REM ***********************************************************************
90 REM
100 !"              First define the structure"\!\!
110 REM
120 INPUT "             Number of ELEMENTS = ",I9     \REM I = ELEMENTS
130 INPUT "             Number of   NODES = ",J9      \REM J = NODES
140 !\!
150 REM
160 REM INITIALIZE
170 REM
180 REM NOTE\ PROGRAM CONTAINS SPACE FOR BENDING DATA
190 REM
200 DIM A(I9,6), E(I9,6), E1(I9,8), F1(6), K1(6,6), M(6), N(J9,9), N1(J9,3)
210 DIM K0(I9,3), S(6), U1(6), F(40)
220 REM
230 REM
240 !"Input data for each NODE:  Degree-of-freedom (DOF), 1=active, 0=nonact."
250 !\!
260 !"     DOF     Applied  Forces  (lb, N)           Coordinates"
270 !" # X Y Z      X         Y         Z        X         Y         Z"
280 FOR J=1 TO J9
290 !%2I,J, \ INPUT1" ",N(J,1) \ INPUT1" ",N(J,2) \ INPUT1" ",N(J,3)
300 !TAB(13), \ INPUT1" ",N(J,4) \ !TAB(23), \ INPUT1" ",N(J,5)
310 !TAB(33), \ INPUT1" ",N(J,6) \ !TAB(43), \ INPUT1" ",N(J,7)
320 !TAB(53), \ INPUT1" ",N(J,8) \ !TAB(63), \ INPUT1" ",N(J,9)
330 INPUT"   OK? (CR/N)",Z$
340 IF Z$="N" THEN 290
350 NEXT \ !\!
360 REM
370 REM  NUMBER THE DEGREES OF FREEDOM FROM 1 TO M
380 FOR J=1 TO J9
390 FOR K=1 TO 3
400 IF N(J,K)=0 THEN 430
410 M=M+1 \ N(J,K)=M
420 F(M)=N(J,K+3)                        \REM APPLIED FORCE VECTOR
430 NEXT
440 NEXT
450 !\!"          There are ",M," active degrees of freedom"\!
460 DIM F0(M), U(M), K(M,M)
470 !
480 !"Input data to define ELEMENT location, stiffness, and initial conditions"
490 !"       Note multipliers for E*A and Temp-Strain"
```

Program Listing 4.2

```
500 !\!
510 !"Element   Located    E*A     Delta-   Temp."
520 !"  No.     Between   *(10^-6)   Zero   Strain"
530 !"           Nodes    (lb, N)   (in,m)  *(10^6)"\!
540 FOR I=1 TO I9
550 !"    ",I,TAB(11),\INPUT1"",E(I,1)\!TAB(14),\INPUT1"",E(I,2)
560 !TAB(20), \ INPUT1"   ",E(I,3)
570 !TAB(30), \ INPUT1"   ",E(I,5)
580 !TAB(40), \ INPUT1"   ",E(I,6)
590 E(I,3)=E(I,3)*10^6 \ E(I,6)=E(I,6)*10^-6 \ E(I,4)=E(I,4)*10^6
600 INPUT"   OK? (CR/N)",Z$
610 IF Z$="N" THEN 550
620 NEXT \!\!\!
630 REM
640 REM *************************************************************************
650 REM              ESTABLISH THE STIFFNESS MATRIX K(M,M)
660 REM
670 FOR U=1 TO M \ FOR V=1 TO M \ K(U,V)=0 \ NEXT \ NEXT
680 FOR I=1 TO I9
690 REM ESTABLISH LOCAL (K)
700 N1=E(I,1) \ N2=E(I,2)            \REM N1 IS NODE AT LOCAL ORIGIN
710 REM
720 FOR K=1 TO 3 \ M(K)=N(N1,K) \ M(K+3)=N(N2,K) \ NEXT
730 X=N(N2,7)-N(N1,7)
740 Y=N(N2,8)-N(N1,8)
750 Z=N(N2,9)-N(N1,9)
760 L=SQRT(X*X+Y*Y+Z*Z)
770 C1=X/L
780 C2=Y/L
790 C3=Z/L
800 S0=SQRT(E(I,3)/L)
810 S(1)=S0*C1 \ K0(I,1)=S(1)
820 S(2)=S0*C2 \ K0(I,2)=S(2)
830 S(3)=S0*C3 \ K0(I,3)=S(3)
840 S(4)=-S(1) \ S(5)=-S(2) \ S(6)=-S(3)
850 REM
860 REM PUT TERMS INTO GLOBAL MATRIX
870 REM
880 FOR U=1 TO 6 \ FOR V=1 TO 6
890 M1=M(U) \ M2=M(V)
900 IF M1=0 THEN 920 \ IF M2=0 THEN 920
910 K(M1,M2)=K(M1,M2)+S(U)*S(V)
920 NEXT \ NEXT
930 REM
940 REM  ESTABLISH LOADING VECTOR F0(M) AND PUT -FORCES INTO OUTPUT
```

**Program Listing 4.2** (*continued*)

```
950 REM
960 F=-S0*(E(I,5)+E(I,6)*L)
970 IF F=0 THEN 1010
980 FOR U=1 TO 6
990 F0(M(U))=F0(M(U))-F*S(U) \ E1(I,U+2)=-F*S(U)
1000 NEXT
1010 NEXT
1020 REM
1030 REM COMBINE FORCE VECTORS
1040 REM
1050 FOR K=1 TO M \ F(K)=F(K)-F0(K) \ NEXT
1060 REM
1070 REM ***************************************************************************
1080 REM        ***** SOLUTION BY GAUSSIAN ELIMINATION *****
1090 REM
1100 REM              TRIANGULARIZE MATRIX
1110 REM
1120 FOR I=1 TO M
1130  FOR J=I+1 TO M
1140   Z=-K(J,I)/K(I,I)
1150    FOR K=I TO M
1160    K(J,K)=K(J,K)+Z*K(I,K)
1170    NEXT
1180    F(J)=F(J)+Z*F(I)
1190  NEXT
1200 NEXT
1210 REM              CALCULATE DISPLACEMENTS
1220 REM
1230 FOR I=M TO 1 STEP-1
1240 Z=0
1250 FOR J=I+1 TO M
1260  Z=Z+U(J)*K(I,J)
1270 NEXT
1280 U(I)=(F(I)-Z)/K(I,I)
1290 NEXT
1300 REM        PUT DISPLACEMENTS IN N1(J9,3) ARRAY
1310 M1=1
1320 FOR J=1 TO J9 \ FOR K=1 TO 3
1330 IF N(J,K)<>M1 THEN 1360
1340 N1(J,K)=U(M1)
1350 M1=M1+1
1360 NEXT \ NEXT
1370 REM
1380 REM ***************************************************************************
```

**Program Listing 4.2** (*continued*)

```
1390 REM          CALC FORCES ON EACH ELEMENT AT BOTH NODES
1400 REM
1410 FOR I=1 TO I9
1420 N1=E(I,1) \ N2=E(I,2)
1430 E1(I,1)=N1 \ E1(I,2)=N2
1440 REM
1450 REM  DETERMINE LOCAL U1(6) VECTOR
1460 REM
1470 FOR K=1 TO 3 \ M1=N(N1,K) \ U1(K)=U(M1) \ NEXT
1480 FOR K=1 TO 3 \ M1=N(N2,K) \ U1(K+3)=U(M1) \ NEXT
1490 REM
1500 REM  GET LOCAL K0
1510 REM
1520 FOR U=1 TO 6 \ FOR V=1 TO 6
1530 S(1)=K0(I,1) \ S(2)=K0(I,2) \ S(3)=K0(I,3)
1540 S(4)=-S(1) \ S(5)=-S(2) \ S(6)=-S(3)
1550 K1(U,V)=S(U)*S(V)
1560 NEXT \ NEXT
1570 REM
1580 REM CALCULATE FORCES\   (F1)=(K0)*(U1)
1590 REM
1600 FOR U=1 TO 6
1610 F1(U)=0
1620 FOR V=1 TO 6
1630 F1(U)=F1(U)+K1(U,V)*U1(V)
1640 NEXT
1650 E1(I,U+2)=E1(I,U+2)+F1(U)
1660 NEXT
1670 REM
1680 NEXT
1690 REM
1700 GOSUB 1950
1710 REM
1720 !\ INPUT "Run other loadings for this structure? (CR/N) ",Z$
1730 IF Z$="N" THEN CHAIN "2.01*"
1740 Z4=Z4+1
1750 !\!
1760!"             Applied Forces"
1770!"   Node         X     Y     Z"
1780 FOR J=1 TO J9
1790 !"    ",J,
1800 INPUT1 "            ",N(J,4) \ INPUT1 "      ", N(J,5)
1810 INPUT "      ",N(J,6)
1820 FOR K=1 TO 3 \ M1=N(J,K)
1830 F(M1)=N(J,K+3)
```

**Program Listing 4.2** (*continued*)

```
1840 NEXT
1850 NEXT
1860 !"Element   Delta-Zero   T-Strain *(10^6)"\!
1870 FOR I=1 TO I9
1880 !%7I,I,\ INPUT1"        ",E(I,5)
1890 !TAB(27),
1900 INPUT1"",E(I,6) \ INPUT"   OK? (CR/N)",Z$
1910 IF Z$="N" THEN 1880
1920 NEXT
1930 GOTO 660
1940 REM
1950 REM ***************************************************************************
1960 REM        OUTPUT SUBROUTINE
1970 Z3=0
1980 REM
1990 !"              Nodal  Displacements"
2000 !"              X           Y           Z"
2010 FOR J=1 TO J9
2020 !"  ",J, TAB(11), %8F5,N1(J,1), TAB(22),N1(J,2),TAB(33),N1(J,3)
2030 IF Z3<15 THEN 2050
2040 Z3=0 \ INPUT"CR to cont",Z$
2050 NEXT
2060 !\!\!
2070 Z3=0
2080 !"              Nodal Forces      (lbs.)"
2090 !"Element     Node      Fx        Fy        Fz"
2100 !
2110 FOR I=1 TO I9
2120 !%4I,I,%13I, E1(I,1), %10F1, E1(I,3), %10F1, E1(I,4), %10F1, E1(I,5)
2130 !%4I,I, %13I, E1(I,2), %10F1, E1(I,6), %10F1, E1(I,7), %10F1, E1(I,8)
2140 IF Z3<15 THEN 2160
2150 Z3=0 \ INPUT "CR to cont",Z$
2160 NEXT
2170 FOR I=1 TO I9 \ FOR K=3 TO 8 \ E1(I,K)=0 \ NEXT \ NEXT
2180 FOR U=1 TO M \ F0(U)=0 \ NEXT
2190 !\!\!\RETURN
```

**Program Listing 4.2** (*continued*)

## 4.10 PROBLEMS

For Probs. 4.1 to 4.10, using the following format, determine and fill in the numbers required for a complete problem solution via the structural analysis programs—i.e., code each problem.

If a computer is available solve for the forces and displacements using the FEM programs.

Prob. No. _____    No. of elements _____    No. of nodes _____

**Data for each *node***

| Node | dof x | y | Applied forces x | y | Coordinates x | y |
|------|-------|---|------------------|---|---------------|---|
| 1 | | | | | | |
| 2 | | | | | | |
| 3 | | | | | | |
| 4 | | | | | | |
| 5 | | | | | | |
| 6 | | | | | | |
| 7 | | | | | | |

**Data for each *element***

| Element No. | Located between nodes | $EA \times 10^{-6}$, lb, N | $\delta_0$, in, m | Temp. strain $\times 10^6$ |
|-------------|-----------------------|----------------------------|-------------------|----------------------------|
| 1 | | | | |
| 2 | | | | |
| 3 | | | | |
| 4 | | | | |
| 5 | | | | |
| 6 | | | | |
| 7 | | | | |

**4.1**

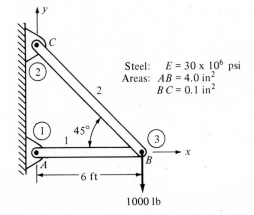

Steel:  $E = 30 \times 10^6$ psi
Areas:  $AB = 4.0$ in$^2$
$BC = 0.1$ in$^2$

6 ft

1000 lb

**Figure P4.1**

**4.2**

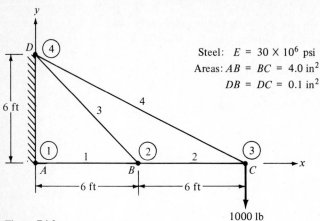

Steel: $E = 30 \times 10^6$ psi
Areas: $AB = BC = 4.0$ in$^2$
$DB = DC = 0.1$ in$^2$

**Figure P4.2**

**4.3**

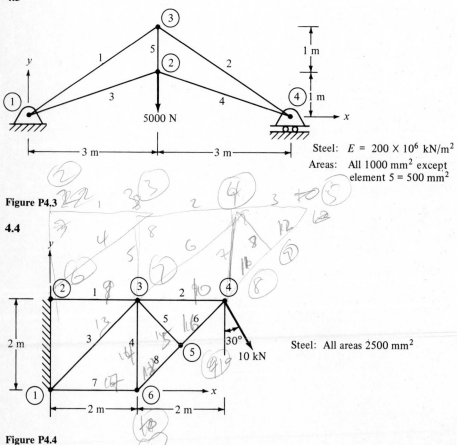

Steel: $E = 200 \times 10^6$ kN/m$^2$
Areas: All 1000 mm$^2$ except
element 5 = 500 mm$^2$

**Figure P4.3**

**4.4**

Steel: All areas 2500 mm$^2$

**Figure P4.4**

**4.5**

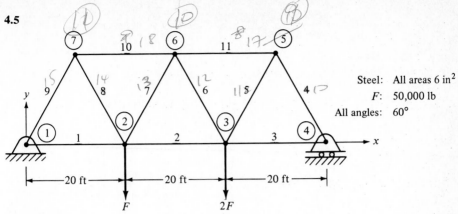

Steel: All areas 6 in$^2$
F: 50,000 lb
All angles: 60°

**Figure P4.5**

**4.6** Same as Prob. 4.5, except node 4 is a fixed support.

**4.7**

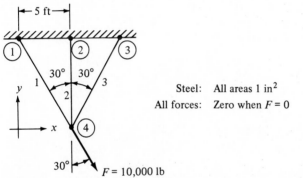

Steel: All areas 1 in$^2$
All forces: Zero when $F = 0$

$F = 10,000$ lb **Figure P4.7**

**4.8** Same geometry as in Prob. 4.7. $F = 0$, but bar 3-4 heated 50°F.

**4.9** Same geometry as in Prob. 4.7. $F = 0$, but rod 2-4 was made 0.050 in too long and had to be forced into place.

**4.10**

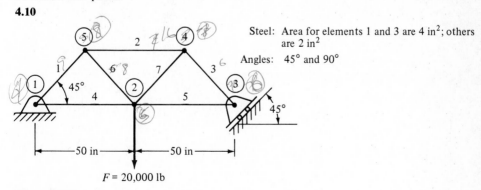

Steel: Area for elements 1 and 3 are 4 in$^2$; others are 2 in$^2$

Angles: 45° and 90°

$F = 20,000$ lb

**Figure P4.10**

**4.11** From the following data, reconstruct the problem.

Number of ELEMENTS = 5
Number of NODES = 4

Input data for each NODE:
For degrees of freedom, 'active'=1; 'nonactive=0

| Node | Deg of free. x | y | Applied forces Fx | Fy | Coordinates x | y |
|---|---|---|---|---|---|---|
| 1 | 0 | 0 | 0 | 0 | 0 | 10 |
| 2 | 1 | 1 | 0 | 0 | 10 | 10 |
| 3 | 1 | 1 | 0 | 0 | 10 | 0 |
| 4 | 0 | 0 | 0 | 0 | 0 | 0 |

| Element No. | Located Between Nodes | E*A *(10^-6) (lbs) | Delta-Zero (in) | Temp. Strain *(10^6) |
|---|---|---|---|---|
| 1 | 1 2 | 30 | 0 | 0 |
| 2 | 2 3 | 30 | 0 | 600 |
| 3 | 3 4 | 30 | 0 | 0 |
| 4 | 1 3 | 30 | 0 | 0 |
| 5 | 4 2 | 30 | -.010 | 0 |

**4.12** From the following data, reconstruct the problem.

| Node | Deg of free. x | y | Applied forces Fx | Fy | Coordinates x | y |
|---|---|---|---|---|---|---|
| 1 | 0 | 0 | 0 | 0 | 0 | 0 |
| 2 | 0 | 0 | 0 | 0 | 100 | 0 |
| 3 | 0 | 0 | 0 | 0 | 200 | 0 |
| 4 | 1 | 1 | 0 | 0 | 50 | -50 |
| 5 | 1 | 1 | 0 | 0 | 150 | -50 |
| 6 | 1 | 1 | 0 | -2000 | 100 | -100 |

Input data to define ELEMENT location, stiffness, and initial conditions
Note multipliers for E*A

| Element No. | Located Between Nodes | E*A *(10^-6) (lbs) | Delta-Zero (in) | Temp. Strain *(10^6) |
|---|---|---|---|---|
| 1 | 1 4 | 30 | 0 | 0 |
| 2 | 4 2 | 30 | 0 | 0 |
| 3 | 2 5 | 30 | 0 | 0 |
| 4 | 5 3 | 30 | 0 | 0 |
| 5 | 4 5 | 60 | 0 | 0 |
| 6 | 4 6 | 10 | 0 | 0 |
| 7 | 5 6 | 10 | 0 | 0 |

For Probs. 4.13 to 4.21, carry out a structural analysis "by hand"; that is,

1. Determine the local element stiffness matrices $[k_1]$
2. Determine the global stiffness matrix $[k]$
3. Establish the force vector $\{F\}$
4. Solve for the displacements $\{u\}$
5. Calculate the forces on each member from the local stiffness matrices $[k_1]$ and $\{u\}$

**4.13** Example 4.1

**4.14** Problem 3.5

**4.15** Problem 3.6

**4.16** Problem 3.7

**4.17** Problem 3.8

**4.18** Problem 3.20

**4.19** Problem 3.21

**4.20** Problem 3.22

**4.21** Problem 3.23

Solve Probs. 4.22 to 4.25 using the structural analysis program for three-dimensions and axial loading.

**4.22** The derrick shown carries a load of 1000 lb. Determine the forces in each member, show them on a sketch of the member, and determine the motion of nodes 3 and 5 for ($a$) $\theta = 0$, ($b$) $\theta = 30$, ($c$) $\theta = 50$.

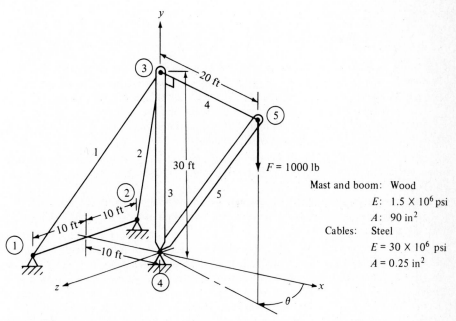

**Figure P4.22**

Note that node 5 is free to swing about the $y$ axis. The computer might not like that!

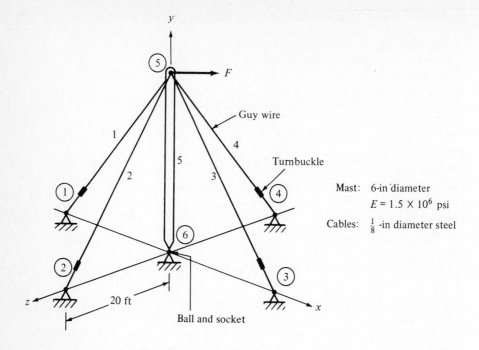

Mast:  6-in diameter
$E = 1.5 \times 10^6$ psi

Cables:  $\frac{1}{8}$ -in diameter steel

**Figure P4.23**

**4.23** A 60-ft wooden mast, shown in the accompanying illustration, is "guyed" with four $\frac{1}{8}$-in-diameter steel wires. Each wire has a "turnbuckle" for adjusting its length for proper tightness. After the wires are just barely tight, each is "shortened" by $\frac{1}{2}$ in.

(a) How large are the initial forces in the wires and the mast?

(b) A horizontal force $F$ is applied at the top of the mast in various directions. How large can $F$ be before one or more wires become slack?

**4.24** A vertical tower (shown in the accompanying illustration without supports) is made from nine pieces of aluminum tubing fastened together. Assume that the joints act like ball and socket joints.

(a) How much will the tower compress (shorten) under a 500-lb vertical load?

(b) Compare the answer to (a) with the compression in an "equivalent" single tube having three times the area.

**4.25** The tower shown in Prob. 4.24 is supported by three guy wires from points 2, 3, and 4. The wires are attached to the ground 15 ft from the tower base at points 6, 7, and 8. The wires are $\frac{1}{8}$-in-diameter steel, and each is shortened $\frac{1}{4}$ in from its zero-force length (using turnbuckles).

(a) If a 100-lb horizontal force is applied at the top, how far does it move?

(b) Determine the maximum stresses in the steel and in the aluminum.

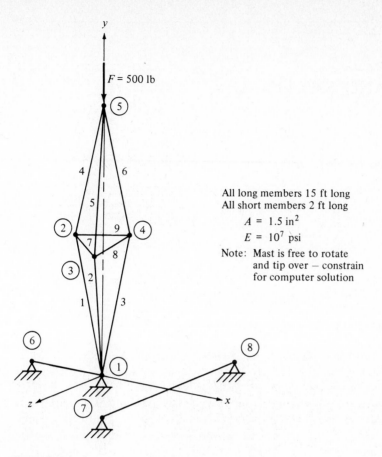

All long members 15 ft long
All short members 2 ft long

$A = 1.5$ in$^2$
$E = 10^7$ psi

Note: Mast is free to rotate
and tip over — constrain
for computer solution

**Figure P4.24**

# FIVE

## STRESS AND STRAIN

### 5.1 OVERVIEW

In Chap. 3 we introduced the concepts of *stress* and *strain*. We defined stress as being a force divided by the area over which it acted, and defined strain in terms of a change in length divided by an original length.

We saw that stress and strain can vary from point to point within a body, so that we must consider the force acting over a very small area, and the change of a very small length. In the limit, we considered stress and strain at a *point* where the area and initial length both approached zero.

We also saw that the tensile test provided information relating tensile (or compressive) stresses and tensile (or compressive) strains—information that could be used to predict the behavior of axially loaded members.

In Chap. 3, we considered a very special set of stresses and strains (although this set has very wide applicability):

The force was always perpendicular to the area.
The change in length was always parallel to the original length.

In this chapter we will relax these requirements and consider a general state of stress and strain. We will see that, in general, we must consider a three-dimensional set of stresses and strains in order to describe fully the condition of force and deformation at any point on or in a body. We will also see that equilibrium and continuity place certain restrictions on the stresses and strains.

In Chap. 3 we saw how the compliance of a structure varied when we considered different sets of *xyz* axes. We saw that there are principal directions of compliance such that a force in a certain direction produces motion only in that direction. In addition to analytical techniques, we saw how the Mohr's circle representation could help in understanding this rather complex concept of tensor transformation.

In this chapter, we will see that both stress and strain, at any point, also vary as we rotate the $xyz$ axes relative to the body. Stress and strain are tensor quantities, and the equations governing rotation of axes are analogous to those for compliance. The Mohr's circle diagrams for stress and strain are essentially identical to that for compliance; all of the same rules apply.

Finally, we will very briefly consider the empirical determination of stress and strain.

## 5.2 STRESS AT A POINT

We have defined stress as force divided by area. We know that a force is a vector quantity having both magnitude and direction. We also know that an area has both magnitude and direction (as specified in terms of the outwardly directed normal vector **n**). Hence, a stress is the *ratio* of two vector quantities; and the ratio of two vectors is *not a* simple vector quantity. We all have some intuitive understanding of vector quantities; it is easy for us to understand that a force vector can be resolved into components. You may be tempted to think of stresses as vectors; resist that temptation; you will generally get wrong answers.

To date we have considered stress where the force vector and the area normal were parallel, i.e., where the force vector was *normal* to the area. We call such a stress a "normal stress," and it is either tensile (+) or compressive (−). Because the force acting on an elemental area can act in *any* direction, we must extend our definition of stress to encompass all such possibilities.

Figure 5.1 shows three elemental areas $dA$ with forces $dF$ applied. The

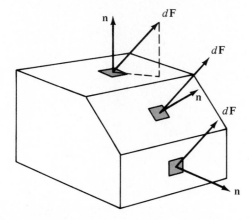

**Figure 5.1** Showing identical force vectors $dF$ that are not parallel to the area normals **n**.

elemental areas $dA$ can be either on the surface of a body, or *inside* the body. We recall that the only way we can consider forces acting on an area within a body is by sectioning (in our imagination) the body through that area such that the force becomes an *external* force. Thus the forces shown could actually be externally applied surface forces, or could be internal forces acting where we have sectioned the body.

In Fig. 5.1, the three force vectors $d\mathbf{F}$ are identical in length and direction. However, the area normals are in different directions. Thus the stresses due to the forces will be different for the three areas.

The simplest way for us to understand the action of an arbitrarily directed force is to resolve it into orthogonal $xyz$ components such that one of the components is normal to the area while the other two are tangential to it. Figure 5.2 shows three force components $dF_x$, $dF_y$, and $dF_z$, acting on the area $dA$ whose normal $\mathbf{n}$ is in the $+x$ direction. The force component normal to the area produces a *normal* stress (tensile or compressive). The tangential components, which exert shearing-type actions, produce *shear* stresses. Thus the total stress on the $x$ face of Fig. 5.2 is replaced by three stress components:

Normal stress:     $\dfrac{dF_x}{dA_x} = \sigma_{xx}$     (as $dA \to 0$)

Shear stress:     $\dfrac{dF_y}{dA_x} = \tau_{xy}$     (as $dA \to 0$)

Shear stress:     $\dfrac{dF_z}{dA_x} = \tau_{xz}$     (as $dA \to 0$)

For each stress component there are two pertinent directions, that of the area normal and that of the force component. The double-subscripting system used here identifies these two directions: The first subscript refers to the area normal (i.e., the $x$ area), and the second subscript refers to the force direction.

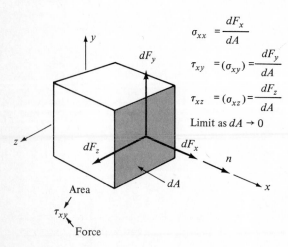

$$\sigma_{xx} = \frac{dF_x}{dA}$$

$$\tau_{xy} = (\sigma_{xy}) = \frac{dF_y}{dA}$$

$$\tau_{xz} = (\sigma_{xz}) = \frac{dF_z}{dA}$$

Limit as $dA \to 0$

**Figure 5.2** The total force $d\mathbf{F}$ is resolved into three components normal and tangential to $dA$.

The *sign* of the stress is determined by the *signs* associated with the area normal and the force component. Area normals and force components can be either positively (+) or negatively (−) directed.

A stress is *positive* when both signs are positive (+ × + = +) or when both are negative (− × − = +). The stress is *negative* when one sign is positive and one negative (+ × − = −, − × + = −). The three stress components in Fig. 5.2 are positive.

Different texts may use different symbols for stress; some use $\sigma$ for both normal and shear stresses. We will use $\sigma$ and $\tau$ to differentiate between normal stress and shear stress.

Figure 5.2 shows *stress* acting at a *point* in a body, on an *area* whose normal is in the $+x$ direction. Clearly we can have many other areas passing through the same point, but oriented in different directions. Fortunately, we can completely define the *state of stress* at a *point* in terms of the stresses acting on any three orthogonal areas—i.e., on the $x$, $y$, and $z$ areas.

Figure 5.3 shows the general description of stress at a point. Each face can have three stresses, one normal and two shear. All of the stresses shown are positive according to the sign convention. We should reiterate that these are stresses at a *point*. In each case, we have taken the ratio of force to area as that area was reduced (in the limit) to zero. Hence, although we show the stresses acting on a cubic element, that element has *zero* volume. In fact, the $+x$ face that you can see is precisely the same area as the $-x$ face that is hidden. Hence the stresses that act on the $+x$ face must be identical to those acting on the $-x$ face. You should note that the sign convention assures that this will be the case.

The nine stress quantities that define the state of stress at a point are conveniently represented as a "stress matrix," which is often labeled $[\sigma_{ij}]$ or merely $\sigma_{ij}$.

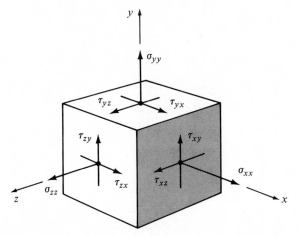

**Figure 5.3** Stress at a *point*. Although shown on a "cube," the cube has zero volume; the length of each side has been reduced to zero.

Direction

$x$    $y$    $z$

$$\text{Face} \quad \begin{array}{c} x \\ y \\ z \end{array} \left[ \begin{array}{ccc} \sigma_{xx} & \tau_{xy} & \tau_{xz} \\ \tau_{yx} & \sigma_{yy} & \tau_{yz} \\ \tau_{zx} & \tau_{zy} & \sigma_{zz} \end{array} \right] = [\sigma_{ij}] \equiv \sigma_{ij}$$

*Note*: In many studies involving three-dimensional effects, matrices such as $[\sigma_{ij}]$ occur. For simplicity of writing, the "indicial" notation is often used. Here, when we write $\sigma_{ij}$, it is understood that both $i$ and $j$ will take on the values 1, 2, and 3, thus forming the entire matrix.

In this book, we will deal explicitly with the individual components of stress and will show them on a sketch such as shown in Fig. 5.3 rather than using the matrix/indicial notation. This will give a better understanding of what is really happening to a piece of material.

## 5.3 PLANE STRESS

In our studies to date, we saw that frequently a real three-dimensional problem could be modeled satisfactorily in two dimensions, and we could simply ignore the third dimension. In dealing with stresses, we have a somewhat analogous situation. In many cases we will be able to study the stresses from a two-dimensional point of view. However, when we consider the *effects* of those stresses on a material, we will have to consider the full three dimensions. This will become clear in Chap. 7.

A two-dimensional state of stress exists whenever all of the stress components associated with a given direction are zero. For instance, if in Fig. 5.3 all of the components having any $z$ subscript are zero, we have a two-dimensional state of stress in the $xy$ plane. This is called "plane stress" and is shown in Fig. 5.4.

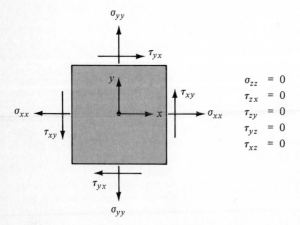

$$\sigma_{zz} = 0$$
$$\tau_{zx} = 0$$
$$\tau_{zy} = 0$$
$$\tau_{yz} = 0$$
$$\tau_{xz} = 0$$

**Figure 5.4** Plane stress (at a point) in the $xy$ plane.

There are many practical examples of plane stress. The tensile test and all of the axially loaded members we have studied are cases of plane stress. Narrow bent beams that we will analyze later are all in plane stress. At the surface of any material, if the surrounding environmental pressure is negligible, there will be no stress on the surface and we will have plane stress. If the surface pressure is negligible, and if the material is relatively "thin," such as sheet metal, not only will we have plane stress at the surface, but in the interior of the sheet as well.

In Secs. 5.4 and 5.5 we will study the requirements that equilibrium places on stress:

1. How stresses can vary *from point to point* within the material
2. How stresses vary *at a point* when we rotate the coordinate axes relative to the material.

The analyses are much simpler for plane stress than for the general case. Accordingly, we will first consider only plane stress, and in the following section extend the analyses to three dimensions.

## 5.4 STRESS EQUILIBRIUM (VARIATIONS FROM POINT TO POINT)

In some instances, such as in the central portion of the tensile test specimen, the stresses are essentially constant and uniform throughout the volume. In most cases, however, the stresses will vary from point to point in the body.

As usual, when conditions of loading vary from point to point, we will *isolate* a differential element and study the requirements of equilibrium.

Figure 5.5 shows a $dx\,dy$ element; the height in the $z$ direction is simply 1.

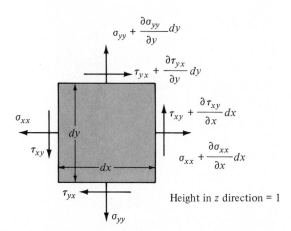

Height in $z$ direction = 1

**Figure 5.5** Isolation of $dx\,dy$ element (unity height in the $z$ direction).

This element looks very similar to the one shown in Fig. 5.3; however, this one has finite (although small) dimensions $dx$ and $dy$, while in Fig. 5.3 the dimensions were zero. Now when we write the stress magnitudes on the four sides, they can vary left to right, and they can vary up and down. Thus, for instance, we have $\sigma_{xx}$ on the left-hand side and $\sigma_{xx} + d\sigma_{xx}$ acting on the right-hand side. In Fig. 5.5, we have allowed for variation of all stresses with $x$ and $y$.

We want to apply equilibrium relations to the element shown in Fig. 5.5. However, equilibrium applies to *forces* and *moments*, not to stresses. Hence, we must multiply the stresses by the areas over which they act to get forces; then we can apply equilibrium. Figure 5.6 shows the same element with the corresponding forces applied. There are four forces in the $x$ direction and four in the $y$ direction.

However, when we apply equilibrium and sum the forces in the $x$ direction, we see that two pairs of forces cancel, and likewise in the $y$ direction. We then have:

For $\Sigma F_x = 0$: $\qquad\qquad \dfrac{\partial \sigma_{xx}}{\partial x} \, dx \, dy + \dfrac{\partial \tau_{yx}}{\partial y} \, dy \, dx = 0$

For $\Sigma F_y = 0$: $\qquad\qquad \dfrac{\partial \sigma_{yy}}{\partial y} \, dy \, dx + \dfrac{\partial \tau_{xy}}{\partial x} \, dx \, dy = 0$

*Note*: In case you have not studied them in calculus, the derivatives above are "partial derivatives." Basically, when taking the derivative with respect to $x$, $y$ is held constant, and vice versa.

Setting the moments about the center of the element equal to zero, only the shear-stress terms appear. Note that the shear-stress forces are multiplied by either $dx/2$ or $dy/2$.

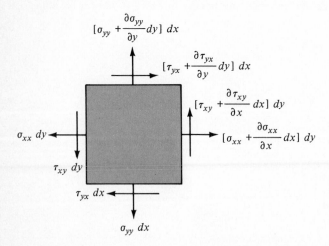

**Figure 5.6** Isolated element $dx \times dy \times 1$, showing the forces due to the stresses.

For $\Sigma M_z = 0$: $$\frac{2\tau_{xy}\,dy\,dx}{2} - \frac{2\tau_{yx}\,dx\,dy}{2} = 0$$

where $(\partial\tau_{xy}/\partial x)\,dx\,dy \ll \tau_{xy}\,dy$.

If we now divide all of the above equilibrium equations by $dx\,dy$, we have the final equilibrium requirements:

$$\frac{\partial\sigma_{xx}}{\partial x} + \frac{\partial\tau_{yx}}{\partial y} = 0$$

$$\frac{\partial\sigma_{yy}}{\partial y} + \frac{\partial\tau_{xy}}{\partial x} = 0 \qquad (5.1)$$

$$\tau_{xy} = \tau_{yx}$$

The lower equation tells us that the shear stress on the $x$ face must equal that on the $y$ face; that is, $\tau_{ij} = \tau_{ji}$ (if we were to represent the stress at a point in matrix form, the matrix would be symmetrical). This greatly simplifies our work because the number of shear-stress variables is cut in half.

The first two of Eqs. (5.1) tell us that if stresses vary, they must do so in a very special way. If a normal stress varies along its own direction, then the shear stress must vary in the perpendicular direction and the sum of the variations must be zero.

For the case of *plane stress*, we can now reduce our writing effort by simplifying the stress subscripting system. There will be no ambiguity if we adopt:

$$\sigma_x = \sigma_{xx}$$

$$\sigma_y = \sigma_{yy}$$

$$\tau = \tau_{xy} = \tau_{yx}$$

Figure 5.7 shows two (zero-area) elements, one with all positive stresses, the

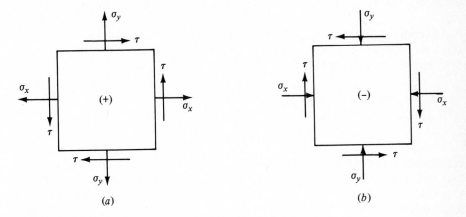

**Figure 5.7** Plane stress at a *point*: (*a*) all stresses positive; (*b*) all stresses negative.

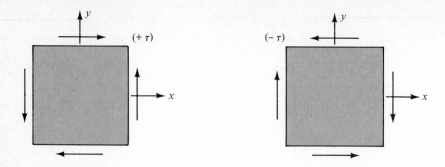

**Figure 5.8** Showing positive and negative shear stresses.

other having all negative stresses. For normal stresses, the sign convention seems "natural." A positive tensile stress tends to elongate the element in the direction of the stress; a negative, compressive stress tends to compress the element.

Shear stresses (and in fact most characteristics associated with shear) do not seem to have a "natural" sign convention. Part of the reason is that we have no words to distinguish positive shear from negative shear. For normal stresses (and strains) we have "tension/compression" and "longer/shorter," words we all know. For shear it is different and you simply will have to accept a set of conventions. Our conventions are completely logical, rigorous, and consistent; but you may feel a bit uncomfortable with them initially.

Using our standard sign convention, involving the sign of the force vector and the sign of the area normal, convince yourself that the stresses shown in Fig. 5.7a are positive while those shown in Fig. 5.7b are negative. You might note that the shear-stress arrowheads always point to two opposite corners of the element. In Fig. 5.8 we can see that if the arrowheads are in the first and third quadrants (using standard rotation convention), then the shear is positive; if the arrowheads are in the second and fourth quadrants, the shear is negative.

**Example 5.1** It has been suggested that the following equations represent the state of stress in a particular structural member. Under what conditions (if any) could the equations represent a *possible* state of stress?

$$\sigma_x = Axy$$

$$\sigma_y = Bxy$$

$$\tau = C(1 - y^2)$$

We are not asked if the equations give the *correct* stresses (we could not answer that without knowing all details of the problem), we are asked if they represent a *possible* set. If the stresses do not satisfy equilibrium, they cannot be a possible set.

All we need do is to take the appropriate partial derivatives of the stresses and see if they satisfy Eqs. (5.1):

$$\frac{\partial \sigma_x}{\partial x} = Ay \qquad \frac{\partial \tau}{\partial y} = -2Cy$$

$$\frac{\partial \sigma_y}{\partial y} = Bx \qquad \frac{\partial \tau}{\partial x} = 0$$

Applying Eqs. (5.1), we find that equilibrium is satisfied only if $A = 2C$ and $B = 0$.         ***

## 5.5 STRESS EQUILIBRIUM (VARIATION WITH ANGLE AT A POINT)

Whenever we study the stresses in a structural member, we must establish a coordinate system. The appropriate coordinate system is generally obvious from the problem itself. We choose a set of coordinates that allows us to calculate stresses due to the loads.

*However*, the stresses relative to those axes may *not* be the "worst" from the point of view of the material under load.

For instance, in the tensile test, where the only loading is along the axis, the appropriate coordinate system is obvious. It would seem reasonable that the axial direction would also be the most important, or "worst," as far as the material is concerned. We will see in Chap. 7 that for some materials the axial direction is the most important; however for the majority of engineering materials, it is *not*.

The problem we want to address here, then, is as follows:

Given the stresses, at a point, relative to one set of coordinate axes, calculate the stresses, at that *same* point, relative to a different (rotated) set of axes.

Fortunately, our old friend "equilibrium" provides all the information we need.

At this point there are two ways we could proceed: We could analyze a general state of plane stress, or we could analyze a special state of stress which can then be related to the general case. It is much easier, and clearer, to start with the special case. Accordingly, we will make some general statements without providing a proof of their correctness, and proceed from there. The truth of the statements will subsequently become obvious.

For any state of stress, at any point, in any material, there *always* exists a set of *principal stress* coordinate axes. Relative to the principal axes, all shear stresses are zero. On the faces of a principal stress element we find only normal stresses—the *principal stresses*.

If you will accept the above, we can proceed to calculate the stresses acting on an element that is rotated relative to the principal element.

When we considered principal axes of compliance, we used subscripts I, II, and III (or 1, 2, and 3) to distinguish the principal axes from a general *xyz* set. We will follow the same system here. Figure 5.9*a* shows a principal stress element in plane stress; i.e., the third principal stress is zero. In Fig. 5.9*b*, stresses are shown acting on an element in the *xyz* coordinate system that is at an angle relative to the principal axes. The rotation has been about the third principal stress axis (III), which corresponds to the *z* axis for this example.

When we vary the angle $\alpha$, we clearly do not change the forces acting on the material, nor do we change the effects of those forces; we only change the direction of the faces on which the stresses will be calculated. The effects of the stresses shown in Fig. 5.9*b must* be identical to the effects of the stresses shown in Fig. 5.9*a*.

Figure 5.10 shows the construction required to assure that the *xyz* stresses are equivalent to the principal stresses. Instead of isolating a square stress element, we select a triangle such that two of the faces are principal faces, while the hypotenuse is either an *x* or *y* face (*x* face as shown). So we see in Fig. 5.10*a* both the known stresses ($\sigma_1$ and $\sigma_2$) together with two of the unknown stresses ($\sigma_x$ and $\tau$). Recall that we cannot apply equilibrium to stresses per se, but must multiply them by areas to obtain forces. Figure 5.10*b* shows the pertinent areas; if the hypotenuse is equal to area 1, then the two sides equal sin $\alpha$ and cos $\alpha$ (the height in the *z* direction being 1). Finally, Fig. 5.10*c* shows the forces which act on the three faces. If we sum the forces in the *x* and *y* directions, we have:

For $\Sigma F_x = 0$: $\qquad \sigma_x = \sigma_1 \cos \alpha \cos \alpha + \sigma_2 \sin \alpha \sin \alpha$

For $\Sigma F_y = 0$: $\qquad \tau = -\sigma_1 \cos \alpha \sin \alpha + \sigma_2 \sin \alpha \cos \alpha$

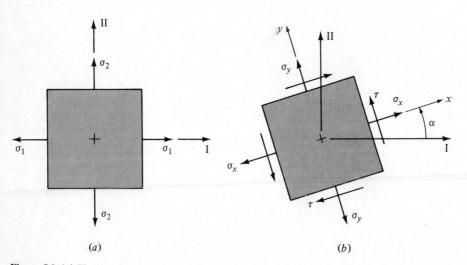

(a)  (b)

**Figure 5.9** (*a*) The *principal stress* element. (*b*) Stresses relative to the *xyz* axes, which are rotated $\alpha$ about the *z*, or III, axis.

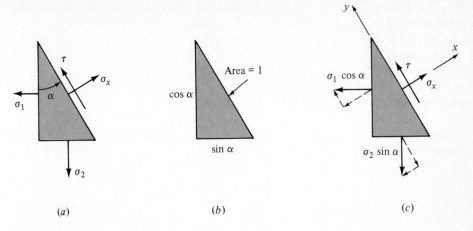

**Figure 5.10** Isolation of a triangular stress element: (*a*) showing the stresses; (*b*) showing the areas; (*c*) showing the forces in equilibrium.

If we isolated an element that included the *y* face, we would obtain an equation for *y*:

$$\sigma_y = \sigma_1 \sin \alpha \sin \alpha + \sigma_2 \cos \alpha \cos \alpha$$

If we compare these equations with those for compliance in Chap. 3 (Sec. 3.7), we see that they are completely analogous. The equations would be identical if we made the following substitutions:

$$C_{11} = \sigma_1 \qquad C_{xx} = \sigma_x$$

$$C_{22} = \sigma_2 \qquad C_{yy} = \sigma_y$$

$$C_{xy} = \tau$$

We can introduce the same double-angle relations that we used then to arrive at the equivalent of Eqs. (3.12), namely:

$$\sigma_x = \frac{\sigma_1 + \sigma_2}{2} + \frac{\sigma_1 - \sigma_2}{2} \cos 2\alpha$$

$$\sigma_y = \frac{\sigma_1 + \sigma_2}{2} - \frac{\sigma_1 - \sigma_2}{2} \cos 2\alpha \qquad (5.2)$$

$$\tau = \qquad -\frac{\sigma_1 - \sigma_2}{2} \sin 2\alpha$$

Equations (5.2) allow us to calculate the stresses relative to an *xy* coordinate system in terms of the principal stresses and the angle between the *xy* axes and the I–II axes. Given three inputs ($\sigma_1$, $\sigma_2$, and $\alpha$), the three equations determine three outputs ($\sigma_x$, $\sigma_y$, and $\tau$). Obviously we could work the problem in reverse; given stresses relative to the *xy* axes, we could compute the principal stresses and the angle of the principal stress element. Thus the earlier statement regarding the existence of principal stresses is verified.

While Eqs. (5.2) are sufficient for manipulating stresses, the Mohr's circle construction gives us a far better understanding of what is really happening. Accordingly, we will emphasize the Mohr's circle approach. In the following example, we will see that we can use the same rules for Mohr's stress circles that we did for Mohr's compliance circles.

**Example 5.2** The principal stress element shown in the accompanying illustration is in plane stress. Plot the Mohr's circle; find the $xy$ axes such that the shear stress is maximum; draw and label the maximum shear-stress element.

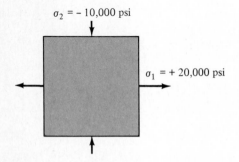

On a Mohr's stress circle, normal stresses will be plotted along the abscissa ($\sigma_1$, $\sigma_2$, $\sigma_x$, $\sigma_y$), and the shear stresses will be plotted vertically. Just as with compliance, we will plot the $x$ face down or up depending on the sign of the shear stress; the $x$ face is plotted below the abscissa if the shear stress is positive and vice versa.

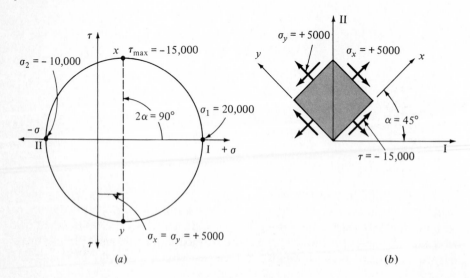

(a)                                                    (b)

The Mohr's circle illustrated is constructed as follows:

1. Plot the two principal stresses on the abscissa (zero shear stress), and draw the circle through them. Unlike our prior work with Mohr's circles, principal stresses can be either positive or negative (tensile or compressive).
2. Note that the maximum shear stress $\tau_{max}$ occurs at the top *and* the bottom of the circle; label the top $x$ and the bottom $y$ (or vice versa). The magnitude of $\tau_{max}$ is obviously equal to the radius, which is 15,000 psi.
3. Draw the second sketch ($b$) with the I–II axes as shown. Now draw the $xy$ axes; because we chose $x$ at the top of the circle, rotate 90° counterclockwise, on the circle, from the I axis to the $x$ axis. We must rotate in the same direction but one-half as far (45°) in real space. Once we know the direction of the $x$ axis, we draw both axes and the stress element.
4. Now label the element with proper stress values. The normal stresses are straightforward. The sign of the shear stress must come from our convention. We chose $x$ upward; therefore the shear stress had to be negative as shown.

It might seem that we arbitrarily established the sign of the shear stress by putting $x$ at the top of the circle instead of the bottom. If you will rework the problem, putting $x$ at the bottom of the circle, the shear stress will then be positive. *However*, the $x$ axis will now be directed clockwise 45° from I and the stress element will be identical. ∗∗∗

**Example 5.3** At a point in plane stress, the stresses are known relative to the set of $xyz$ coordinates, as shown in the accompanying illustration. Determine the principal stress magnitudes and directions. Show a properly labeled and oriented principal stress element.

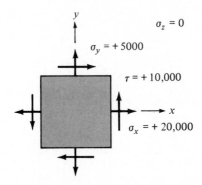

The approach is the same used in the previous example except that we will first plot points representing the $x$ and $y$ faces, and then draw the circle to establish the principal stress points as shown below:

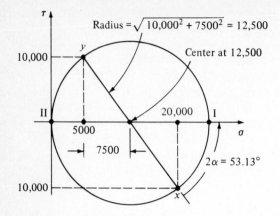

1. Plot $x$ at 20,000 to the right and 10,000 down (positive shear, plot $x$ down).
   Plot $y$ at 5000 to the right and 10,000 up.
2. Draw a line from $x$ to $y$ locating the center of the circle at 12,500.
3. Calculate the radius from the triangles (12,500 psi) and draw the circle which
   establishes I and II.
4. Calculate the angle between the $x$ axis and the I axis ($2\alpha = 53.13°$).
5. Draw the properly oriented element as shown below.

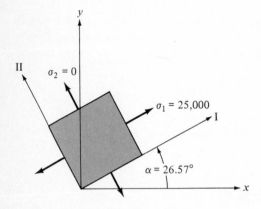

        ***

    In the above examples, both of which were plane stress problems, we were
rotating the axes about the $z$ axis, which was also the III axis.

    *Whenever* we use the Mohr's circle for stresses, we can rotate *only* about a
principal stress axis. Hence we must know one principal axis in order to use
this method. Fortunately, we most often do know a principal direction—from
symmetry, from a surface that is not loaded, or from a surface having only
normal stress applied.

## 5.6 THREE-DIMENSIONAL STATE OF STRESS

### 5.6.1 Equilibrium Equations

We saw that whenever stress varies from point to point, in a system under plane stress, equilibrium places restrictions on just how the stresses might vary. We obtained relationships among the partial derivatives of the stresses.

If now we consider the general, three-dimensional case, we will have precisely the same restrictions except now there will be more terms. When we derived the equilibrium relations among the stresses, we isolated a small element $dx \times dy \times 1$. We assumed that the only forces acting on the element were those due to the stresses acting on the surface of the element. In general, there may also be forces acting on the mass within the element, forces due to external fields of gravity, magnetism, or acceleration.

These forces, which act directly on the *body* of the element, as opposed to acting on the surface of the element, are called "body forces." We will simply denote them as $X$, $Y$, or $Z$: the body forces *per unit volume*, acting in the $x$, $y$, and $z$ directions, respectively. When these are incorporated into the three-dimensional relations, we have:

$$\frac{\partial \sigma_{xx}}{\partial x} + \frac{\partial \tau_{yx}}{\partial y} + \frac{\partial \tau_{zx}}{\partial z} + X = 0$$

$$\frac{\partial \tau_{xy}}{\partial x} + \frac{\partial \sigma_{yy}}{\partial y} + \frac{\partial \tau_{zy}}{\partial z} + Y = 0 \qquad (5.3)$$

$$\frac{\partial \tau_{xz}}{\partial x} + \frac{\partial \tau_{yz}}{\partial y} + \frac{\partial \tau_{zz}}{\partial z} + Z = 0$$

### 5.6.2 Mohr's Circle for Three Dimensions

When we considered plane stress at a point, and rotated the coordinate axes through that point, we saw that equilibrium provided a set of transformation equations. The Mohr's circle representation of those equations made it very easy to see how stress varied with rotation of axes. Let us now consider the three-dimensional state of stress at a point.

Looking at any of the Mohr's circles we have considered so far, we observe that when we have a circle passing through the principal stress points I and II, the circle involves rotation about the III axis. That is, if we look down along the III axis, we can observe rotation about that axis, in the I-II plane as shown in Fig. 5.11.

If we think about it for a moment, we realize that even if there had been a stress on the III face, our equations for rotation in the I-II plane would still hold—a force on the III face would not have any components in the $x$, $y$, I, or II directions. Therefore, as long as we consider rotations about one principal axis, we can use Mohr's circle to determine stresses in the plane of the other two principal axes.

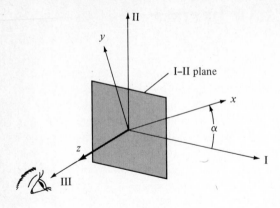

**Figure 5.11** "Looking down" the III axis, we can observe rotation ($\alpha$) in the I-II plane.

If we can observe rotations in the I-II plane when we look down the III axis, we can look down the I axis to observe rotation in the II-III plane, etc. Figure 5.12 shows that we can obtain three separate Mohr's circles by looking down the three principal axes in succession.

We can superpose the three Mohr's circles shown in Fig. 5.12 to obtain a single plot, as shown in Fig. 5.13; the three circles obviously pass through the three principal stress points I, II, and III.

Given the three principal stresses, we can plot the three circles, and can calculate stresses when we rotate the axes about any *one* principal axis at a time. We cannot (using this construction) consider the general case involving simultaneous rotation about two or more principal axes. We can, however, make a very useful, general statement—one we will not prove:

The stresses on *any* area, at *any* rotation, when plotted on the three-dimensional Mohr's circle diagram shown in Fig. 5.13, will be represented by a point either *on* one of the three circles, or within the shaded area shown in Fig. 5.13. There are *never* stresses outside the large circle, or within the smaller circles.

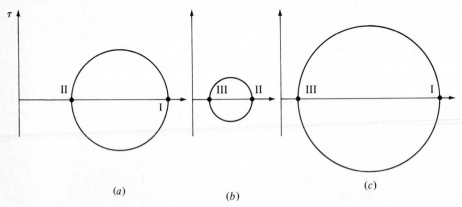

**Figure 5.12** Mohr's circle representation of stress: (*a*) looking down the III axis; (*b*) looking down the I axis; (*c*) looking down the II axis.

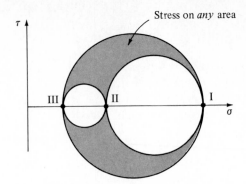

Stress on *any* area

III    II    I

**Figure 5.13** Superposition of the three Mohr's circles shown in Fig. 5.12.

It is indeed fortunate that in most instances we know, a priori, at least one principal stress direction. We can then look down that axis to determine the other two. Thus we can construct the three-dimensional Mohr's stress diagram shown in Fig. 5.13.

Clearly, we can calculate the general three-dimensional case from equilibrium relations (see S.P. Timoshenko and J.N. Goodier, *Theory of Elasticity*, 3d ed., McGraw-Hill, New York, 1970). However, this author, in many years of involvement with stress and strain, has never had to make that calculation.

**Example 5.4** The state of stress at a point is shown relative to the *xyz* axes in the accompanying illustration. There are no stresses on the *z* face; therefore, the *z* direction is a principal direction. Determine the magnitude of the maximum shear stress, and show it on a properly oriented stress element.

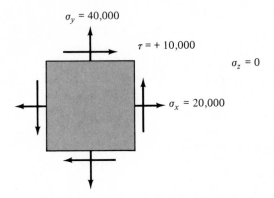

$\sigma_y = 40,000$

$\tau = +10,000$

$\sigma_z = 0$

$\sigma_x = 20,000$

As usual, we start by plotting the *x* and *y* faces, as shown. *x* is plotted down because the shear stress is positive. We then draw the solid circle through *x* and *y*. The circle is centered halfway from *x* to *y*; the angle $2\alpha$ is 45°, and the principal stresses are

$$\sigma_1 = 30,000 + 14,142 = 44,142 \text{ psi}$$

$$\sigma_2 = 30,000 - 14,142 = 15,858 \text{ psi}$$

The third principal stress, being zero, is plotted at the origin. We can now draw the other two circles (shown dashed) through the principal stresses.

It is clear that the maximum shear stress is not on the solid circle, but is at the top and bottom of the *largest* circle passing through I and III. The magnitude is clear (see the drawing below):

$$\tau_{max} = \frac{\sigma_1 - \sigma_3}{2} = 22,071 \text{ psi}$$

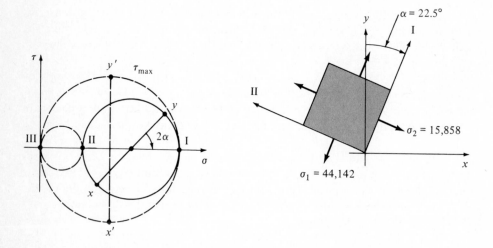

Now let us try to draw the element on which $\tau_{max}$ will appear. This is a bit complicated, and is made more difficult because this page is printed in two dimensions; however, we will do our best. This geometry is the most complex that we will ask you to consider in this text.

The maximum shear stress is plotted on the circle passing through I and III. We see this circle when "looking down" the II axis, as shown in the illustration below. If we "looked down" the II axis, we would see the I axis pointing up, and the III axis pointing to our right. On this I-III plane, we can draw the maximum shear-stress element.

As before, we can arbitrarily label the axes associated with the maximum shear-stress element as $x'$ and $y'$, $x'$ being arbitrarily at the bottom of the Mohr's circle. To go from the I axis to the $x'$ axis, we rotate clockwise 90° on the Mohr's circle; on the I-III plane, we must rotate 45°, as shown. So we establish the $x'y'$ axes on the I-III plane, and draw the element. Because $x'$ is plotted "down," the shear stress is positive.

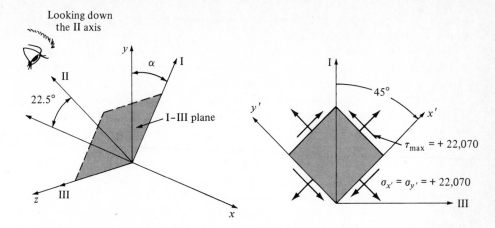

We will see in Chap. 7 that the maximum shear-stress magnitude is frequently a critical quantity (we do not often have to worry about direction!). When considering plane stress, it is very easy to forget that the largest circle *may* pass through *zero*. In Example 5.4, if we had plotted only the solid circle, and taken $\tau_{max}$ to be its radius, we would have made a serious error. This is an instance when "zero" can be a very important quantity.

## 5.7 STRAIN AT A POINT

Although stress and strain are quite different quantities, we will see certain similarities between the way in which we proceed to analyze strain and the way in which we analyzed stress. First, we will define strain at a point, then we will consider the two-dimensional case of plane strain. We will determine how strain may vary from point to point, and how it varies with rotation of axes.

Strain, in general, is associated with motion. Strain at a point must be associated with motion of that point. However, if every point on a body undergoes the *same* motion, the body merely translates and is not strained at all. Thus the motion, or displacement, of a single point cannot define the strain at that point. Strain involves the *relative motion* between two or more points. It is only when two points undergo different motions that we have deformation, or strain. In this book, we will consider only small deformations and small strains (strains of less than a few percent).

In Chap. 3 we defined strain ($\epsilon$) as the change in length divided by the original length (as the original length approached zero). Just as the tensile (compressive) stresses were very special cases, so the tensile (compressive) strains are also special cases. In a more general sense, we must consider the overall distortion of a volume of material (as the dimensions of the volume approach zero).

One of the best ways to visualize strain is to consider the change in shape and size of a minute cubic element. It is also instructive to consider the sphere that is contained within the cube.

Just as we did with stress, we will consider the two-dimensional case for strain first. In this instance, there is *zero strain* in the *z* direction; we call this "plane strain."

Figure 5.14*a* shows a square in the *xy* plane with an inscribed circle. In Fig. 5.14*b* through *e*, the element has been distorted in various ways. In Fig. 5.14*b* and *c*, the element is lengthened in one direction and shortened in the other; the circle has become elongated, or elliptical, with major and minor axes in the *x* and *y* directions. In Fig. 5.14*d* the element has become larger in both directions such that the square remains a square and the circle remains a circle; the element has changed size, but not shape. In Fig. 5.14*e*, the square has become a rhombus of equal area. The circle has become elliptical, but now the major and minor axes are inclined.

If we concentrate on the circles shown in Fig. 5.14, we see that the deformations in Fig. 5.14*b*, *c*, and *e* are very similar except that they occur at different angles. In fact the geometrical deformation, as shown by the circle becoming elliptical, could be the same in all three cases. But the perceived deformation, in terms of the change in shape of the square, is quite different. If we had chosen a different set of *xy* axes, the deformations of the square would again be different.

Consider now the deformation relative to the *xy* axes. Figure 5.15 shows two vectors, *OB* and *OA*, in the *x* and *y* directions. These could represent two sides of the original square shown in Fig. 5.14. Let the original length of these vectors be $L_0$. Following deformation, point *A* has moved to *A'*, and *B* to *B'*. Each point has undergone a displacement in the *x* direction and one in the *y* direction, as represented by the short displacement vectors. In each case, one displacement vector is parallel to the original vector, and one is at right angles to it.

We have used a double-subscript system for the displacement vectors that should appear familiar to you: The first subscript defines the direction of the original vector, while the second defines the direction of the displacement.

The displacements with like subscripts (*xx* or *yy*) produce *normal* strains of extension or compression—the same type of strains we saw in Chap. 3.

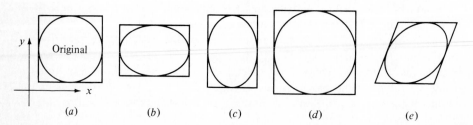

(a)    (b)    (c)    (d)    (e)

**Figure 5.14** Showing how an originally square element can be distorted into various sizes and shapes.

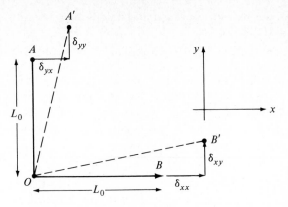

Figure 5.15 Strain defined in terms of the motion of the ends of two orthogonal vectors.

Normal strain (small):
$$\epsilon_{xx} = \frac{\delta_{xx}}{L_0}$$
$$\epsilon_{yy} = \frac{\delta_{yy}}{L_0}$$
as $L_0 \to 0$ (5.4)

The displacements having unlike subscripts ($xy$ or $yx$) produce *shear*-type deformation of the material and/or gross rotation of the material. We will define shear strain and rotation as follows:

Shear strain (small):
$$\gamma_{xy} = \gamma_{yx} = \frac{\delta_{xy}}{L_0} + \frac{\delta_{yx}}{L_0}$$
as $L_0 \to 0$ (5.5)

Rotation:
$$\omega = \frac{\delta_{xy}/L_0 - \delta_{yx}/L_0}{2}$$

(In this text we will not concern ourselves with rotation.)

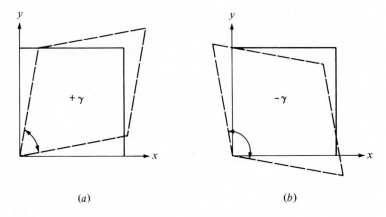

(a)                                    (b)

**Figure 5.16** Showing (a) positive shear strain and (b) negative shear strain.

Just as with stresses, we will let the sign of the strain be determined by the signs associated with the double subscripts.

For small strains (and we are concerned primarily with very small strains), we can see from Fig. 5.15 that the shear strain is equal to the *change* in the original right angle *BOA*. As shown in Fig. 5.16, when the angle between the *x* and *y* axes becomes smaller, the shear strain is positive; when it grows, shear is negative. If you compare Fig. 5.8 with Fig. 5.16, you will see that positive shear stress produces positive shear strain, just as positive normal stress produces positive normal strain.

## 5.8 COMPATIBILITY OF STRAIN

When we considered stress at a point, we found that equilibrium determined how stresses could vary from point to point and how they could vary with rotation of axes. When we consider strains, they must be geometrically "compatible." The only requirement for geometric compatibility is that we do not produce voids, or place two masses in a volume that can hold only one; i.e., we must have continuity.

When we look for a way to express geometric compatibility in terms of strains, it is not clear just how to do so. However, it is very easy to express the compatibility requirements in terms of the overall *motions* that produced the strains.

If we consider a body in the *xyz* coordinate system, we can define the motion of every point ($x$, $y$, $z$) in terms of the components of displacement in the *x*, *y*, and *z* directions. The displacement components are generally labeled *u*, *v*, and *w*.

$$u(x, y, z) = \text{displacement of point } x, y, z \text{ in the } x \text{ direction}$$

$$v(x, y, z) = \text{displacement of point } x, y, z \text{ in the } y \text{ direction}$$

$$w(x, y, z) = \text{displacement of point } x, y, z \text{ in the } z \text{ direction}$$

Compatibility requires that the displacements *u*, *v*, and *w* be *continuous* functions of *x*, *y*, and *z*. Any discontinuity will represent either fracture or evaporation of mass!

In order that the compatibility requirements be of use, we must formally relate the displacements *u*, *v*, and *w* to the strains in the material. Returning to the two-dimensional case, Fig. 5.17 shows two vectors *OA* and *OB* of initial length *dy* and *dx*. After deformation, the vectors have moved to the primed positions.

In terms of the displacement functions for *u* and *v*, we see that point *O* has moved *u* to the right and *v* upwards. If *A* and *B* also moved *u* and *v*, there would be no deformation, only translation.

However, point *B* moves to the right not only *u*, but in addition it moves an amount equal to the variation in *u* associated with the length *dx* (the partial

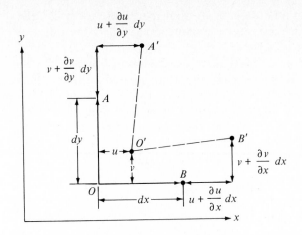

**Figure 5.17** Geometry to determine strains in terms of the displacement functions.

derivative term). If we compare Figs. 5.17 and 5.15, we can immediately write the strains in terms of the displacements:

$$\epsilon_{xx} = \frac{\partial u}{\partial x}$$

$$\epsilon_{yy} = \frac{\partial v}{\partial y} \tag{5.6}$$

$$\gamma_{xy} = \frac{\partial v}{\partial x} + \frac{\partial u}{\partial y}$$

Thus, *if* the displacement equations for $u$, $v$, and $w$ are continuous, then the strains [as given by Eqs. (5.6)] are compatible. This does not mean that they are correct, just that they are possible.

From Eqs. (5.6), we can calculate three strain quantities from only two initial equations of motion ($u$ and $v$). The three strains cannot be independent, but must be related. If we take the proper partial derivatives we can readily show that:

$$\frac{\partial^2 \epsilon_{xx}}{\partial y^2} + \frac{\partial^2 \epsilon_{yy}}{\partial x^2} = \frac{\partial^2 \gamma_{xy}}{\partial x\, \partial y} \tag{5.7}$$

## 5.9 ROTATION OF STRAIN AXES

If (for plane strain) we know the displacement functions $u(x, y)$, and $v(x, y)$ relative to the $xyz$ coordinate system, it is simply a matter of geometric manipulation to determine the displacement functions $u'$ and $v'$ relative to an $x'y'z$ set of axes that are rotated an angle $\alpha$ about the $z$ axis. We can then take the partial derivatives of the new functions with respect to the new axes [as in Eqs. (5.6)] to determine the strains in the new coordinate system. It should not surprise you that the resulting relationships are similar to those for stress and compliance—namely, the strains are tensor quantities that follow the rules we have established for the Mohr's circle.

At any point within a deformed body there will be a set of *principal strain axes* (I-II-III). Relative to this set of axes, there are no shear strains, only normal strains.

A cubic *principal strain element* can undergo a change of dimension, but it will remain rectangular. Again, if we consider the rotation of axes about one of the principal axes, the Mohr's circle construction is valid. The equations for strain relative to the $xy$ axes when rotated about the III (or $z$) axis are:

$$\epsilon_{xx} = \frac{\epsilon_1 + \epsilon_2}{2} + \frac{\epsilon_1 - \epsilon_2}{2} \cos 2\alpha$$

$$\epsilon_{yy} = \frac{\epsilon_1 + \epsilon_2}{2} - \frac{\epsilon_1 - \epsilon_2}{2} \cos 2\alpha \tag{5.8}$$

$$\frac{\gamma_{xy}}{2} = \qquad -\frac{\epsilon_1 - \epsilon_2}{2} \sin 2\alpha$$

Equations (5.8) are *almost* the same as those for the rotation of axes for stress; the difference is that the third equation is for *half* of the shear strain rather than the full value. When using Eqs. (5.8), or their Mohr's circle representation, it is very easy to forget that factor of 2. Just try to remember that there is *something* different about the Mohr's circle for strain. Given this one qualification (the factor of 2), we can treat strain rotations precisely as we did stresses. The sign convention for shear is the same, so we plot the Mohr's circle in exactly the same way.

When we deal with plane strain, in a manner analogous to stress, we can drop the double-subscript system without ambiguity:

$$\epsilon_x = \epsilon_{xx} \qquad \epsilon_y = \epsilon_{yy} \qquad \gamma = \gamma_{xy}$$

**Example 5.5** Consider the state of plane strain at a point given by:

$$\epsilon_x = 800 \times 10^{-6}$$

$$\epsilon_y = 200 \times 10^{-6}$$

$$\gamma = -600 \times 10^{-6}$$

Draw the Mohr's circle diagram, then draw two properly oriented and labeled strain elements, the principal strain element, and one showing the maximum shear strain in the $xy$ plane. Draw the distorted element.

To construct the Mohr's circle, plot $x$ $800 \times 10^{-6}$ to the right and $300 \times 10^{-6}$ upward (half the shear strain; negative shear, plot $x$ upward). Plot $y$ at the opposite end of a diameter, and draw the circle. Calculate the principal strains from the triangles.

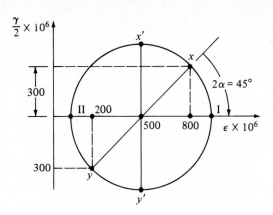

$$\epsilon_{1,2} = [+500 \pm 300\sqrt{2}] \times 10^{-6}$$
$$\epsilon_1 = 924 \times 10^{-6}$$
$$\epsilon_2 = 75.7 \times 10^{-6}$$
$$\left(\frac{\gamma}{2}\right)_{max} = 424 \times 10^{-6}$$
$$\gamma_{max} = 848 \times 10^{-6}$$

To draw the principal strain element on the $xy$ plane, locate the I axis by rotating 22.5° (half of 45°) clockwise from the $x$ axis. Draw the distorted element as shown (exaggerated).

The maximum shear strain axes are labeled $x'$ and $y'$; rotate counter-clockwise half of 45° from the $x$ axis to the $x'$ axis. Because we arbitrarily put $x'$ *up* at the top of the circle, the shear strain is negative, and the angle between the $x'$ and the $y'$ axes must increase (see the second accompanying illustration).

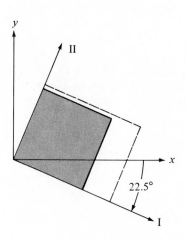

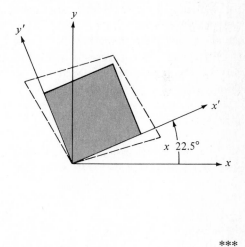

\*\*\*

## 5.10 THREE-DIMENSIONAL STATE OF STRAIN

The generalization from plane strain to the three-dimensional case is similar to that for stress. The only requirement has already been stipulated, namely, that all three displacements ($u$, $v$, $w$) be continuous functions of $x$, $y$, and $z$.

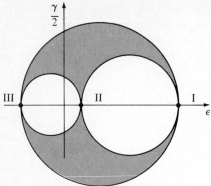

**Figure 5.18** Mohr's circle for the three-dimensional state of strain.

In terms of the displacement functions, the six components of strain are

$$\epsilon_{xx} = \frac{\partial u}{\partial x} \qquad \gamma_{xy} = \frac{\partial u}{\partial y} + \frac{\partial v}{\partial x}$$

$$\epsilon_{yy} = \frac{\partial v}{\partial y} \qquad \gamma_{yz} = \frac{\partial v}{\partial z} + \frac{\partial w}{\partial y} \qquad (5.9)$$

$$\epsilon_{zz} = \frac{\partial w}{\partial z} \qquad \gamma_{zx} = \frac{\partial w}{\partial x} + \frac{\partial u}{\partial z}$$

We now have six strain quantities derived from the original three independent motion equations; thus there must be three relations among the six strains. These are obtained from Eq. (5.7) by merely cycling the *xyz* subscripts.

For rotation of axes, as with stresses, there are three Mohr's circles corresponding to rotation about each of the principal axes. The strains associated with *all* other rotations are within the shaded area shown in Fig. 5.18.

Again, your only problem will be to remember the factor of 2 on the shear-strain axis.

## 5.11 MEASUREMENT OF STRAIN

Whenever we deal with a *complex* component, our analyses are at best approximate, and we must adopt a suitable degree of humility relative to the precision of our predictions.

The only way we will really know how a part will behave is to test it to failure. Because a part may be subject in service to a wide range of loading conditions, we should make numerous tests to failure under various conditions. Clearly, such a process can become prohibitive in terms of time and money.

When we test critical or expensive parts, we would like to extract as much information as possible from a single test, information which can be used to lend credence (or doubt) to our analyses. In this way we increase the accuracy of our predictions and our confidence in them.

Regarding the subject matter of this text, we would like to be able to measure the stresses and the strains at all points on and within a body when it is loaded. Simply put, this cannot be done. In general, we can make observations only at the outside of a part. We then must deduce, from the external observations, the internal conditions.

Furthermore, it is generally impossible to measure stress directly; we simply have no physical means for doing so. What we can do, however, is to measure the displacements and deformations due to the loading, and from these deduce the stresses. The fascinating field called "Experimental Stress Analysis" really involves *strain* measurement followed by stress deduction.

Many methods and instruments have been developed for the measurement of strain. A generic problem is that for most "engineering materials," the elastic modulus is high, and the strains to be measured are very small. For highly elastic materials such as rubber, we could observe the change in shape of a circle scribed on the surface, but this would be difficult to do for a piece of steel. We must use methods which are inherently sensitive to very small displacements or strains.

In this section, we will very briefly discuss a number of the more important methods for measurement of strain.

### 5.11.1 Optical Methods

Because the wavelength of light is small and reproducible, there are a number of optical methods for strain measurement:

Holography provides us with the potential for observing simultaneously the displacements of numerous points on the surface of a body. However, the method is not yet developed well enough for our purposes.

Photoelasticity is a well-developed method particularly suited to plane stress problems. The part is modeled in an optically birefringent material (typically a plastic), and is loaded while polarized light is passed through it. When properly done, optical lines, or fringes, can be observed and interpreted regarding strain.

Reflective photoelasticity involves the use of a layer or birefringent material applied to the surface of a part. Polarized light is reflected from the part and interpreted in terms of the surface strains.

Three-dimensional photoelasticity can be carried out using various "freezing" and sectioning techniques, but is very cumbersome.

### 5.11.2 Brittle Lacquer

Brittle materials are inherently very sensitive to strain. If strained very much, they break (in tension). This is exploited in the "brittle lacquer" method of strain indication. Basically, a thin layer of a very brittle lacquer is sprayed onto the surface of the part. After the lacquer has hardened, the part is loaded. If the part is carefully observed, the first cracks that appear in the lacquer will indicate the point of maximum tensile strain. As loading is continued, one can record the relative loads at which cracking occurs in various areas of the coating. This gives some information about relative strain levels. This is a simple test which gives good qualitative results. In addition, the direction of the principal strains can be inferred from the direction of the cracks.

### 5.11.3 Strain Gages

It is not surprising in today's world that one of the most used methods for strain measurement involves sensitive electronic measurement techniques. Any conductor or semiconductor will undergo a change of resistance if it is stretched or compressed, semiconductors being perhaps 50 to 200 times as sensitive as conductors.

"Strain gage" is the generic name applied to a variety of devices used for the electrical measurement of strain. They are used by the millions, not only as a means of indicating strain as part of the design/test procedure, but also as the sensing "transducer" in many instruments for measuring force, pressure, acceleration, etc. Because strain gages are so widely used, and because they provide good quantitative results, we will describe them in some detail.

Figure 5.19 shows schematically three types of gages made for measuring strain in the $y$ direction. The length of gages ($l$) is typically 0.06 to 0.5 in, and they are simply "glued" onto the surface to be tested:

1. Wire gages use a length of wire that is about 0.001 in in diameter. For such a small diameter, the ratio of surface area to cross-sectional area is high, and any glue is capable of transmitting enough force to make the wire stretch when the substrate is stretched. The wire is formed into a flat coil and glued between layers of insulating paper.
2. In foil gages, which are the most widely used today, the "coil" is photo-etched from a thin (0.0001-in) layer of metal foil sandwiched between layers of plastic.
3. Semiconductor gages use a very thin piece of P- or N-type semiconducting material with leads attached, which is secured between layers of plastic.

When the metal gages shown in Fig. 5.19 are stretched in the $y$ direction, their resistance increases due primarily to increased length and reduced area (Poisson effect). Gage sensitivity is expressed in terms of resistance change divided by initial resistance. The "gage factor" is the ratio of this relative resistance change to the axial strain. For metal gages, the gage factor is on the order of 2 to 3.

Semiconductor gages experience a much greater resistance change due to the change in electron mobility with strain. Their gage factors range from 50 to 300. This high sensitivity is accompanied by relatively large, undesirable temperature effects.

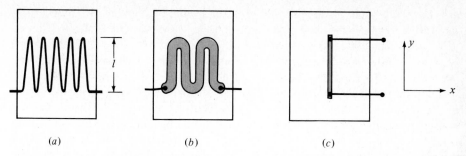

(a)          (b)          (c)

**Figure 5.19** Typical strain gages: (a) wire gage; (b) foil gage; (c) semiconductor gage.

Without going into great detail, we can say that when we use conventional electronics, the resolution of the gages is approximately $10^{-6}$ to $10^{-5}$ (strain) for metal gages and $10^{-8}$ to $10^{-7}$ (strain) for semiconductor gages. With great care, these numbers can be reduced by a factor of 10.

Let us see what this all means: Figure 5.20 shows a gage attached to the side of a steel rod having a cross-sectional area of $0.2 \text{ in}^2$. When a force $F$ is applied, the gage stretches. Assuming a resolution of $10^{-6}$, a stress of 30 psi will produce a reading; this requires only 6 lb!

When the 6 lb is applied, if the gage was originally 0.1 in long, it will have lengthened $0.1 \times 10^{-6} = 10^{-7}$ in! This is approximately 10 atomic spacings of the iron in the rod! With the semiconductor gages, using care with the electronics, we can resolve about 1/100 of an atomic spacing!

For most applications, strain gages are wired into a "Wheatstone Bridge" configuration, a configuration of four resistors, any or all of which can be strain gages. With proper gage placement and wiring, many undesirable effects can be eliminated from the output.

### 5.11.4 Strain Rosettes

Of particular interest in the context of stress and strain are gages made especially to measure the total state of strain. Generally, gages are cemented onto a surface that has zero stress acting on it, i.e., on a surface that is a principal stress face. Generally, a principal stress direction will correspond to a principal strain direction. Hence, we should be able to use the Mohr's circle construction, which permits rotation about a principal strain axis, to determine the principal strains in the plane of the surface.

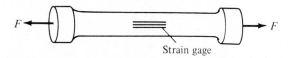

Strain gage

**Figure 5.20** Strain gage attached to a steel rod.

We know that to specify the state of strain in a plane, we need three quantities: typically two normal strains and the associated shear strain. There is no way we can determine three quantities from one strain gage; in fact, we require an independent measurement for each quantity we wish to determine. We need three quantities; therefore we need three strain gages.

Special strain gages are made for this purpose: Three gages are assembled into a single unit which is called "strain rosette." Figure 5.21 shows the two common strain rosettes; the 45° rosette where the gages are at 45° to each other, and the 60° rosette. If we obtain strain measurements from the three gages, we can then determine the state of strain (in the plane of the surface).

Consider first the analysis of data from a 45° rosette where we have determined the three strains $\epsilon_a$, $\epsilon_b$, and $\epsilon_c$ from the rosette shown in Fig. 5.21. The appropriate Mohr's circle construction is shown in Fig. 5.22. Because the gages are 45° apart in real space, the points representing the $a$, $b$, and $c$ directions are 90° apart on the Mohr's circle. The process of constructing the

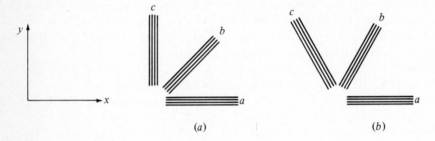

**Figure 5.21** Strain gage rosettes: 45° and 60°.

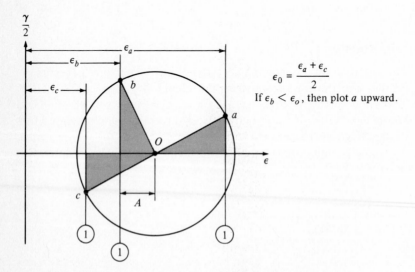

$$\epsilon_0 = \frac{\epsilon_a + \epsilon_c}{2}$$

If $\epsilon_b < \epsilon_o$, then plot $a$ upward.

**Figure 5.22** Mohr's circle construction for a 45° strain gage rosette.

Mohr's circle is as follows:

1. Lay off the strains $\epsilon_a$, $\epsilon_b$, and $\epsilon_c$ along the abscissa, and draw in three vertical lines (labeled ①) through these points.
2. The problem is simple geometry: Construct a circle, intersecting the vertical lines at $a$, $b$, and $c$, such that points $a$, $b$, and $c$ are 90° apart and in the proper rotational sequence.
3. Because $a$ and $c$ are at opposite ends of a diameter, the center of the circle is located half-way between them at point $O$.
4. The only question remaining is how far $a$ and $c$ should be above or below the axis. Looking at Fig. 5.22, we see three identical triangles (shaded); clearly the distance up (or down) to $a$ is exactly equal to the horizontal distance $A$, measured from the center of the circle to $b$.
5. Assuming that $a$, $b$, and $c$ are always 45° counterclockwise from each other as in Fig. 5.21, then, if $b$ is to the left of the center, plot $a$ upwards.

If you prefer to use equations with our usual sign convention, we can write for the 45° rosette:

$$\epsilon_x = \epsilon_a$$

$$\epsilon_y = \epsilon_c \tag{5.10}$$

$$\frac{\gamma}{2} = \epsilon_b - \frac{\epsilon_a + \epsilon_c}{2}$$

For a 60° rosette, where the gages are arranged as shown in Fig. 5.21, the construction is as follows:

1. The center of the circle is at the average reading:

$$\epsilon_0 = \frac{\epsilon_a + \epsilon_b + \epsilon_c}{3}$$

2. Calculate the appropriate strains:

$$\epsilon_x = \epsilon_a$$

$$\epsilon_y = -\frac{\epsilon_a}{3} + \frac{2\epsilon_b}{3} + \frac{2\epsilon_c}{3} \tag{5.11}$$

$$\frac{\gamma}{2} = \frac{\epsilon_b - \epsilon_c}{\sqrt{3}}$$

## 5.12 COMPUTER EXAMPLE—MOHR'S CIRCLE ANALYSIS

The calculations associated with the Mohr's circle certainly are not at all complex. However, they can become tedious if they must be repeated many times. Program Listing 5.1 shows a simple program for calculating the principal stresses, the direction of the principal stress axes, and the maximum shear

```
10 !"              MOHR'S CIRCLE for STRESS"\!
20 REM
30 !"Input stresses (or compliance, strain, or area-moments of inertia"
40 !" For strain, input HALF the shear-strain for Tau-xy"\!
50 INPUT"    Sigma-x = ", X \!
60 INPUT"    Sigma-y = ", Y \!
70 INPUT"    Tau-xy  = ", T \!
80 C=(X+Y)/2                       \REM  CENTER OF CIRCLE
90 D=(X-Y)/2
100 R=(D*D+T*T)^.5                 \REM  RADIUS OF CIRCLE
110 IF D<>0 THEN 140
120 A=45
130 GOTO 150
140 A= ABS(28.6479*ATN(T/D))
150 IF D<0 THEN A=90-A
160 IF T<0 THEN A=-A
170!\!
180 !"        Sigma-1 = ", C+R \!
190 !"        Sigma-2 = ", C-R \!
200 !"        Alpha   = ", A \!
210 !"        Tau-max = ", R \!
220 !"        Positive alpha means that the I-axis is"
230 !"        counterclockwise (alpha) from the x-axis." \!\!
240 !\!
250 INPUT"   Do you want the stresses at other values of alpha? (CR/N)",Z$
260 !\!
270 IF Z$="N" THEN END
280 INPUT"   Alpha = ", B1 \!
290 B=(2*B1-2*A)/57.2958
300 T=-R*SIN(B) \ X=C+R*COS(B) \ Y=C-R*COS(B)
310 !"        Sigma-x' = ",X \!
320 !"        Sigma-y' = ",Y \!
330 !"        Tau-x'y' = ",T \!
340 !\!
350 GOTO 250
```

**Program Listing 5.1**

stress. Also shown as an example is Program Output 5.1. The program in addition will compute the stresses at any angle to the original axes (the program does not consider the third principal stress).

Although the program is written for Mohr's stress circle, it can be used for compliance and strain just as well (be sure to input *half* of the shear strain). In Chap. 8 we will see that this program can also be used to determine certain characteristics of a beam cross section.

MOHR'S CIRCLE for STRESS

Input stresses (or compliance, strain, or area-moments of inertia
 For strain, input HALF the shear-strain for Tau-xy

　　Sigma-x = 25000

　　Sigma-y = -10000

　　Tau-xy  = -10000

　　　　Sigma-1 =  27655.641

　　　　Sigma-2 =  -12655.641

　　　　Alpha   =  -14.872446

　　　　Tau-max =  20155.641

　　　　Positive alpha means that the I-axis is
　　　　counterclockwise (alpha) from the x-axis.

Do you want the stresses at other values of alpha? (CR/N)

Alpha = 45

　　　　Sigma-x' =  -2499.9869

　　　　Sigma-y' =  17499.987

　　　　Tau-x'y' =  -17500.004

Do you want the stresses at other values of alpha? (CR/N)N

**Program Output 5.1**

## 5.13 PROBLEMS

For Probs. 5.1 to 5.6: In the text, we have considered stress and strain relative to the *xyz* coordinate system. For devices that are "circular" in shape, it may be easier to use the cylindrical coordinates *rθz* shown below.

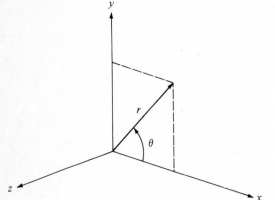

**Figure P5.1–5.6**

The displacements in the *r*, *θ*, and *z* directions are usually called *u*, *v*, and *w*, just as in the rectangular coordinate system.

When we are concerned with stress and strain at a point, there is no difference between the two systems. However, when we move from point to point, the associated equations are different.

**5.1** Show the cylindrical coordinate element that is equivalent to that shown in Fig. 5.3. Show all positive stresses.

**5.2** If there are zero stresses on the *z* face, and the stresses are symmetrical about the *z* axis, show that the coordinate directions are also the principal stress directions, and that equilibrium gives:

$$\frac{\partial \sigma_r}{\partial r} + \frac{\sigma_r - \sigma_\theta}{r} = 0$$

**5.3** Consider a condition of plane stress where there is zero stress on the *z* face. Draw elements that are equivalent to those shown in Figs. 5.5 and 5.6. Show that the equilibrium relations are:

$$\frac{\partial \sigma_{rr}}{\partial r} + \frac{1}{r}\frac{\partial \tau_{r\theta}}{\partial \theta} + \frac{\sigma_{rr} - \sigma_{\theta\theta}}{r} = 0$$

$$\frac{\partial \tau_{r\theta}}{\partial r} + \frac{1}{r}\frac{\partial \sigma_{\theta\theta}}{\partial \theta} + 2\frac{\tau_{r\theta}}{r} = 0$$

**5.4** Show that the equilibrium relations, for a general state of stress in cylindrical coordinates, are:

$$\frac{\partial \sigma_{rr}}{\partial r} + \frac{1}{r}\frac{\partial \tau_{r\theta}}{\partial \theta} + \frac{\partial \tau_{zr}}{\partial z} + \frac{\sigma_{rr} - \sigma_{\theta\theta}}{r} = 0$$

$$\frac{\partial \tau_{r\theta}}{\partial r} + \frac{1}{r}\frac{\partial \sigma_{\theta\theta}}{\partial \theta} + \frac{\partial \tau_{\theta z}}{\partial z} + \quad 2\frac{\tau_{r\theta}}{r} = 0$$

$$\frac{\partial \tau_{zr}}{\partial r} + \frac{1}{r}\frac{\partial \tau_{\theta z}}{\partial \theta} + \frac{\partial \sigma_{zz}}{\partial z} + \quad \frac{\tau_{zr}}{r} = 0$$

**5.5** If the deformations are symmetrical about the $z$ axis, and there is zero strain in the $z$ direction (plane strain), then each point must move in a purely radial direction ($v = w = 0$). For this condition, show that the strains are:

$$\epsilon_r = \frac{\partial u}{\partial r} \qquad \epsilon_\theta = \frac{u}{r} \qquad \gamma_{r\theta} = 0$$

**5.6** Show that for plane strain ($\epsilon_z = 0$) the strains are:

$$\epsilon_r = \frac{\partial u}{\partial r} \qquad \epsilon_\theta = \frac{1}{r}\frac{\partial v}{\partial \theta} + \frac{u}{r} \qquad \gamma_{r\theta} = \frac{\partial v}{\partial r} + \frac{1}{r}\frac{\partial u}{\partial \theta} - \frac{v}{r}$$

**5.7** For the cases shown in the accompanying illustration, draw the Mohr's stress circles (all three), showing the principal stresses and the maximum shear stress.

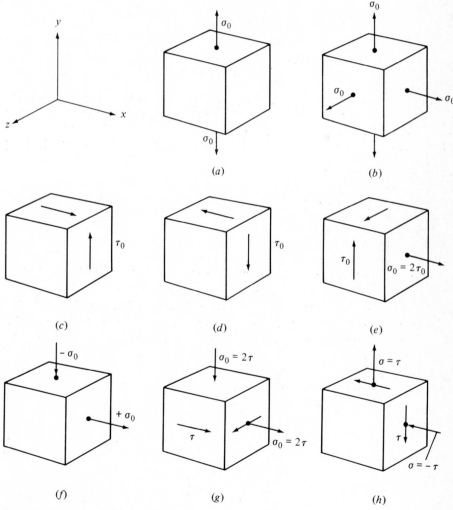

**Figure P5.7**

**5.8** The following cases are all *plane* stress, where $\sigma_z = 0$. For each case, draw the three Mohr's circles (on a single diagram), determine the principal stresses and directions, and show a properly oriented principal stress element (in the $xy$ plane).

| Case | $\sigma_x$, psi | $\sigma_y$, psi | $\tau_{xy}$, psi |
|------|-----------------|-----------------|------------------|
| (a) | +8,000 | 0 | +3,000 |
| (b) | +75,000 | −15,000 | −30,000 |
| (c) | +30,000 | −30,000 | 0 |
| (d) | −6,000 | +4,000 | −6,000 |
| (e) | −6,000 | −20,000 | +12,000 |

**5.9** Carry out the same procedures as in Prob. 5.8 for the following cases:

| Case | $\sigma_x$ | $\sigma_y$ | $\tau_{xy}$ |
|------|------------|------------|-------------|
| (a) | +500 MPa | 0 MPa | +300 MPa |
| (b) | +300 MPa | −120 MPa | −160 MPa |
| (c) | +700 MN/m² | −700 MN/m² | 0 MN/m² |
| (d) | −200 MN/m² | +40 MN/m² | −120 MN/m² |
| (e) | −300 MPa | −750 MPa | −450 MPa |

**5.10** In Chap. 3, we saw that the hoop stress due to internal pressure in a cylinder was $\sigma_\theta = pr/t$ when the cylinder was "thin" ($t/r \ll 1$). Consider now the "closed" thin-walled cylindrical tank shown in the accompanying illustration. Show that the radial, axial, and tangential directions are the principal stress directions. Draw the Mohr's stress circles, and determine the maximum shear stress.

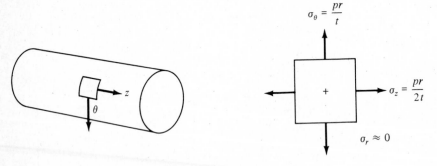

**Figure P5.10**

**5.11** A thin-walled cylinder, of radius $r$ and thickness $t$, is fitted with freely sliding pistons at each end as shown in the accompanying illustration. Determine the principal stresses and the maximum shear stress.

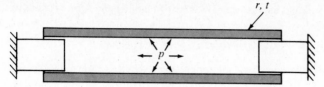

**Figure P5.11**

**5.12** For the cylinder discussed in Prob. 5.10 show in a sketch the plane on which the maximum shear stress acts.

**5.13** For a thin-walled *sphere*, of radius $r$ and thickness $t$, subject to internal pressure $p$, determine the principal stresses, their directions, and the maximum shear stress.

**5.14** A high-pressure tank has a long cylindrical portion, 10 in in diameter, and $\frac{1}{2}$ in thick. When it is filled with air at 2000 psi, what is the maximum tensile stress and the maximum shear stress?

**5.15** A U-shaped pressure tube (see the accompanying illustration) is made from three lengths of pipe ($A$, $B$, and $C$) joined with 90° elbows and sealed with end caps. The pipe is 1 in in diameter, with a thickness of 0.08 in. If the assembly is pressurized to 500 psi, what are the principal stresses and the maximum shear stress in member $B$?

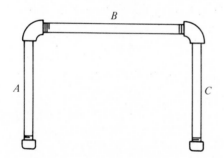

**Figure P.5.15**

**5.16** The cases below are all in plane strain; the strain in the $z$ direction is zero.

1. Draw the $xy$ strain element showing the deformation greatly exaggerated.
2. Draw the Mohr's strain circle, and determine the principal strains.
3. Sketch the deformed principal strain element and the maximum shear-strain element.

| Case | $\epsilon_x$ | $\epsilon_y$ | $\gamma_{xy}$ |
|------|------|------|------|
| ($a$) | +0.000800 | +0.000200 | −0.001200 |
| ($b$) | −0.001600 | +0.000400 | +0.001600 |

The following are in units of "microstrain," i.e., $10^{-6}$:

| | | | |
|------|------|------|------|
| ($c$) | +300 | −500 | −600 |
| ($d$) | −850 | +150 | −1000 |

**5.17** The principal strains at a point are given below. Sketch a deformed element showing the maximum shear strain.

$$\epsilon_1 = 0.001000 \qquad \epsilon_2 = 0.000500 \qquad \epsilon_3 = -0.000500$$

**5.18** The data below were taken from a 45° strain gage rosette, shown in the accompanying illustration. Draw the Mohr's strain circle, determine the principal strains, and show a properly oriented principal strain element.

| Case | $\epsilon_a$ | $\epsilon_b$ | $\epsilon_c$ | |
|------|------|------|------|------|
| (a) | +1000 | +600 | +200 | $\mu$ strain |
| (b) | +1000 | +600 | −400 | $\mu$ strain |
| (c) | −500 | −1000 | +800 | $\mu$ strain |
| (d) | +800 | 0 | 0 | $\mu$ strain |

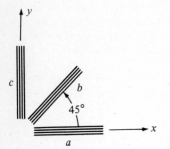

**Figure P5.18**

**5.19** Carry out the same procedures as in Prob. 5.18, but do so for the 60° rosette shown in the accompanying illustration.

| Case | $\epsilon_a$ | $\epsilon_b$ | $\epsilon_c$ | |
|------|------|------|------|------|
| (a) | 0 | +800 | +800 | $\mu$ strain |
| (b) | +1500 | +600 | −200 | $\mu$ strain |
| (c) | +866 | 0 | −866 | $\mu$ strain |
| (d) | +1200 | +300 | +1000 | $\mu$ strain |

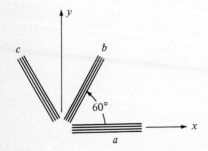

**Figure P5.19**

**5.20** Long tubes, to be used for high-pressure closed-end pressure vessels, can be fabricated from sheet metal, formed into the proper shape, and welded. Two configurations are shown in the accompanying illustration: a rolled tube with a straight longitudinal weld, and one with a helical weld at an angle.

If the maximum allowable tensile stress in the weld is limited to 75 percent of the maximum allowable in the sheet metal, which configuration is preferable?

For the helical weld case, what helix angle $\alpha$ would make the weld stress 75 percent of that in the sheet?

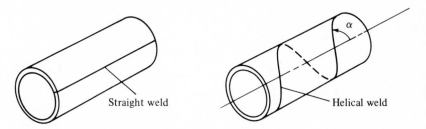

Straight weld    Helical weld

**Figure P5.20**

**5.21** Pressure vessels can be fabricated from fiber-reinforced plastics, in which the fibers, in tension, carry the pressure-induced stresses in the axial and tangential directions, while the plastic seals the cylinder and transfers the pressure loading to the fibers.

Consider the case where the fibers are wound at an angle $\alpha$ as shown in the accompanying illustration. If $\alpha = 0$, the fibers can carry the axial stresses but not the tangential; for $\alpha = 90°$, we have the opposite situation. There is an optimum angle $\alpha$ such that both stresses are properly carried.

Analyze the small element shown in the accompanying illustration to determine the optimum value of $\alpha$. Do *not* attempt to use Mohr's circle; the fibers can carry a force in one direction only.

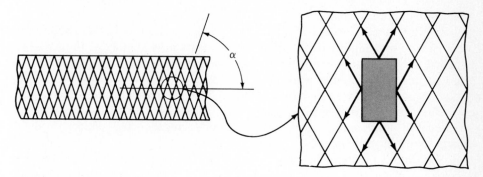

**Figure P5.21**

# STRESS-STRAIN-TEMPERATURE
# RELATIONS

## 6.1 OVERVIEW

In Chap. 5, we considered stress and strain from a purely theoretical point of view; it made no difference what material was being strained under the influence of stress as long as the material could be considered a "continuum," and the strains were "small." Given these two stipulations, the manipulations in Chap. 5 followed from equilibrium and geometry. They were independent of the material involved.

In this chapter we will see how stress and strain are related for various real materials. In addition to the strain produced by stress, we will include the strain due to temperature change.

Implicit in the concept of strain depending upon stress is the assumption that time is not important, i.e., that strain occurs immediately as the stress is applied and does not change with time. For the great portion of engineering design, this assumption is warranted and gives good results. However, if the "rate of loading" is very high (as in impact), or if the temperature is high, then we must consider time effects. The effects of a high rate of loading can be treated via *wave mechanics*. We will briefly consider the effects of high temperature in this chapter.

We should mention, in passing, that materials exposed to a high flux of nuclear radiation can undergo marked changes in behavior; embrittlement and volume growth being common changes.

The actual relationships between stress, temperature, time, and strain are generally very complex—complex to the point that simple approximations are almost always used in practice. In Chap. 7, we will consider the *limits* imposed on loading due to inherent material limitations (yield, fracture, etc.). Again, the actual limitations are so complex that we must resort to simple approximations. Accordingly, it seems appropriate, in an introductory text, to consider only the simplest relationships. Hopefully, at the expense of some depth, we can make the subject understandable.

We know that all ductile materials can undergo *plastic* strains that are large compared to the elastic strains. However, most devices are designed to operate well within the elastic range. Hence, for normal temperatures we will consider only the *elastic* stress-strain relations.

If we research the subject, we find that the stress-strain-temperature relations are *nonlinear* to a greater or lesser extent. We will consider (as do most others) only the linear approximation.

Most real materials, once they have been processed into a device, are not isotropic. That is, their properties are different in different directions. In most metals, the degree of anisotropy is small, and the assumption of isotropy is reasonable. At the other extreme, many of our fiber-reinforced materials are *highly* anisotropic. In this text, we will consider primarily the isotropic case, and will briefly treat "orthotropy," the most common deviation from isotropy.

Within the context of the above limitations, let us proceed to explore the stress-strain-temperature-time relations. We might note here that the relations between stress, strain, and temperature are frequently called the "constitutive relations."

## 6.2 IDEAL MATERIALS

For the purpose of analysis, various "ideal materials" have been defined. This does not mean that they are ideally suited to any specific use, but that they are ideally suited to simple analysis. Figure 6.1 shows the tensile stress-strain curves for four such ideal materials, all being independent of time.

We note the following about the materials shown in Fig. 6.1:

The "ideal elastic" material is linear without limit, having the same modulus $E$ in all directions.

The "ideal plastic" material has no elastic deformation, but flows plastically at a constant "flow stress" which equals the yield stress.

The "ideal elastic, perfectly plastic" material combines the above two deformations, being linear elastic up to the yield point and exhibiting a constant flow stress thereafter.

The "ideal elastic, linear strain-hardening" material combines linear elasticity with a linearly increasing flow stress in the plastic region.

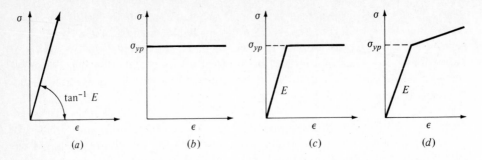

**Figure 6.1** "Ideal materials": (*a*) ideal elastic; (*b*) ideal plastic; (*c*) ideal elastic, perfectly plastic; (*d*) ideal elastic, strain-hardening plastic.

For most of this chapter we will consider an ideal elastic, isotropic material, a material that has linear stress-strain relations without limit. In Chap. 7 we will introduce the limits we must impose on the stresses, limits which are a function of both the material and the design.

In this chapter we will also briefly consider an elastic, orthotropic material and a material that "creeps" at elevated temperature.

## 6.3 STRESS-STRAIN RELATIONS

### 6.3.1 Linear Materials

For a completely general case of stress and strain, we have six components of stress and six components of strain. For a linear elastic material, each component of strain could be influenced by each component of stress. To describe such a relationship could require the 36 coefficients shown in the matrix below:

| | $\sigma_{xx}$ | $\sigma_{yy}$ | $\sigma_{zz}$ | $\tau_{xy}$ | $\tau_{yz}$ | $\tau_{zx}$ |
|---|---|---|---|---|---|---|
| $\epsilon_{xx} =$ | * | * | * | * | * | * |
| $\epsilon_{yy} =$ | * | * | * | * | * | * |
| $\epsilon_{zz} =$ | * | * | * | * | * | * |
| $\gamma_{xy} =$ | * | * | * | * | * | * |
| $\gamma_{yz} =$ | * | * | * | * | * | * |
| $\gamma_{zx} =$ | * | * | * | * | * | * |

You should note that the coefficients of this matrix have units of strain divided by stress (in$^2$/lb, m$^2$/N), and represent the *compliance* of the material rather than the stiffness.

Based on our previous studies of compliance matrices, you should not be surprised to learn that the matrix is symmetrical (for linear materials), so that *only* 21 coefficients are really required. For a fully anisotropic material, we would have to deal with all 21 coefficients, a situation we will avoid!

For an *isotropic* material, we can use symmetry to eliminate many terms from the matrix. Because the materials we will consider are linear elastic, we can superpose the effects of the various stress components to obtain the effects due to a combination of stresses. For the same reason, we can consider the effects of the stresses independently.

First let us ask, can a *shear stress* cause a *normal strain*? Figure 6.2 shows an element subject only to $\tau_{xy}$. Clearly, if we rotate the element about the 45° axis $AA$ as shown, the stresses will look exactly the same. However, if the element had undergone a normal strain in the $x$ direction (Fig. 6.2b), rotation about $AA$ would not give the same deformation picture.

Because we have isotropy, the same stress pattern must produce the same strain pattern. Consequently, shear stress *cannot* produce normal strain. This observation reduces six of the 21 coefficients to zero:

|  | $\sigma_{xx}$ | $\sigma_{yy}$ | $\sigma_{zz}$ | $\tau_{xy}$ | $\tau_{yz}$ | $\tau_{zx}$ |
|---|---|---|---|---|---|---|
| $\epsilon_{xx} =$ | * | * | * | 0 | 0 | 0 |
| $\epsilon_{yy} =$ |  | * | * | 0 | 0 | 0 |
| $\epsilon_{zz} =$ |  |  | * | 0 | 0 | 0 |
| $\gamma_{xy} =$ |  |  |  | * | * | * |
| $\gamma_{yz} =$ |  |  |  |  | * | * |
| $\gamma_{zx} =$ |  |  |  |  |  | * |

Now ask: Can a shear stress in one plane ($\tau_{xy}$) cause shear strain in another plane? In Fig. 6.3 we see the same rotation about the 45° axis $AA$. If the shear stress produced a shear strain in the $xz$ plane as shown, a different shape would result after rotation about $AA$. The shear strain ($\gamma_{xz}$) must be zero for the strain element to look the same after rotation. This observation removes three

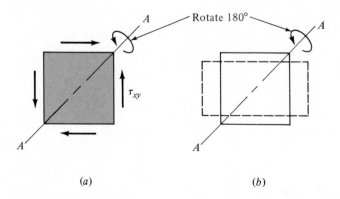

*(a)*          *(b)*

**Figure 6.2** Showing that a shear stress cannot produce a normal strain in a linear isotropic material.

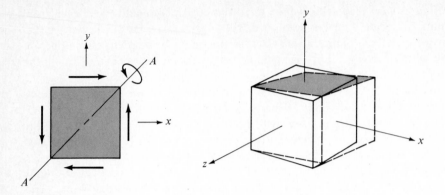

**Figure 6.3** Showing that $\tau_{xy}$ cannot produce $\gamma_{xz}$ in a linear isotropic material.

more coefficients:

|  | $\sigma_{xx}$ | $\sigma_{yy}$ | $\sigma_{zz}$ | $\tau_{xy}$ | $\tau_{yz}$ | $\tau_{zx}$ |
|---|---|---|---|---|---|---|
| $\epsilon_{xx} =$ | * | * | * | 0 | 0 | 0 |
| $\epsilon_{yy} =$ |  | * | * | 0 | 0 | 0 |
| $\epsilon_{zz} =$ |  |  | * | 0 | 0 | 0 |
| $\gamma_{xy} =$ |  |  |  | * | 0 | 0 |
| $\gamma_{yz} =$ |  |  |  |  | * | 0 |
| $\gamma_{zx} =$ |  |  |  |  |  | * |

The obvious question now is: Can a normal stress in one direction produce normal strain in another direction? The answer is obviously *yes* because we are familiar with the Poisson effect in the tensile test.

We know that when we apply a positive normal stress in the $x$ direction ($\sigma_{xx}$), we also have negative normal strains in the $y$ and $z$ directions.

$$\text{Due to } \sigma_{xx}: \quad \epsilon_{xx} = \frac{\sigma_{xx}}{E}$$

$$\epsilon_{yy} = -\frac{\nu\sigma_{xx}}{E}$$

$$\epsilon_{zz} = -\frac{\nu\sigma_{xx}}{E}$$

We also know that if we apply a shear stress $\tau_{xy}$, we must obtain a shear strain $\gamma_{xy}$, thus:

$$\gamma_{xy} = \frac{\tau_{xy}}{G}$$

where $G$ (the shear modulus) relates elastic shear stress and strain.

We can now write the complete set of elastic stress-strain equations for a linear isotropic material:

$$\epsilon_{xx} = \frac{1}{E}\left[\sigma_{xx} - \nu(\sigma_{yy} + \sigma_{zz})\right]$$

$$\epsilon_{yy} = \frac{1}{E}\left[\sigma_{yy} - \nu(\sigma_{zz} + \sigma_{xx})\right]$$

$$\epsilon_{zz} = \frac{1}{E}\left[\sigma_{zz} - \nu(\sigma_{xx} + \sigma_{yy})\right]$$

$$\gamma_{xy} = \frac{\tau_{xy}}{G}$$

$$\gamma_{yz} = \frac{\tau_{yz}}{G}$$

$$\gamma_{zx} = \frac{\tau_{zx}}{G}$$

$$(6.1)$$

### 6.3.2 Nonlinear Materials

Although we will not consider nonlinear materials in any detail, we should note here that nonlinearity can void the symmetry arguments used above.

For instance, consider a material where the tensile modulus ($E$) is much greater than the compressive modulus. This would be the case with a fabric or many of our foams. If we subject this material to shear, the Mohr's circle shows us that in effect we have applied tension and compression at 90° to each other. The material will elongate slightly because of the tension, but will compress substantially because of the compression. The net result will be a decrease in volume—i.e., shear can produce normal strain (this effect allows us to "wring out" a piece of cloth by twisting it).

In a similar way, particulate materials such as sand, being much stiffer in compression than in tension, will undergo "dilatation" (increase in volume) when they are sheared. These effects are considered in detail in texts on soil mechanics.

## 6.4 THE ELASTIC CONSTANTS

It would appear that (for linear isotropic materials) three elastic constants ($E$, $G$, $\nu$) are required (from experiment) if we wish to calculate strains as functions of stresses. The tensile test, you will recall, gave us two of the three, namely $E$ and $\nu$. Fortunately, the three constants are not independent, and we can compute $G$ in terms of $E$ and $\nu$.

Consider the two states of stress and strain shown in Fig. 6.4. In Fig. 6.4*a*, equal and opposite principal stresses of magnitude $\sigma_0$ produce equal and opposite principal strains ($\epsilon_1$ and $\epsilon_2$), as shown. In Fig. 6.4*b*, an element at 45° is under "pure shear" $\tau$, which produces the shear strain $\gamma$.

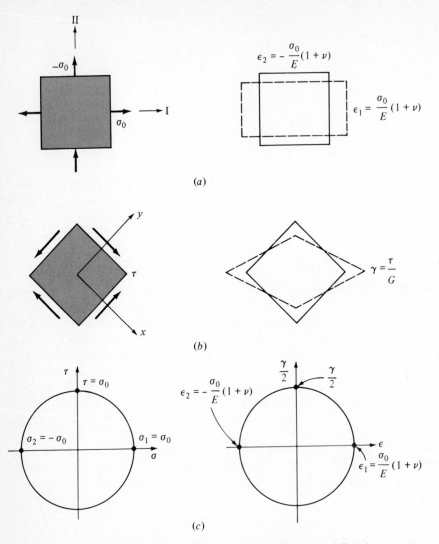

**Figure 6.4** Stress, strain, and Mohr's circle for calculating $G$ in terms of $E$ and $\nu$.

The two states of stress will be identical if $\tau = \sigma_0$ as seen on the Mohr's circle for stress. If the stresses are identical, the strains must also be identical; and from the Mohr's strain circle we see that:

$$\frac{\gamma}{2} = \frac{\sigma_0}{E}(1 + \nu)$$

Thus:

$$G = \frac{\tau}{\gamma} = \frac{\sigma_0}{2[(\sigma_0/E)(1 + \nu)]} = \frac{E}{2(1 + \nu)} \tag{6.2}$$

A fourth elastic constant that is often used is called the "bulk modulus" ($B$), and relates a hydrostatic pressure ($p$) to the resulting fractional change in volume ($\Delta V/V$):

$$B = \frac{-p}{\Delta V/V} = \frac{E}{3(1-2\nu)} \tag{6.3}$$

Although $B$ is often given as a positive number, you must realize that a positive pressure (negative stress) will produce a decrease in volume.

## 6.5 TEMPERATURE EFFECTS

In Chap. 4 we noted that when an axial member is heated, it becomes longer because of thermal expansion. We let $\alpha$ represent the *linear coefficient* of thermal expansion. This does not mean that thermal expansion is linear, it means that $\alpha$ represents the linear approximation to an inherently nonlinear effect. We now want to include thermal expansion, or better, "thermal strain" in the stress-strain relations.

An increase (or decrease) in temperature at a point is a scalar quantity; it has no direction. Or we might say that it is the same in all directions. This obviously means that, for isotropic material, the thermal strain must also be the same in all directions; and it must be a normal strain.

The change in length, per unit initial length, due to temperature is simply $\alpha \, \Delta T$. Clearly, this equals the thermal strain. All we need do, then, is to add an $\alpha \, \Delta T$ term to each of the normal strains in Eqs. (6.1).

If we heat an isotropic body slowly, so that the temperature is essentially uniform at any instant, and if the body is not constrained in any way, then it will simply grow in all directions. Every dimension will have the same fractional increase, and there will be no stresses in the body.

On the other hand, if a body is heated rapidly, or if it is constrained, it can change shape as well as size, and "thermal stresses" may result.

Of course, there are other temperature effects besides thermal strain. Many of the physical properties change radically as we approach the melting temperature of a material. When the temperature exceeds about 0.4 times the absolute melting temperature, substantial changes can be expected.

## 6.6 THE EQUATIONS OF ELASTICITY

In Chap. 3, as set forth in Eqs. (3.12), we noted that all problems in mechanics involve the interaction of three sets of relations:

Equilibrium relations
Conditions of geometric fit
Force-deformation relations

In Chap. 3, we studied these relations relative to macroscopic bodies. In Chaps.

4 and 5, we have concerned ourselves with conditions at a single point and how those conditions must vary from point to point.

We find that we now have exactly the same three sets of relations, except that they now apply on a microscopic basis. For reference purposes we assemble the three sets of relations:

1. Equilibrium equations [Eqs. (5.3)]:

$$\frac{\partial \sigma_{xx}}{\partial x} + \frac{\partial \tau_{yx}}{\partial y} + \frac{\partial \tau_{zx}}{\partial z} + X = 0$$

$$\frac{\partial \tau_{xy}}{\partial x} + \frac{\partial \sigma_{yy}}{\partial y} + \frac{\partial \tau_{zy}}{\partial z} + Y = 0 \qquad (6.4)$$

$$\frac{\partial \tau_{xz}}{\partial x} + \frac{\partial \tau_{yz}}{\partial y} + \frac{\partial \sigma_{zz}}{\partial z} + Z = 0$$

2. Compatibility of strain [Eqs. (5.7) and (5.9)]:

$$\epsilon_{xx} = \frac{\partial u}{\partial x} \qquad \gamma_{xy} = \frac{\partial u}{\partial y} + \frac{\partial v}{\partial x}$$

$$\epsilon_{yy} = \frac{\partial v}{\partial y} \qquad \gamma_{yz} = \frac{\partial v}{\partial z} + \frac{\partial w}{\partial y}$$

$$\epsilon_{zz} = \frac{\partial w}{\partial z} \qquad \gamma_{zx} = \frac{\partial w}{\partial x} + \frac{\partial u}{\partial z}$$

$$\frac{\partial^2 \epsilon_{xx}}{\partial y^2} + \frac{\partial^2 \epsilon_{yy}}{\partial x^2} = \frac{\partial^2 \gamma_{xy}}{\partial x \, \partial y} \qquad (6.5)$$

$$\frac{\partial^2 \epsilon_{yy}}{\partial z^2} + \frac{\partial^2 \epsilon_{zz}}{\partial y^2} = \frac{\partial^2 \gamma_{yz}}{\partial y \, \partial z}$$

$$\frac{\partial \epsilon_{zz}}{\partial x^2} + \frac{\partial \epsilon_{xx}}{\partial z^2} = \frac{\partial \gamma_{zx}}{\partial x \, \partial z}$$

3. Stress-strain-temperature relations:

$$\epsilon_{xx} = \frac{1}{E} [\sigma_{xx} - \nu(\sigma_{yy} + \sigma_{zz})] + \alpha \, \Delta T$$

$$\epsilon_{yy} = \frac{1}{E} [\sigma_{yy} - \nu(\sigma_{zz} + \sigma_{xx})] + \alpha \, \Delta T$$

$$\epsilon_{zz} = \frac{1}{E} [\sigma_{zz} - \nu(\sigma_{xx} + \sigma_{yy})] + \alpha \, \Delta T$$

$$\gamma_{xy} = \frac{\tau_{xy}}{G} \qquad (6.6)$$

$$\gamma_{yz} = \frac{\tau_{yz}}{G} \qquad G = \frac{E}{2(1 + \nu)}$$

$$\gamma_{zx} = \frac{\tau_{zx}}{G}$$

The three sets of equations assembled above provide the complete basis for the theory of elasticity for linear isotropic materials. For more complex materials, only the relations in Eqs. (6.6) must be changed.

In the past, a great deal of effort has gone into finding solutions of the above equations for specific geometries, loads, and constraints. It should be apparent that this is not a simple task!

In general, one cannot find a satisfactory solution because of the mathematical complexity; and approximate solutions must be used. Either the problem can be replaced by a simpler, approximately correct problem, or the above equations can be satisfied only approximately, or both.

The most recent efforts have involved use of the finite-element method (FEM), together with large computational capability, to obtain solutions that are approximately correct.

Unlike the computerized structural analyses that we have considered in this book, where we obtain solutions that are essentially "exact," most stress analysis via the finite-element method is only approximate. While we can completely specify the state of (for instance) a beam in terms of the motions of the nodes at its ends, complete specification of a general problem requires an infinite number of nodes. This is clearly out of the question, and a finite set of nodes is used. The above equations are then satisfied at the nodes, but not necessarily in between.

**Example 6.1** Consider a 1-cm cube of aluminum placed between "rigid" (nonmoving) steel jaws, and then heated 100°C. Ignore friction between the cube and the jaws and determine the stresses acting on the cube in the $x$, $y$, and $z$ directions:

1. When the jaws constrain motion in the $x$ direction only
2. When both $x$ and $y$ motions are constrained
3. When the cube is enveloped in three sets of jaws that effectively restrain motion in all directions

For the aluminum, let $E = 70$ GPa; $\alpha = 23 \times 10^{-6}/°C$, and $\nu = 0.3$. In part ($a$) of the accompanying illustration (page 198) we know:

$$\left.\begin{array}{l} \sigma_y = 0 \\[2mm] \sigma_z = 0 \end{array}\right\} \quad \text{(free surfaces)}$$

$$\epsilon_x = 0 \quad \text{(motion constrained)}$$

From the equation for $\epsilon_x$:

$$\epsilon_x = 0 = \frac{1}{E}[\sigma_x - (0+0)] + \alpha T$$

$$\sigma_x = -E\alpha\,\Delta T = -161\,\text{MPa}$$

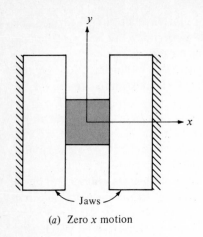

(a) Zero x motion

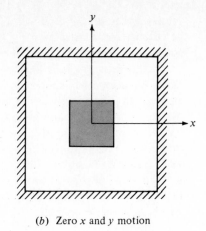

(b) Zero x and y motion

In part (b) we have:

$$\sigma_z = 0 \qquad \text{(free surface)}$$

$$\left.\begin{array}{l} \epsilon_x = 0 \\ \epsilon_y = 0 \end{array}\right\} \quad \text{(motions constrained)}$$

$$\epsilon_x = 0 = \frac{1}{E}[\sigma_x - \nu(\sigma_y + 0)] + \alpha\,\Delta T$$

From symmetry $$\sigma_x = \sigma_y$$

Therefore $$\sigma_x = -\frac{E\alpha\,\Delta T}{1-\nu} = -230\,\text{MPa}$$

In part (c), all motions are restrained and all surface stresses are equal such that:

$$\sigma_x = \sigma_y = \sigma_z = \frac{-E\alpha\,\Delta T}{1-2\nu} = -403\,\text{MPa} \qquad\qquad ***$$

## 6.7 ORTHOTROPY

All of the above relations hold for a linear *isotropic* material. Many of our materials of construction are well approximated as isotropic. However, there are very important exceptions.

If we think about it, we will realize that in most structural, load-bearing elements, the material is stressed much more in some directions than in others. It would seem intuitively appropriate in such a case to use a material that is "stronger" in the directions that require the greatest strength. It would seem to be a more "efficient" use of the material.

In numerous instances, nature has been very efficient in its "use" of materials. A tree is a beautiful example: The structure of wood is thoroughly consistent with the types of loads normally imposed. However, wood is not isotropic; it fits a classification called "orthotropic."

Figure 6.5 shows an idealized piece of wood with the "grain" running in the $x$ direction. We can rotate the piece 180° about the $x$ axis, the $y$ axis, or the $z$ axis, and it will appear just the same; if rotated about other axes, it will not appear the same. This is the definition of an orthotropic material: a material having three orthogonal axes of 180° rotational symmetry. We have seen "principal axes" for stress, strain, and compliance. Now we have "principal axes" for a material.

For an orthotropic material, the stress-strain relations are simplest when referred to the principal material directions, and are given below (where the $xyz$ axes *must* be the principal axes for the material):

$$
\begin{array}{c}
\begin{array}{ccccccc} \sigma_{xx} & \sigma_{yy} & \sigma_{zz} & \tau_{xy} & \tau_{yz} & \tau_{zx} & \Delta T \end{array} \\
\begin{array}{c} \epsilon_{xx} = \\ \epsilon_{yy} = \\ \epsilon_{zz} = \\ \gamma_{xy} = \\ \gamma_{yz} = \\ \gamma_{zx} = \end{array}
\left[
\begin{array}{ccccccc}
S_{11} & S_{12} & S_{13} & & & & \alpha_x \\
S_{21} & S_{22} & S_{23} & & & & \alpha_y \\
S_{13} & S_{23} & S_{33} & & & & \alpha_z \\
& & & S_{44} & & & \\
& & & & S_{55} & & \\
& & & & & S_{66} &
\end{array}
\right]
\end{array}
\qquad (6.7)
$$

The nine stress coefficients and the three thermal coefficients must be determined empirically or through analysis. If the stress-strain relations are considered relative to nonprincipal material axes, then we find that a shear stress produces normal strains and vice versa. Clearly, it is best to always use the principal material axes.

Man-made materials that are orthotropic include fiber-reinforced plastics, cloth, metal-reinforced concrete, etc. When we design the material to suit the need, as we do with many composites, we can frequently make the use of Eq. (6.7) considerably easier. We can *build* the structure such that the principal stress axes and the principal material axes coincide. This, in effect, removes the shear-stress terms from Eq. (6.7).

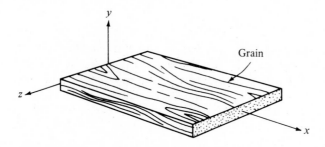

**Figure 6.5** Wood, having directional characteristics, is an example of an orthotropic material.

## 6.8 CREEP

As we have noted several times previously, materials behave differently at elevated temperatures; and "differently" generally means worse, not better. However, many processes and devices are functionally more efficient if operated at high temperatures; hence there is strong motivation to develop materials for high-temperature applications.

Some materials, such as lead and many polymers, behave in a "high-temperature" fashion even at room temperature.

As a generality, when the temperature of a material exceeds about 0.4 times its absolute melting temperature, the thermal energy of the atoms becomes sufficient to "assist" the applied stresses in producing deformation, and the behavior of the material changes significantly: The elastic modulus decreases, the yield stress decreases, and deformation is no longer independent of time.

A specimen under tension can slowly grow longer and eventually fracture at stresses far below the yield point. The time-dependent deformations are permanent, nonrecoverable, plastic deformations. This process, occurring very slowly, is called "creep."

Whenever we wish to consider time-dependent effects on an engineering structure, we are faced with a problem—namely, we might want to predict the effects over a lifetime of perhaps 20 years, but we cannot run a test for 20 years to get the appropriate data! So we are faced with the problem of predicting long-term behavior based on relatively short-term test data.

There are several components of creep deformation. However, if we restrict ourselves to the practical condition of relatively long life, one component predominates—creep which occurs at a relatively constant rate of strain. Empirically the creep strain rate depends on both the temperature and the stress, and is material-dependent. An approximate relation is:

$$\frac{d\epsilon}{dt} = A\sigma^m e^{-h/kT} \tag{6.8}$$

where $A$ = constant
$m$ = stress exponent, typically 4 to 8
$h$ = thermal activation energy
$k$ = Boltzmann's constant
$T$ = absolute temperature

We will not delve further into creep phenomena here; you have been alerted to the entirely different behavior that takes place at elevated temperatures, a behavior that can easily become the predominant one.

## 6.9 COMPUTER EXAMPLE

In Chap. 5 we saw how to determine the principal stresses at the surface of a machine element through the use of strain rosettes. In this chapter, we have related stress and strain. So now we can determine the stresses acting through strain measurement.

Program Listing 6.1 shows a program for computing the principal stresses and strains from data from a 45° rosette. Program Output 6.1 gives the output. The significance of "Tau-max" and "Safety-factor" will become apparent in Chap. 7.

```
20 !"                45-Degree Strain Rosette"\!
30 !"                Zero stress on z-face"
40 !"                Poisson's ratio = .3"
50 !"                A to B to C counterclockwise (45-degrees)"\!
60 REM
70 !" Input material parameters (or zeros)"\!
80 E9=1
90 INPUT " Tensile Yield Stress, (10^3 psi,  MPa) = ",S0 \!
100 INPUT " Modulus of Elasticity, E, (10^6 psi,  GPa) =",E \ !
110 IF E>0 THEN 130
120 E=30 \ E9=0
130 !" Input strains A, B, and C, in micro-strain (10^-6)"\!
140 INPUT "Strain A = ",A
150 INPUT "       B = ",B
160 INPUT "       C = ",C
170  C0= (A+C)/2              \REM   Center of Circle
180 D= (A-C)/2
190 G= ABS(B-C0)
200 R= SQRT(D*D+G*G)          \REM   Radius of Circle
210 E1 = (C0+R)               \REM   Principal Strains
220 E2 = (C0-R)
230 D1 = E1*E \ D2 = E2*E
240 S1 = (D1+.3*D2)/(1-.09)   \REM   Principal Stresses
250 S2 = D2+.3*S1
260 E3 = -.3*(S1+S2)/E
270 S3 = 0
280 D1 = ABS(S1-S2)
290 D2 = ABS(S2-S3)
300 D3 = ABS(S3-S1)
310 IF D2>D1 THEN D1=D2
320 IF D3>D1 THEN D1=D3       \REM   Maximum diameter Mohr's Circle
330 !
340 IF E9=0 THEN 360
350 !"Principal Stresses are: ",S1/1000,"    ",S2/1000,"   and zero"\!
360 !"Principal Strains are : ",E1,"    ",E2,"    ",E3 \!
370 IF S0=0 THEN 410
380 IF E9=0 THEN 410
```

**Program Listing 6.1**

```
390 !"    Tau-max = ",%10F2,D1/2000,"    (ksi, MPa)"
400 !"    Safety-factor = ",%5F2,1000*S0/D1
410 !\!
420 INPUT" Other Rosette Data?   (Y/CR) ",Z$
430 IF Z$="Y" THEN 10
```

**Program Listing 6.1** (*continued*)

```
            45-Degree Strain Rosette

              Zero stress on z-face
              Poisson's ratio = .3
              A to B to C counterclockwise (45-degrees)

Input material parameters (or zeros)

Tensile Yield Stress, (10^3 psi,  MPa) = 60

Modulus of Elasticity, E, (10^6 psi,  GPa) =30

Input strains A, B, and C, in micro-strain (10^-6)

Strain A = 100
       B = 200
       C = -150

Principal Stresses are:  4.8683579    -7.0112151    and zero

Principal Strains are :  232.39075    -282.39075    21.428572

    Tau-max =     5.94    (ksi, MPa)
    Safety-factor = 5.05
```

**Program Output 6.1**

## 6.10 PROBLEMS

**6.1** An isotropic material is compressed by a piston of area $A$ within a "rigid" chamber (see accompanying illustration). Find the motion of the piston when the force $F$ is applied (neglect friction).

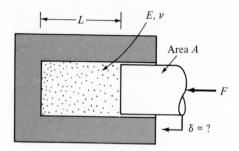

Figure P6.1

**6.2** Using Fig. 6.4, show that $G = E/[2(1 + v)]$  [Eq. (6.2)].

**6.3** Show that the bulk modulus $B = E[3(1 - 2v)]$  [Eq. (6.3)].

**6.4** Poisson's ratio $v$ ranges from 0 to 0.5 for various materials. From Prob. 6.3 it is seen that when $v = 0.5$, then the material is "incompressible." Show that if you could develop a material with $v > 0.5$, you would have solved all of our energy-shortage problems.

**6.5** The slab of metal shown in the accompanying illustration is placed between a pair of "rigid" platens which do not permit motion in the $y$ direction. When the forces $F$ are applied, determine the contact stresses between the slab and the platen. State your assumptions.

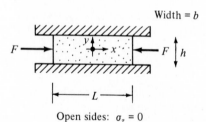

Width = $b$

Open sides: $\sigma_z = 0$          **Figure P6.5**

**6.6** Consider the stress-strain-temperature relationships in Eqs. (6.6). The strains are given in terms of the stresses and the temperature change. Determine the inverse relation: the stresses in terms of the strains and temperature change.

**6.7** Determine the dimensionless change in volume associated with a change in temperature $\Delta V/V = f(\Delta T)$.

**6.8** The following conditions represent the state of stress in a sheet of steel lying in the $xy$ plane. There are no forces acting on the surfaces of the sheet ($\sigma_z = 0$; plane stress).

Determine the principal strains in the $xy$ plane for the cases indicated and show them on a properly oriented element. Also determine the strain in the $z$ direction.

$$E = 30 \times 10^6 \text{ psi (206.8 GPa)} \qquad v = 0.3$$

| Case | $\sigma_x$ | $\sigma_y$ | $\tau_{xy}$ | |
|------|-----------|-----------|------------|-----|
| (a) | +90,000 | +45,000 | −30,000 | psi |
| (b) | −25,000 | +25,000 | 0 | psi |
| (c) | −30,000 | +20,000 | +25,000 | psi |
| (d) | +700 | −250 | +300 | MPa |
| (e) | −500 | −500 | +500 | MPa |

**6.9** The following represents the state of strain in the same steel sheet described in Prob. 6.8. For the cases indicated determine the magnitude and direction of the principal stresses and show them on a properly oriented element. Also determine the strain in the $z$ direction.

| Case | $\epsilon_x$ | $\epsilon_y$ | $\gamma_{xy}$ |
|------|-------------|-------------|--------------|
| (a) | +0.000950 | −0.000325 | +0.000750 |
| (b) | +0.000450 | +0.001250 | −0.000800 |
| (c) | −0.001300 | −0.000500 | +0.000400 |

**6.10** In the flat sheet of Prob. 6.8 it is known that:

$$\sigma_x = 450 \qquad \tau_{xy} = 150 \,(\text{MPa}) \qquad \epsilon_z = -0.000400$$

Determine the principal stresses.

**6.11** A 45° strain rosette is glued to the free surface of a structural steel element. The following strains are indicated under load. Determine the normal stress in the direction of gage $a$.

$$\epsilon_a = 0.001250 \qquad \epsilon_b = 0.000750 \qquad \epsilon_c = -0.000375$$

**6.12** For a 45° strain rosette glued to the free surface of a structural steel element, find the principal stresses and the maximum shear stress when the following strains are indicated under load:

$$\epsilon_a = -0.000450 \qquad \epsilon_b = -0.000125 \qquad \epsilon_c = 0.000900$$

**6.13** A 60° strain rosette glued to the surface of an aluminum member indicates the following strains under load. Determine the maximum shear stress.

$$\epsilon_a = 0.000650 \qquad \epsilon_b = 0.000500 \qquad \epsilon_c = -0.000300$$

**6.14** A cylindrical steel pressure vessel has a radius of 50 mm and a thickness of 2 mm. If the tank is pressurized to 50 atm, how much does the diameter increase?

**6.15** A strain gage is glued to the outside of a cylindrical pressure vessel such that it indicates strain in the axial direction. When a microstrain of 100 is read, how great is the internal pressure?

$$r = 5 \,\text{in} \qquad t = 0.2 \,\text{in} \qquad E = 20 \times 10^6 \,\text{psi} \qquad \nu = 0.25$$

**6.16** The same as Prob. 6.15 except that as the pressure is increased, the temperature of the tank also increases. When a microstrain of 100 is indicated, the temperature has increased 50°F. How great is the pressure then? $\alpha = 6.5 \times 10^{-6} \,°\text{F}^{-1}$.

**6.17** A strain gage is glued to the outside of a spherical pressure vessel. Determine a relationship for the internal pressure $(p)$ in terms of the strain indicated $(\epsilon)$, and $r$, $t$, $E$, and $\nu$.

**6.18** A pressure cylinder is subject to an internal pressure $p$ and an axial load $F$. Strain gages in the axial and tangential directions indicate the following. How large are $p$ and $F$?

$$
\begin{aligned}
&\text{Axial strain} && = 0.000650 && E = 10^7 \text{ psi} \\
&\text{Tangential strain} && = 0.000875 && \nu = 0.3 \\
& && && r = 1 \text{ in} \\
& && && t = 0.1 \text{ in}
\end{aligned}
$$

**6.19** A 45° strain rosette is glued to the outside of a cylindrical pressure tank. The tank is subject to pressure $p$, axial force $F$, and possible temperature changes $\Delta T$. Determine $p$ and $F$ under the following conditions:

Assume that the gages are "compensated" for temperature changes such that they indicate the true, total strain.

$$
\begin{aligned}
&\text{Gage } a \text{ (axial strain)} && = 0.000050 && E = 10^7 \text{ psi} \\
&\text{Gage } b && = 0.000375 && \nu = 0.3 \\
&\text{Gage } c \text{ (tangential)} && = 0.000700 && \alpha = 10^{-5}\,°\text{F}^{-1} \\
& && && r = 2.5 \text{ in} \\
& && && t = 0.2 \text{ in} \\
& && && \Delta T = +200°\text{F}
\end{aligned}
$$

**Probs. 6.20 to 6.23:** Composite materials are frequently made with strong, high modulus fibers embedded in a polymer matrix, as shown in the accompanying illustration. Typical fibers are glass or carbon filaments; typical matrix materials are polyester and epoxy resins. The simplest structure to analyze consists of a uniform dispersion of parallel fibers in the matrix. When sheets of such material are made, they are orthotropic with the principal axes, as shown. The (isotropic) properties of the two constituent materials are:

$$
\text{Fiber: } E_f,\ \nu_f \qquad \text{Resin: } E_r,\ \nu_r
$$

The "volume fraction" of the fiber is $V_f$ (the ratio of the fiber volume to the total volume). Clearly, $V_r = 1 - V_f$. For the purposes of analysis, consider the model shown where the composite consists of alternating slabs of fiber and resin.

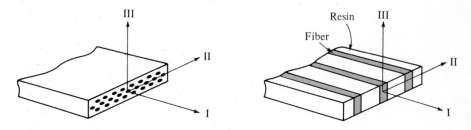

**Figure P6.20 to 6.23**

**6.20** Find the "equivalent" modulus in the I direction [$1/S_{11}$ in Eq. (6.7)]. Show that for many cases, where the fibers are much stiffer than the resins, this modulus can be approximated by $E_{11} \cong E_f V_f$.

**6.21** Find the "equivalent" modulus in the II direction [$1/S_{22}$ in Eq. (6.7)]. Show that frequently $E_{22} \cong E_r/V_r$.

**6.22** Show that when pulled in the I direction, the strain in the II direction (contraction) due to the strain in the I direction is given by:

$$\frac{S_{12}}{S_{11}} = -(\nu_f V_f + \nu_r V_r)$$

**6.23** Consider now a composite material made from layers of the composite just illustrated and discussed. Alternating layers have fibers running in the I and II directions, such that the properties in the I and II directions are identical. Determine the equivalent modulus in the I and II directions. Show that it frequently can be approximated by:

$$E_{11} \cong \frac{E_f V_f}{2}$$

Having identical properties in two principal directions, is the composite isotropic in that plane? No.

# STRESS LIMITS IN DESIGN

## 7.1 OVERVIEW

We have devoted Chaps. 5 and 6 to a discussion of stress and strain, and we have seen how to manipulate both and how they are related. Most of the rest of this book is concerned with determining the magnitude of the stresses and strains for various structural elements, under various conditions of load.

In this chapter we ask: How large can we permit the stresses to be? Or conversely: How large must a part be to withstand a given set of loads? What are the overall conditions or limits that will determine the size and material for a part?

Design limits are based on *avoiding failure* of the part to perform its desired function. Because different parts must satisfy different functional requirements, the conditions which limit load-carrying ability may be quite different for different elements. As an example, compare the design limits for the floor of a house with those for the wing of an airplane.

If we were to determine the size of the wooden beams in a home such that they simply did not break, we would not be very happy with them; they would be too "springy." Walking across the room would be like walking out on a diving board.

Obviously, we should be concerned with the maximum "deflection" that we, as individuals, find acceptable. This level will be rather subjective, and different people will give different answers. In fact, the same people may give different answers depending on whether they are paying for the floor or not!

An airplane wing structure is clearly different. If you look out an airplane window and watch the wing during turbulent weather, you will see large deflections; in fact you may wish that they were smaller. However, you know that the important issue is that of "structural integrity," not deflection.

We want to be assured that the wing will remain intact. We want to be assured that no matter what the pilot and the weather do, that wing will continue to act like a good and proper wing. In fact, we really want to be assured that the wing will *never* fail under *any* conditions. Now that is a pretty tall order; who knows what the "worst" conditions might be?

Engineers who are responsible for the design of airplane wing structures *must* know, with some degree of certainty, what the "worst" conditions are likely to be. It takes great patience and dedication for many years to assemble enough test data and failure analyses to be able to predict the "worst" case. The general procedure is to develop statistical data which allow us to say how *frequently* a given condition is likely to be encountered—once every 1000 hours, or once every 10,000 hours, etc.

In the first part of this chapter we will introduce two concepts associated with the fact that our stress analyses are never precise: the concepts of "stress concentration" and "safety factor." These both *must* be considered in the design of any structural element.

In the second part of the chapter we will consider the various *modes* of failure and how they are handled analytically for different materials. These will include yielding of ductile materials, fracture of brittle materials, and fracture by "fatigue."

## 7.2 STRESS CONCENTRATION

When we first considered the tensile test, we saw that the central portion of the specimen was very carefully made, with large "fillets" or "radii" blending the actual gage section into the ends which had to be gripped. If the specimen had been made with *abrupt* changes in geometry, the test data could have been quite different.

Whenever we have an *abrupt* change in the geometry of a part, the stresses near that change will be higher than the average values you might calculate. Figure 7.1 shows a flat metal sheet with a force $F$ of 100 lb applied. The area at the right end ($C$) is half the area at the left ($A$); thus we would expect the stress at $A$ to be 500 psi and that at $C$ to be 1000 psi.

For purposes of discussion, we have drawn lines on the specimen shown in Fig. 7.1 that can be thought of as "lines of force" (the material between each pair of lines carrying an equal portion of the load $F$). At ends $A$ and $C$, the lines are uniformly spaced, representing a uniform force distribution and therefore a uniform stress distribution. However, at $B$, where the cross section changes, the lines are crowded together, or "concentrated" at the corner.

Not only is the stress at $C$ greater than that at $A$, but the maximum stress at $B$ is greater than at either $A$ or $C$. The ratio of the maximum stress at $B$ to the average stress at $C$ is called the "stress-concentration factor" $K$.

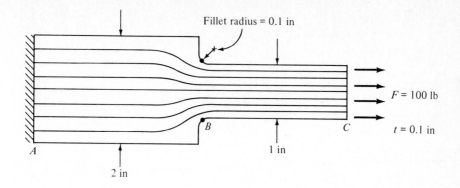

**Figure 7.1** Showing how "lines of force" are concentrated near a change of cross section.

For the case shown, where the radius of the fillet is 0.1 in, the maximum stress at $B$ is about 2000 psi, the stress-concentration factor $K$ being about 2.

In all structural designs we try to minimize the magnitude of the stress concentrations and to use materials that are less sensitive to stress concentration. As a generality, which will be made more specific later in this chapter, we can say that:

Even in "good" designs, stress concentrations of 2 to 4 are common.
Brittle fracture and fatigue fracture are both *highly* sensitive to stress concentration.
Ductile yield is little affected by $K$, but ductile fracture may be changed to "brittle" in a "notch-sensitive" material.

Theoretically, the magnitude of a stress concentration can approach infinity when we have a very sharp crack in a material. Obviously, the stresses cannot approach infinity for any real material; so we define a *theoretical* stress-concentration factor $K_t$. This represents the ratio of the maximum stress to the mean stress for a theoretical material that is linear and remains elastic (does not yield or fracture).

There are many tabulations of $K_t$ in the handbooks and literature, values obtained by both analytical and empirical methods. Figure 7.2 shows typical curves of $K_t$ for tension, bending, and twisting. We see that as the fillet radius approaches zero, $K_t$ goes to infinity.

Figure 7.3 shows an elliptical hole in a large (relative to the hole) plate under tensile stress. The axes of the ellipse are parallel and normal to the stress axis. The value of $K_t$ (determined analytically through the theory of elasticity) is simply:

$$K_t = 1 + \frac{2a}{b} \qquad \text{(for an elliptical hole)} \qquad (7.1)$$

For a circular hole, where $a = b$ and the hole is small relative to the plate, $K_t = 3$. No matter how small the hole is, $K_t$ still equals 3!

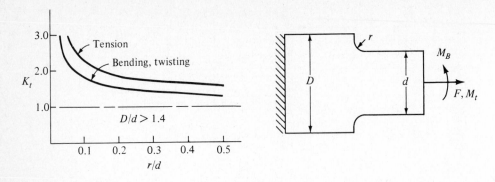

**Figure 7.2** Approximate stress-concentration factors $K_t$ for tension of a plate, bending of a beam, and twisting of a shaft.

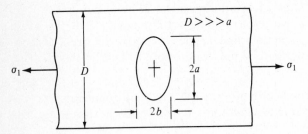

**Figure 7.3** Elliptical hole in a large plate.

When we have a circular or elliptical hole under tension, the maximum tensile stress (at the side of the hole) can be three or more times the mean applied stress.

At the ends of the hole, where you would expect zero stress, there will be a *compressive* stress. For the circular hole, the magnitude of the compressive stress equals the mean tensile stress. This effect, where stress sign is changed, is important in brittle fracture; you may apply compressive stresses, but at the ends of a properly oriented crack there will be tension.

For ellipses with large aspect ratios, the value of $K_t$ can be expressed in terms of the major axis $a$, and the radius of curvature $\rho$ as:

$$K_t = 2\sqrt{\frac{a}{\rho}} \quad \text{(for sharp elliptical holes)} \qquad (7.2)$$

Thus, whenever we have a "sharp" crack in a part, the radius at the tip of the crack is exceedingly small, and $K_t$ is exceedingly large.

A common "temporary fix," when a part has a crack in it, is to drill a hole at the tip of the crack, thus increasing the radius of curvature and reducing $K_t$.

## 7.3 SAFETY FACTOR

Presumably you have come to the conclusion that our analytic procedures are only approximate. This is not because this is just an introductory text, but because there are many imponderables and many possibilities for error—and not just errors of analysis.

As we said earlier, our object is to avoid failure. Suppose, however, that a part *has* failed in service, and we are asked: *Why*? "Error" as such can come from three distinctly different sources, any or all of which can cause failure:

1. *Error in design*: We the designers or the design analysts may have been a bit too optimistic: Maybe we ignored some loads; maybe our equations did not apply or were not properly applied; maybe we overestimated the intelligence of the user; maybe (heaven forbid) we slipped a decimal point.
2. *Error in manufacture*: When a device involves heavily stressed members, the effective strength of the members can be greatly reduced through improper manufacture and assembly: Maybe the *wrong* material was used; maybe the heat treatment was not as specified; maybe the surface finish was not as good as called for; maybe a part was "out of tolerance"; maybe the surface was damaged during machining; maybe the threads were not lubricated at assembly; or perhaps the bolts were not properly tightened.
3. *Error in use*: As we all know, we can damage almost anything if we try hard enough, and sometimes we do so accidentally: We went too fast; we lost control; we fell asleep; we were not watching the gages; the power went off; the computer crashed; *he* was taking a coffee break; *she* forgot to turn the machine off; you failed to lubricate it, etc.

Any of the above can happen: Nothing is designed perfectly; nothing is made perfectly; and nothing is used perfectly. When failure does occur, and we try to determine the cause, we can usually examine the design; we can usually examine the failed parts for manufacturing deficiencies; but we cannot usually determine how the device was used (or misused). In serious cases, this can give rise to considerable differences of opinion, differences which frequently end in court.

In an effort to account for all the above possibilities, we design every part with a *safety factor*. Simply put, the safety factor (SF) is the ratio of the load that we think the part can withstand to the load we expect it to experience. The safety factor can be applied by increasing the design loads beyond those actually expected, or by designing to stress levels below those that the material actually can withstand (frequently called "design stresses").

$$\text{Safety factor} = \text{SF} = \text{failure load/design load}$$

$$= \text{failure stress/design stress} \qquad (7.3)$$

It is difficult to determine an appropriate value for the safety factor. In general, we should use larger values when:

1. The possible consequences of failure are high in terms of life or cost.
2. There are large uncertainties in the design analyses.

Values of SF generally range from a low of about 1.5 to 5 or more. When the incentives to reduce structural weight are great (as in aircraft and spacecraft), there is an obvious conflict. Safety dictates a large SF, while performance requires a small value. The *only* resolution involves reduction of uncertainty. Because of extreme care and diligence in design, test, manufacture, and use, the aircraft industry is able to maintain very enviable safety records while using safety factors as low as 1.5.

We might note that the safety factor is frequently called the "ignorance factor." This is not to imply that engineers are ignorant, but to help instill in them humility, caution, and care. An engineer is *responsible* for his or her design decisions, both ethically and legally. Try to learn from the mistakes of others rather than making your own.

## 7.4 DUCTILE YIELD

When we studied the tensile test, we saw that all ductile materials have a more or less well defined yield point which marks the end of elastic deformation and the onset of plastic deformation. If a member is loaded in simple tension, it will yield when the stress reaches the yield stress $\sigma_{yp}$. However, we are now able to consider a general state of stress, where we may have applied various stresses in different directions. We now must learn how to predict the onset of yielding when we have "combined stresses."

We know that in the completely general case, we can represent the state of stress at a point in terms of the three principal stresses $\sigma_1$, $\sigma_2$, $\sigma_3$, or graphically in terms of the three Mohr's stress circles. What combinations of these stresses will produce yielding?

Let us first make some observations about the mechanism of plastic flow in metals, and then introduce "yield criteria" for combined stresses. We will consider only isotropic materials.

### 7.4.1 Plastic Flow Mechanism

If we take very careful measurements during a tensile test of a metal, we find that during the elastic extension, the volume of the specimen increases. The atoms are pulled away from their equilibrium positions, and the interatomic spacing *increases* in proportion to the strain in the axial direction and *decreases* according to Poisson's ratio in the transverse directions.

However, as we move into the plastic regime, the volume does not continue to increase in proportion to the axial strain. If there is little strain hardening, the volume remains essentially constant, and Poisson's ratio becomes $\frac{1}{2}$. The interatomic spacing no longer increases, but remains constant.

The mechanism of plastic flow that satisfies the above observations is the mechanism of *shear*. In essence, yield and plastic flow involve shearing or sliding of atomic layers relative to each other. Figure 7.4 shows schematically how layers of material can "slip" over each other to produce large shear strains without changes in volume. Although the mechanism actually produces shear strain $\gamma$, if we rotated our axes 45° we would see large tensile strain along $AB$ (remember Mohr's circle for strain?).

In the case of an isotropic tensile specimen, we can think of slip taking place on many, many planes at 45° to the axis. Thus we observe what appears to be a uniform extension of the material in the axial direction; the actual mechanism, however, involves slip on a great many discrete planes. In fact, because engineering metals are polycrystalline, and individual crystals have preferred slip planes and directions, the total slipping mechanism is most complex.

Furthermore, if we estimate the stress required to slip one layer of a crystal over another, we find it is higher, by several orders of magnitude, than we observe. So the mechanism cannot involve slipping of an entire layer. Rather, it involves the migration of imperfections within the crystal which result in what we call "slip."

Figure 7.5 shows how an inchworm moves by advancing its tail, thus

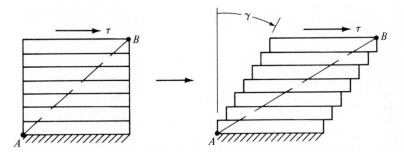

**Figure 7.4** Showing how "slip" between layers can produce plastic strain.

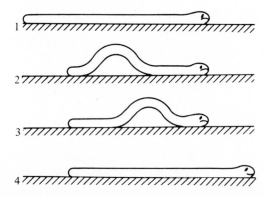

**Figure 7.5** Motion of an inchworm due to a travelling wave. *(After Orowan.)*

producing a "wave" which moves forward. At no time does the entire worm slide; it moves forward incrementally. So it is with plastic flow: The imperfections in the crystal structure, which are called "dislocations," move throughout the material, causing incremental gross motion.

Anyone interested in the details of plastic flow will find an extensive, and fascinating, literature on dislocation theory. We will not attempt any detail here: Suffice it to say that plastic flow occurs via dislocation motion; strain hardening is due primarily to blocking and "stacking up" of dislocations; and hardening through heat treatment involves blocking of dislocation motion via well-dispersed hard particles.

### 7.4.2 Yield Criteria

Because plastic flow and yield are essentially shear phenomena, it is not surprising that they are controlled primarily by the maximum shear stress at any point. The "maximum shear stress" or "Tresca" yield criterion simply states that yield will occur when the maximum shear stress reaches a critical value $\tau_{yp}$. The value of $\tau_{yp}$ must, of course, be determined by experiment for each material.

$\tau_{yp}$ is usually determined from the standard tensile test. Figure 7.6 shows the Mohr's stress circles for the tensile test at the instant of yield. The magnitude of the maximum shear stress is simply half of the tensile yield stress; that is, yield occurs when

$$\tau_{max} = \tau_{yp} = \frac{\sigma_{yp}}{2}$$

or
$$\frac{\sigma_1 - \sigma_3}{2} = \frac{\sigma_{yp}}{2} \tag{7.4}$$

$$\sigma_1 - \sigma_3 = \sigma_{yp}$$

We should note that this is a "criterion," not a "law." It is reasonable to

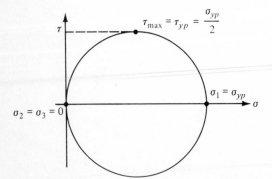

**Figure 7.6** Mohr's stress circles for a tensile test at the instant of yield.

expect that $\tau_{max}$ would control, and experiment gives reasonable (but not exact) agreement.

The other often used criterion for yield and subsequent flow is the simplest symmetrical function of the principal stresses that might be considered. Various physical interpretations have subsequently been given for the function, but are not particularly informative. The "von Mises yield criterion" states: Yield will occur when the sum of the squares of the diameters of the three Mohr's circles reaches a critical value. Again, the critical value is determined in terms of the tensile yield stress $\sigma_{yp}$. In equation form, yield, according to the von Mises yield criterion, occurs when

$$(\sigma_1 - \sigma_2)^2 + (\sigma_2 - \sigma_3)^2 + (\sigma_3 - \sigma_1)^2 = C = 2\sigma_{yp}^2 \qquad (7.5)$$

Either criterion gives satisfactory results; neither is precise. For some states of stress they will give identical results; for others (when the intermediate principal stress is not zero), the Tresca criterion will give a more conservative result. The maximum difference is approximately 15 percent.

For general use (as opposed to advanced study), the Tresca, maximum shear stress, criterion is preferable:

It appeals from a "mechanistic" point of view (shear stress should control shear processes).
It is the more conservative criterion.
It is easier to use.

We now have two criteria to tell us if a given state of stress should cause yielding in any specific material (if we know the tensile yield stress). In the earlier sections of this chapter we discussed stress concentration and the safety factor. How do these apply?

Stress concentrations produce higher stresses in the vicinity of geometrical variations; at a distance from the cause, there is little effect. High stress-concentration factors are associated with small radii of curvature, so the volume subject to high stress concentration is small.

In a ductile material, if $K_t$ times the mean stress exceeds the yield stress, then local yielding occurs in the very small volume affected by $K_t$, but there is little effect elsewhere; and the effect of the yielding flow is to decrease the stress concentration.

Hence, the effect of stress concentration on gross yielding can be ignored.

However, in cyclic loading (fatigue), the small, localized, plastic deformations can accumulate and eventually lead to the formation of a "fatigue crack" and failure.

Also (as will be discussed later), in materials that are "notch-sensitive," the plastic flow at the stress concentration can lead to a brittle fracture (not good).

The application of a safety factor in yield is quite straightforward. We simply divide the actual yield stress by the safety factor to obtain the design stress which can be used in either criterion.

**Example 7.1** A pressure cylinder is made from steel having a tensile yield stress of 50,000 psi. The tank is 10 in in diameter, and the steel is 0.1 in thick. What should be the maximum applied pressure $p$ if we desire a safety factor of 4.0? Consider failure to be yield of the cylindrical portion of the tank.

The appropriate stress element and Mohr's circles are shown in the accompanying illustration. Note that the radial stress, normal to the cylinder surface, is $-p$ inside the tank and 0 outside. From the Mohr's circle, we see that the maximum shear stress will be slightly larger at the inside of the tank.

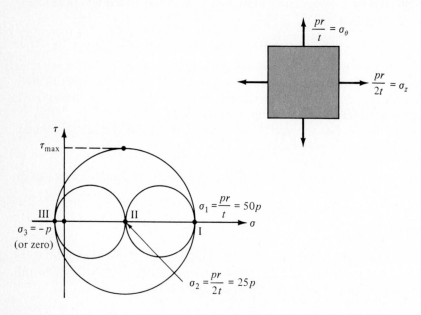

We can apply the two yield criteria directly:

Tresca:
$$\sigma_1 - \sigma_3 = \frac{50,000}{4} = 12,500$$
$$50p - (-p) = 12,500$$
$$p = 245 \text{ psi}$$

von Mises:
$$(\sigma_1 - \sigma_2)^2 + (\sigma_2 - \sigma_3)^2 + (\sigma_3 - \sigma_1)^2 = 2\left(\frac{50,000}{4}\right)^2$$
$$(25p)^2 + (26p)^2 + (-51p)^2 = 3.125 \times 10^8$$
$$3902p^2 = 3.125 \times 10^8$$
$$p^2 = 80,087$$
$$p = 283 \text{ psi}$$

We see that the von Mises criterion permits about 16 percent higher pressure—i.e., Tresca is more conservative. We also see that for "thin-walled" cylinders, we would not have introduced a significant error if we had considered the outside of the tank, which had zero radial stress (the Tresca criterion would have permitted 250 psi).

\*\*\*

## 7.5 BRITTLE FRACTURE

We can define a "brittle" material as one that undergoes little or no plastic deformation prior to fracture; that is, it is essentially elastic until it fractures.

Because it will not yield significantly, any brittle material is inherently very susceptible to the effects of stress concentrations; the stress concentrations cannot be relieved through plastic flow, and have their full theoretical value, $K_t$. For all practical purposes we can say:

> Whenever we experience "brittle fracture," it is caused by tensile stress at a stress concentration. Even if the part is loaded in axial compression, failure will be due to tensile stresses at the tip of a properly oriented (internal) stress concentration.

There are two types of criteria for brittle fracture: In one case, the maximum tensile stress at the tip of the "worst" stress concentration is equated to the cleavage strength of the material. For the other criterion, the elastic energy released, when a crack grows, is equated to the energy required to produce new surface and any associated local plastic strain. The criteria predict much greater strength in compression than in tension, a prediction that is in good qualitative agreement with experiment. However, these criteria generally are not conservative for compressive loading.

Because we prefer a conservative criterion, the brittle-fracture criterion due to Mohr (of Mohr's circle fame) is physically appealing and practically useful. It is assumed that data are available giving the fracture strength in tension ($\sigma_{f_t}$) and in compression ($\sigma_{f_c}$). Mohr's stress circles for the tension and compression tests are drawn as shown in Fig. 7.7. Lines are drawn tangent to the test data circles, and it is assumed (as a criterion) that fracture will not occur if the actual Mohr's circles remain inside the solid lines.

When we use Mohr's criterion, we assume implicitly that the distribution of stress concentrations in our part is the same as it was in the test specimens. The active stress concentrations within a material are generally due to flaws or inclusions of other phases that are statistically distributed. So it is reasonable to expect that a large part *could* have larger flaws than a small test specimen.

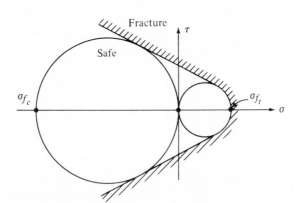

**Figure 7.7** Mohr's criterion for brittle fracture ($K_t = 1$; SF = 1).

Thus there is a "size effect" operative for brittle materials, in which large parts tend to be relatively weaker than small parts.

While the test data reasonably account for the stress concentrations inherent in the material, they do not take account of any gross concentrations due to notches, etc. Thus, for a brittle material:

> The permissible design stresses, for use with the Mohr criterion, must be divided by the theoretical stress-concentration factor and by the safety factor.

Although we blithely discuss "brittle" materials and "ductile" materials, the distinction is not always clear. Under certain states of stress or temperature, normally ductile materials can fail in a brittle fashion, and normally brittle materials can undergo plastic deformation.

We noted in Chap. 3 that in the presence of large hydrostatic compression, a normally brittle material can yield and flow plastically. A qualitative explanation of these effects is based on the Mohr's stress circles shown in Fig. 7.8.

We have drawn upper and lower boundaries at $\tau = \tau_{yp}$, the yield shear stress for that material. We have also drawn the brittle-fracture boundary through $\sigma_{f_t}$, the tensile stress that would cause brittle fracture at the "worst" stress concentration. A material behaves in a brittle or ductile fashion depending on which boundary is reached first.

If the loading produces a Mohr's circle such as $A$ (the largest of the three), yielding occurs; for circle $B$, we would have brittle fracture. We can now make three general statements:

"Hydrostatic pressure" will move the Mohr's stress circles to the left, producing more ductile behavior.

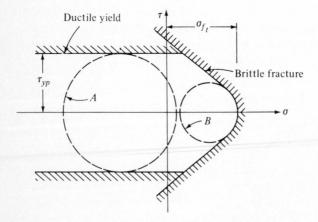

**Figure 7.8** Plastic-flow and brittle-fracture boundaries.

"Hydrostatic tension" will move the Mohr's stress circles to the right toward a
brittle behavior.

Raising the yield shear stress ($\tau_{yp}$) or lowering the tensile fracture stress ($\sigma_{f_t}$)
will increase the probability of brittle failure.

From a design point of view, making a material behave with more ductility
is almost always good. However, when we expect ductile behavior and suffer
brittle fracture, that is invariably *bad*, frequently catastrophically so. (You may
have heard of large ships, made of "ductile" steel, that have broken in half in a
brittle fashion.) There are two main effects that can lead to unexpected brittle
behavior:

1. Large-value stress concentrations in a marginally ductile material, particularly notch-type concentrations, can lead to brittle failure. When yielding
   occurs, the high rate of strain can raise $\tau_{yp}$, and the stress concentration can
   give hydrostatic tension, reducing $\sigma_{f_t}$. Both effects promote brittle behavior.
   Materials which are sensitive to this effect are called (not surprisingly)
   "notch-sensitive." This effect can be clearly demonstrated in polyvinyl-
   chloride (PVC).
2. Reduced temperature, in many metals and polymers, can lead to an increase
   in $\tau_{yp}$ relative to $\sigma_{f_t}$ such that the metal fractures instead of yielding. This
   effect can be quite abrupt, and is referred to as the "ductile-brittle-
   transition temperature."

Because brittle failures tend to be catastrophic, and because new, higher-
strength materials often are susceptible to brittle fracture, we must be particularly careful to avoid the unexpected.


## 7.6 DUCTILE FRACTURE

We will have very little to say about *ductile fracture* for two reasons:

It is an exceedingly complex process that remains a subject of research.

From a practical, design point of view, ductile fracture is not particularly
important. Long before a part will fracture in ductile fashion, it will have
yielded. This generally constitutes "failure" of a part.

In metals, ductile fracture involves the growth and coalescence of voids
within the material, much like "bubbles" growing in a liquid. Voids tend to
initiate at imperfections or around hard-particle inclusions in the material.
Dislocations which stack up at such a location result in void growth.

As you might intuitively expect, voids (like bubbles) tend to grow faster in
the presence of hydrostatic tension than in the presence of hydrostatic com-

pression. Thus, as the Mohr's stress circles move from the tensile side of the diagram to the compressive side, not only can a brittle material become ductile, but a ductile material becomes even more ductile (exhibits greater strain prior to fracture).

In the latter stages of a tensile test of ductile metal, voids start to form in the area of greatest necking. Because of the geometry of necking, where the lines of force must bend outward away from the axis, hydrostatic tension is produced near the specimen axis. Accordingly, voids form first near the axis of the specimen. These can be observed in an "almost" fractured specimen.


## 7.7 FATIGUE FAILURE

If we consider today's world of engineering structures, and ask *what* gives rise to unexpected failure, we find that *fatigue* is probably the worst culprit. Corrosion and wear are probably more costly in the aggregate, but fatigue failures, like brittle failures, tend to be more disastrous.

Just like the other design limitations we are considering, fatigue is a very complex process, one which is not fully understood and which is currently the subject of considerable research. Although fatigue is most important, we cannot provide a thoroughly satisfactory analytical treatment. The approach we will use is a variation, of the classical *S-N method* (*S* for stress; *N* for number of cycles to failure). We should note here, however, that the more advanced methods of "fracture mechanics" are replacing the *S-N* method for advanced structural design. (See J. A. Collins, *Failure of Materials in Mechanical Design*, Wiley, New York, 1981.)

When in doubt, and the consequences of failure are great, there is no substitute for full-scale testing of the actual device. There is a large and growing literature on fatigue, and you can find test data for a wide variety of materials and test conditions.

Fatigue occurs over a period of time, through cyclic loading at stresses below those that would cause immediate failure. A part that may have been in service for hours, months, or years will suddenly break, frequently causing severe problems: A wheel falls off of a car; an engine falls from an airplane, etc. Fatigue failures occur when least expected, generally giving no warning.

You all know that if you bend a wire back and forth a few times, it will break. In such a case, you have bent the wire beyond yield each time, and in some way there has been an accumulation of damage which led to fracture. On the other hand, if you bent the wire back and forth such that the maximum bending stress was perhaps three-quarters of the yield stress, it might take thousands or millions of cycles before failure; in fact, failure might never occur.

In this section, first we will discuss qualitatively the mechanism of fatigue, and then introduce data and criteria for handling various situations.

### 7.7.1 Fatigue Mechanism

Consider the fatigue of a tensile test specimen where we now alternately load the specimen in tension and then compression, as shown in Fig. 7.9a. The *alternating* stress ($\sigma_a$) can be applied equally above and below zero, or there can be a *mean* stress ($\sigma_m$) as well (Fig. 7.9b). For now, let $\sigma_m = 0$.

If, for the alternating stress, we apply a stress equal to the ultimate stress ($\sigma_{ult}$), then the specimen will fail during the first cycle (by definition). If we reduce the stress $\sigma_a$, more cycles will be required for failure. If we make many tests, we can produce curves like that shown in Fig. 7.10 where the number of cycles to failure ($N$) is plotted vs. the applied alternating stress $\sigma_a$. (This curve is often called the *S-N curve*, and is plotted on log-log or semilog coordinates.)

The mechanism of fatigue involves two stages: crack initiation, and crack growth to failure.

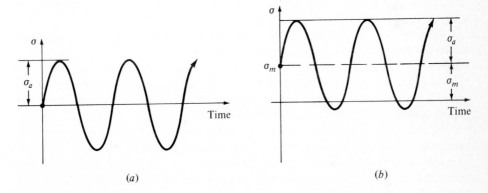

$(a)$ $(b)$

**Figure 7.9** Fatigue loading: ($a$) without mean stress and ($b$) with mean stress.

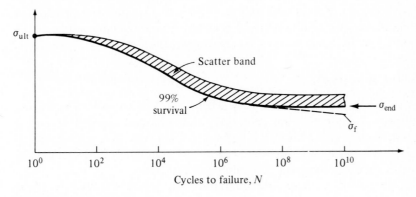

**Figure 7.10** Typical fatigue test curve.

Even at quite low applied stresses, there will be some localized plastic flow at the "worst" stress concentrations. As the number of cycles grows, this plastic flow "accumulates" in the form of a minute crack having a large stress concentration at its tip.

The crack then grows slowly as the stress continues to be cycled. Eventually, the "fatigue crack" will be so large that the remaining area cannot support the load, and it fails, frequently in brittle fashion because of the large stress concentration associated with the fatigue crack. In "high-cycle fatigue" where millions of cycles are generally required to produce failure, it is quite common for the crack to cover more than half the cross-sectional area before failure finally occurs.

### 7.7.2 Fatigue Test Data

Because of uncertainty regarding the "worst" stress concentration and the nature of fatigue crack development, there is a great deal of scatter in fatigue data. If numerous repeat tests are made, the number of cycles to failure can vary by as much as a factor of 10.

In order to give proper statistical treatment to fatigue data, many tests are required. Because the tests are inherently "long-time" tests, this becomes very expensive. A typical "scatter band" is shown in Fig. 7.10. Given sufficient data for analysis, we can plot curves having various statistical significance. A 99 percent survival curve simply means that under conditions below that curve, 99 out of 100 specimens tested will probably survive.

In the past, most fatigue tests were conducted using a rotating cylindrical specimen, with a very smooth test section, subject to a bending moment. During each revolution, the beam elements were subject to an alternating stress. More recently, more tests have been made on a push-pull machine that can apply any combination of mean and alternating loads. The rotating beam machines are the fastest, typically running up to 10,000 rpm. However, even at 10,000 rpm, it takes a while to accumulate $10^8$ cycles (about 1 week); so for 50 data points (which are not all that many), either one has many machines or it takes a long time. The effects of test speed are usually small enough to be ignored; the effects of temperature and environment are not.

### 7.7.3 Stress Concentration

The fatigue crack initiation phase is very sensitive to stress concentrations and they must be taken into account. We saw earlier that the theoretical stress-concentration factor $K_t$ is known for many geometries. We applied $K_t$ in predictions of brittle fracture. Now, however, the crack initiation is caused by plastic flow. Should we apply the full value of $K_t$ or not?

Empirically we find that some materials are more sensitive to stress concentration than others. Accordingly, a different factor is used which can vary from 1 to $K_t$, and which is called "fatigue-strength reduction factor," or $K_f$

for short. If $K_f$ is not known for a particular material-geometry combination, use the full value $K_t$.

### 7.7.4 Endurance Limit—Fatigue Strength

When fatigue test data are plotted, we find two different types of materials:

1. *Steel and titanium*: The *S-N* curve for ferrous materials and titanium becomes horizontal for cycles above $10^6$ to $10^7$. That is, there is a stress below which fatigue failure will not occur. This limiting stress, below which the part should endure forever, is called the "endurance limit," ($\sigma_{end}$) or the "fatigue limit." For clarity we will use the term "endurance limit." As a rough rule of thumb, when other data are not available, $\sigma_{end} = 0.4\sigma_{ult}$ (for ferrous metals).
2. *Other materials*: The *S-N* curves do not level off at high cycles, but continue downward at a very slight slope. Accordingly, instead of an endurance limit, we can specify a "fatigue strength" ($\sigma_f$) at a specific number of cycles (typically $10^7$ to $10^8$). The rule of thumb for these materials is that $\sigma_f = 0.25\sigma_{ult}$ at about $10^8$ cycles.

*Note*: There is some evidence that for steel and titanium occasional very high loads may cause the *S-N* curve to continue sloping downward beyond $10^6$ cycles.

### 7.7.5 Variable-Stress Levels

The discussion above relates to fatigue when the loading is essentially sinusoidal. The majority of real devices do not undergo a nice uniform load over their lifetime, so we must relate constant-load data to variable-load conditions. The simplest method (and the one that is used) involves the assumption that damage at one load level can be added to damage at another level, and that each load level will "use up" a fraction of the total life. This fraction equals the ratio of the actual number of cycles ($n_i$) to the number of cycles to fracture ($N_i$) at that load. The criterion (sometimes called "Miner's rule") can be expressed in the following way: Fracture due to $i$ different load levels occurs when:

$$\sum_i \frac{n_i}{N_i} = 1 \qquad (7.6)$$

### 7.7.6 Effects of Mean Stress

In many instances, the fatigue alternating stress is added to a mean stress as shown in Fig. 7.9. A typical case would be a bolt, tightened to a mean tensile load, and subject to an alternating load (head bolts or connecting-rod bolts in an internal combustion engine, for example). Test data show conclusively that a tensile mean load reduces fatigue life, whereas a compressive load can have either no effect or a beneficial effect. To be conservative, we will assume no effect due to compression, but substantial detriment due to tension.

We have seen that fatigue tests are expensive even when the mean stress is zero; now we have the possibility of an entire family of fatigue curves at different mean-stress levels. It is clear that it would be most desirable if we could predict behavior in the presence of mean stress from tests at zero mean stress.

Again, there are no "laws" to guide our analysis, only criteria that have been proposed and tested. Figure 7.11 is a plot of alternating stress $\sigma_a$ vs. mean stress $\sigma_m$. We can draw curves that represent constant fatigue life (including infinite life) as shown.

Two points on the curves are clear: At $\sigma_m = 0$, the maximum alternating component is simply the endurance limit (fatigue strength) that we have been considering. At the other extreme, if the mean stress equals the ultimate stress for the material, it will fracture without any alternating stress. The Gerber criterion is a cubic equation joining the two points; the Goodman criterion joins them with a straight line. Most data fall between these lines. The Soderberg criterion, which is more conservative, uses the yield stress as the terminus of the line.

Also shown in Fig. 7.11 is a line drawn at 45° from the yield stress. Along this line, the sum of the mean and alternating stresses equals the yield stress. Clearly we must operate below this line to avoid gross yield.

We will use the Goodman line because it is simple and easy to recall. Expressed analytically:

$$\frac{\sigma_a}{\sigma_{\text{end}}} = 1 - \frac{\sigma_m}{\sigma_{\text{ult}}} \tag{7.7}$$

It should be clear that Eq. (7.7) is for failure of smooth specimens. The effects

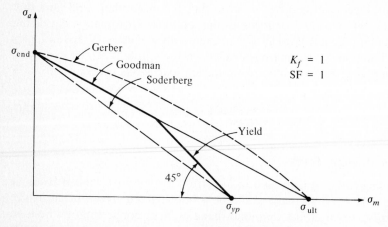

**Figure 7.11** Criteria for the effect of mean stress in fatigue.

of stress concentration and safety factor are not included. Obviously these must be considered, but you will find that various authors will treat these effects in different ways. The following treatment, which is not the "normal" one, is based on extensive work of Snow and Langer (see A. L. Snow and B. F. Langer, "Low Cycle Fatigue of Large-Diameter Bolts," *Trans. ASME B*, pp. 53–61, Feb. 1967).

We saw earlier that stress concentration has a negligible effect on the gross yielding or fracture of a ductile material. This is because the localized yielding near the stress concentration effectively limits the maximum local stress level. Gross yielding cannot occur until the entire cross section reaches the yield condition.

Relative to fatigue, however, the *crack initiation* process *is* a local phenomenon, and any stress concentration can be very detrimental (as expressed by the fatigue strength reduction factor $K_f$). Therefore, when the mean-stress level is zero, we must apply the full value of $K_f$ in reducing the permissible alternating stress.

When we have a relatively small mean stress $\sigma_m$, and we have a stress concentration, the mean stress at the stress concentration is increased by $K_t$. Thus the failure line on the Goodman diagram must be moved towards the origin. Figure 7.12 shows the original Goodman line $OO'$ moved down to $AA'$ (strictly speaking, the ultimate stress should be divided by $K_t$ as noted above; however, for simplicity in calculation we will divide by $K_f$). Both the endurance limit and the ultimate stress have been divided by $K_f$.

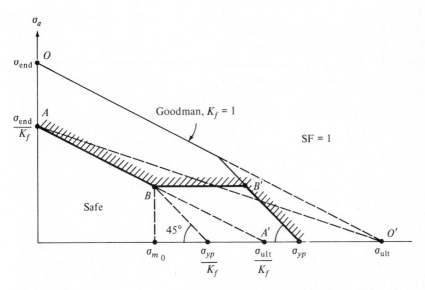

**Figure 7.12** Showing the effects of stress concentration on the Goodman diagram (as expressed in terms of $K_f$).

As we continue to increase the mean stress, a point will be reached when the material at the stress concentration yields. Beyond this point, a further increase in the mean stress can have no further effect (the stress simply cannot exceed the yield stress). This is indicated by the horizontal line $BB'$ in Fig. 7.12. Point $B$ is determined by the onset of local yielding when $K_f(\sigma_m + \sigma_a) = \sigma_{yp}$ (again using $K_f$ in place of $K_t$). Point $B'$ is determined by gross yielding. The $45°$ line drawn from $\sigma_m = \sigma_{yp}/K_f$ gives the condition for local yielding. The critical value of $\sigma_m$ at point $B$ $(\sigma_{m_0})$ is given by:

$$\sigma_{m_0} = \frac{\sigma_{\text{ult}}}{K_f}\left(\frac{\sigma_{yp}/\sigma_{\text{end}} - 1}{\sigma_{\text{ult}}/\sigma_{\text{end}} - 1}\right) \tag{7.8}$$

We can now express the failure criteria for the two regimes:

For $\sigma_m < \sigma_{m_0}$:
$$\sigma_a = \frac{\sigma_{\text{end}}}{K_f}\left(1 - \frac{K_f\sigma_m}{\sigma_{\text{ult}}}\right) \tag{7.9}$$

For $\sigma_m > \sigma_{m_0}$:
$$\sigma_a = \frac{\sigma_{\text{end}}}{K_f}\left(\frac{1 - \sigma_{yp}/\sigma_{\text{ult}}}{1 - \sigma_{\text{end}}/\sigma_{\text{ult}}}\right) \tag{7.10}$$

Point $B'$, determined by the condition that $\sigma_a + \sigma_m = \sigma_{yp}$, is given by:

At $B'$:
$$\frac{\sigma_m}{\sigma_{yp}} = 1 - \frac{\sigma_{\text{end}}}{K_f\sigma_{yp}}\left(\frac{1 - \sigma_{yp}/\sigma_{\text{ult}}}{1 - \sigma_{\text{end}}/\sigma_{\text{ult}}}\right) \tag{7.11}$$

In practice, the maximum mean stress is generally kept below two-thirds of the yield stress.

In order to use Fig. 7.12 for design purposes, we still must apply a safety factor. It is conceivable that different safety factors should be applied to the mean and alternating stresses—sometimes one of them is known with some precision and a smaller SF is warranted.

In many bolt applications, there is less than perfect control of the mean tensile preload. Under such conditions, it is prudent to assume that the mean stress will exceed $\sigma_{m_0}$, and that operation will be in the horizontal portion $BB'$ shown in Fig. 7.12. In essence we assume the maximum detrimental effect of the mean stress. In this case, we need apply a SF to the alternating stress only.

Also shown in Fig. 7.12 is the line $AO'$; this is the "commonly" used failure line discussed in many texts. It is clearly *not* a conservative criterion at moderate values of mean stress.

### 7.7.7 Combined Stress

This discussion of fatigue is relevant to uniaxial cyclic loading: tension and compression. We are frequently concerned with fatigue in torsion or other nonaxial loading, but the methods for treating these cases are not well developed.

For fatigue crack initiation, which involves plastic flow, a Tresca (maximum-shear-stress) criterion is suitable. For crack growth, which is controlled by the maximum tensile stress, a maximum (+) normal stress criterion is probably best. When we are concerned with infinite life, a crack should not form, and the Tresca criterion can be used.

**Example 7.2** A typical high-strength steel used for bolts is AISI 4340 heat-treated to give the following properties:

$$\sigma_{ult} = 140,000 \text{ psi}$$

$$\sigma_{yp} = 115,000 \text{ psi}$$

$$\sigma_{end} = 60,000 \text{ psi}$$

For a bolt, assume that $K_f = 4.0$ at the root of the threads (empirically, $K_f$ varies from 2.5 and 4 for screw threads).

1. Draw the Goodman diagram showing the effect of $K_f$.
2. If a bolt is prestressed to 75,000 psi, and we desire SF = 2.0, how large can $\sigma_a$ be?
3. If we plan to operate at a mean stress of 10,000 psi, and an alternating stress of 5000 psi, how large is the SF?

The Goodman diagram is shown below; the points for $\sigma_{end}$ and $\sigma_{ult}$ are joined by the upper line. Next, the line $AA'$ is drawn through $\sigma_{end}/K_f$ and $\sigma_{ult}/K_f$. Point $B$ can be determined from Eqs. (7.8) and (7.10) to be at 24,063 and 4688. $B'$ is at a mean stress of 110,300 psi. $B$ and $B'$ can also be determined by drawing 45° lines from $\sigma_{yp}/K_f$ and $\sigma_{yp}$.

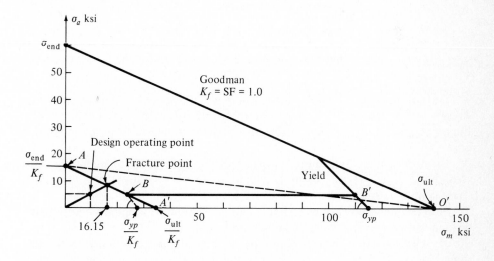

If the bolt is prestressed to 75,000 psi of tension, we are operating in the region between $B$ and $B'$ where the effect of $\sigma_m$ is limited by yielding. The maximum allowable value of $\sigma_a$ is 4688 psi; if we desire SF = 2.0, we must limit $\sigma_a$ to 4688/2 = 2344 psi.

For question 3 above, the design operating point is shown ($\sigma_a = 5000$, $\sigma_m = 10{,}000$). Assuming that we require the same SF for both mean and alternating loads, we simply draw a line, through the origin and the operating point, to the failure line. The intersection determines the SF. From the geometry, we find that the intersection is at $\sigma_m = 16{,}150$ psi. Thus the SF = 16,150/10,000 = 1.6.

Again we have shown the conventional failure criterion as a dashed line from $A$ to $\sigma_{\text{ult}}$. It is not conservative.                                   ***

### 7.7.8 Surface and Environmental Conditions

The fatigue process is very sensitive to the physical condition of the surface and to the atmospheric environment. Crack initiation is always hastened by the presence of tension or stress concentration. Fatigue failure is generally hastened by corrosive environments, even mild ones. The effects can be dramatic (or catastrophic) as shown below.

For purposes of comparison and exposition, consider the effects of a variety of surface conditions on the endurance limit of a fictitious steel (all of these effects might not apply to any single material):

| Surface treatment | $\sigma_{\text{end}}$, psi |
| --- | --- |
| Normal endurance limit | 50,000 |
| Gentle surface grind | 65,000 |
| Electrodischarge machining | 40,000 |
| Shot peen | 75,000 |
| Poor chrome plate | 35,000 |
| In salt-water spray | 7,500 |
| Fretting conditions | 15,000 |

Surface grinding, properly done, will leave a layer of compressive residual stresses at the surface which considerably improves the fatigue resistance. Improper grinding, where the surface is overheated, will reduce the endurance limit.

Electrodischarge machining always leaves a thin layer of metal that has resolidified. This "recast" layer is usually under tensile stress and may have stress-concentrating cracks.

Shot peening and various rolling techniques are used to impart compressive residual stresses to specific areas to retard fatigue.

An electroplated layer is generally in tension, and frequently will show small cracks. Both of these are detrimental from the point of view of fatigue, and can be severe.

Corrosion and fatigue are synergistic; the combined damage is worse than the sum of the individual damages. If conditions might be corrosive, be very careful in selecting the proper material.

"Fretting" occurs when two surfaces, clamped together under load, undergo slight sliding motion. The obvious solution is to clamp the surfaces more tightly; this will stop the fretting of the two surfaces, but will generally increase the tension in some other member. This problem is typical of heavily loaded bolts.

**Example 7.3** Determine the minimum diameter of a bolt for the following conditions:

| | |
|---|---|
| Mean load | 20,000 lb |
| Alternating load | 5,000 lb |
| Ultimate stress | 130,000 psi |
| Yield stress | 95,000 psi |
| Endurance limit | 60,000 psi |
| $K_f = 4.0$    SF = 2.0 | |

The Goodman diagram shown has been adjusted for $K_f$. Also shown is the "load line" representing the possible combinations of $\sigma_a$ and $\sigma_m$ that will satisfy the condition that the mean load be four times the alternating load. The operating point is where the two lines cross, at $C$. Thus, the maximum allowable alternating stress is 7500 psi [from Eq. (7.10)].

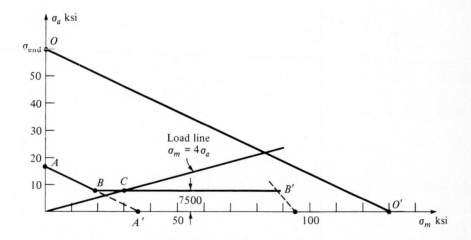

For SF = 2, we must limit the actual alternating stress to 3750 psi when the load is 5000 lb. The required bolt area is 1.33 in²; the diameter at the root of the threads must be greater than 1.30 in (the minimum bolt size would be 1.5″–12). ***

It should be amply clear from the foregoing discussions that fatigue can *greatly* reduce the permissible stress levels. A material may have a yield stress of 100,000 psi, but we may be able to load it to only 2500 psi of alternating stress. Corrosion of fretting can reduce the strength even further. *Always* be alert and on the look out for possible fatigue failure.

## 7.8 COMPUTER EXAMPLE

When carrying out the calculations associated with fatigue in the presence of a mean stress, there are ample opportunities for error. Also, there are alternative solutions depending on the value of the mean stress. This is just the type of problem that is well suited to the computer.

Program Listing 7.1 shows a program for carrying out fatigue calculations. It is longer than most of the prior examples because there are several routes through the program depending on what you wish to determine. Program Output 7.1 gives the printout for an example using the program.

```
10 !"  Fatigue Calculations for Combined Alternating and Mean Stress"\!
20 !"   Input Material Data for Smooth Specimen (Kf = SF = 1)" \!
30 INPUT"     Endurance Limit     (ksi, MPa) = ",S2
40 INPUT"     Tensile Yield Stress (ksi, MPa) = ",S3
50 INPUT"     Ultimate Stress     (ksi, MPa) = ",S4
60 !
70 INPUT"     Fatigue Strength Reduction Factor, Kf = ",K0
80 REM
90 IF S4>S2 THEN 120
100 !\!"Ultimate MUST be greater than the endurance limit"\!
110 GOTO 30
120   S5 = (S4/K0)*((S3/S2-1)/(S4/S2-1))     \REM Sigma-Mo
130   S6 = (S2/K0)*((1-S3/S4)/(1-S2/S4))     \REM Sigma-Ao
140   S8 = S3*(1-(S2/(K0*S3)))*((1-S3/S4)/(1-S2/S4)))  \REM  Sigma-MB'
150   REM
160 !\!
170 !"  Select Mode of Operation:"
180 !"    1. Given Stresses, Calculate Safety Factors"
190 !"    2. Given Safety Factors, Calculate Stresses" \ !
200 INPUT"      Type Choice (1 or 2) ", C1 \ !
210 IF C1 = 2 THEN 460
220 REM
230 !"  Input actual planned (design) operating stresses:" \!
240 INPUT"     Mean Stress,     Sigma-m, (ksi, MPa) = ",S0
250 IF S0<0 THEN S0=0
260 INPUT"     Alternating Stress, Sigma-a, (ksi, MPa) = ",S1
270 IF S0<.9*S8 THEN 310
```

**Program Listing 7.1**

```
280 !
290 !"Mean Stress Too Great ---- Reduce It" \!
300 GOTO 230
310 !
320 R1 = S6/S5                  \REM  Break-point Stress Ratio
330 IF S0>0 THEN 350
340 S7=S2/K0 \ GOTO 420
350 R2 = S1/S0                  \REM  Actual stress Ratio
360 IF R2>R1 THEN 410
370 !"Safety Factor = ",%5F2,S6/S1 \!
380 INPUT"     Try Other Design Stresses?  (CR/N) ",Z$ \!
390 IF Z$<>"N" THEN 230
400 END
410 S7 = (S2/K0)/(1+S2/(R2*S4))    \REM Allowable Alt. Stress
420 !"Safety Factor = ",%5F2,S7/S1 \!
430 GOTO 380
440 REM
450 REM
460 !"   Input Safety Factors:" \!
470 INPUT"        Mean Stress Safety Factor        = ",F1
480 INPUT"        Alternating Stress Safety Factor = ",F2 \!
490 !"Select Operating Mode:" \!
500 !"      1. Given Mean Stress, Calculate Alternating Stress"
510 !"      2. Given Alternating Stress, Calculate Mean Stress" \!
520 INPUT"         Type Choice, 1 or 2 ",C2 \ !
530 IF C2=2 THEN 690
540 INPUT"      Design Mean Stress (ksi, MPa) = ",S0 \!
550 IF S0<0 THEN S0=0
560 IF S0<.9*S8 THEN 600
570 !
580 !"Mean Stress Too Large --- Reduce It." \!
590 GOTO 540
600 IF S0<=S5/F1 THEN 650
610 !"Allowable Design Alternating Stress (ksi, MPa) = ",%10F3,S6/F2\!
620 INPUT "    Try Other Mean Stresses?  (CR/N) ",Z$ \!
630 IF Z$<>"N" THEN 540
640 END
650 S1 = (S2/K0)*(1-K0*S0*F1/S4)   \REM  Allowable Alt Stress
660 !"Allowable Design Alternating Stress (ksi, MPa) = ",%10F3,S1/F2 \!
670 GOTO 620
680 REM
690 INPUT"      Design Alternating Stress (ksi, MPa) = ",S1 \ !
700 IF S1*F2>S6 THEN 750
710 !"Allowable Design Mean Stress (ksi, MPa) = ",%10F3,.9*S8/F1 \!
720 INPUT"    Try Other Alternating Stresses?  (CR/N) ",Z$ \!
```

**Program Listing 7.1** (*continued*)

```
730 IF Z$<>"N" THEN 690
740 END
750 S0 = (S4/K0)*(1-K0*S1*F2/S2)      \REM Allowable Mean Stress
760 IF S0>0 THEN 790
770 !" Alternating Stress is Too HIGH --- Reduce it." \!
780 GOTO 690
790 !"Allowable Design Mean Stress (ksi, MPa) = ",%10F2, S0/F1 \!
800 GOTO 720
```

**Program Listing 7.1** (*continued*)


```
    Fatigue Calculations for Combined Alternating and Mean Stress

    Input Material Data for Smooth Specimen (Kf = SF = 1)

        Endurance Limit       (ksi, MPa) = 50
        Tensile Yield Stress (ksi, MPa) = 80
        Ultimate Stress       (ksi, MPa) = 110

        Fatigue Strength Reduction Factor, Kf = 3.5



    Select Mode of Operation:
      1.  Given Stresses, Calculate Safety Factors
      2.  Given Safety Factors, Calculate Stresses

          Type Choice (1 or 2) 1

    Input actual planned (design) operating stresses:

        Mean Stress,         Sigma-m, (ksi, MPa) = 15
        Alternating Stress, Sigma-a, (ksi, MPa) = 2.5

Safety Factor =  2.86

        Try Other Design Stresses?   (CR/N)
```

**(Same input data as above)**

```
  Select Mode of Operation:
    1.  Given Stresses, Calculate Safety Factors
    2.  Given Safety Factors, Calculate Stresses

        Type Choice (1 or 2) 2
```

**Program Output 7.1**

Input Safety Factors:

```
Mean Stress Safety Factor      = 2.86
Alternating Stress Safety Factor = 2.86
```

Select Operating Mode:

```
1. Given Mean Stress, Calculate Alternating Stress
2. Given Alternating Stress, Calculate Mean Stress

   Type Choice, 1 or 2 1

   Design Mean Stress (ksi, MPa) = 15
```

Allowable Design Alternating Stress (ksi, MPa) =      2.498

```
   Try Other Mean Stresses?  (CR/N)
```

**Program Output 7.1** (*continued*)


## 7.9 PROBLEMS

**7.1** A cylindrical pressure vessel is made from steel having a yield stress in tension of $\sigma_{yp} = 80,000$ psi. In addition to internal pressure of 3000 psi, it is subject to an axial tensile load of 10,000 lb. If radius $r = 2$ in and thickness $t = 0.1$ in, how large is the safety factor (relative to yielding):

(a) According to the Tresca criterion

(b) According to the von Mises criterion

**7.2** Design a spherical pressure vessel to contain 1 cubic foot ($ft^3$) of gas at 5000 psi. We require a safety factor of 4, and the material will have a minimum tensile yield stress of 90,000 psi (design based on yield).

**7.3** The state of stress at a critical location in a member is as shown below. If the tensile yield stress is 40,000 psi, how large is the safety factor?

$$\sigma_x = 13,000 \qquad \sigma_y = -4000 \qquad \sigma_z = 0 \qquad \tau_{xy} = 7000 \qquad \text{(all in psi)}$$

**7.4** 45° strain gage rosettes are attached to the surface of a steel structural element at various locations. The material has a yield stress in tension of 625 MPa. How large is the SF at each location?

| Location | $\epsilon_a$ | $\epsilon_b$ | $\epsilon_c$ |
|---|---|---|---|
| a | 0.001250 | 0.000300 | −0.000750 |
| b | 0.000900 | 0.000650 | 0.001100 |
| c | −0.000050 | −0.000950 | −0.001350 |

**7.5** The ceramic component shown is made from material having a fracture stress in tension of 12,000 psi (brittle). In order to ensure a safety factor of 3, what is the maximum load $F$ that you would allow?

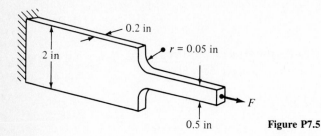

Figure P7.5

**7.6** A sheet of glass has a tensile fracture stress of 20,000 psi. A very small, elliptically shaped groove is scratched into one edge as shown. At what value of $\sigma_x$ would you expect the sheet to fracture?

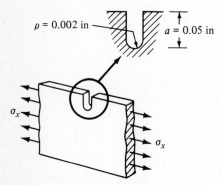

Figure P7.6

**7.7** A tensile member is made from a Kevlar-Epoxy composite. The volume fraction of the fiber is 45 percent. Estimate the tensile fracture stress of the composite.

| | | |
|---|---|---|
| Fiber: | $E = 19 \times 10^6$ psi | $\sigma_{fract} = 500,000$ psi |
| Matrix: | $E = 500,000$ psi | $\sigma_{fract} = 10,000$ psi |

*Note*: Composite fracture strengths are invariably lower than predicted by simple analysis.

**Probs. 7.8 to 7.10:** Composite metal structures are frequently made by "cladding" one metal with another. Typical is aluminum-cladded steel: The steel provides low-cost strength, while the aluminum provides corrosion resistance. Consider an "alclad" sheet, where the thickness of the aluminum is small relative to that of the steel (see the accompanying illustration). The properties are given below in tabular form:

|  | Steel | Aluminum |
|---|---|---|
| $E$ | $30 \times 10^6$ psi | $10 \times 10^6$ psi |
| $\nu$ | 0.3 | 0.33 |
| $\sigma_{yp}$ | 50,000 psi | 10,000 psi |
| $\alpha$ | $6 \times 10^{-6}\,^\circ\mathrm{F}^{-1}$ | $13 \times 10^{-6}\,^\circ\mathrm{F}^{-1}$ |

Assume that the bond between the materials is "sufficiently" strong.

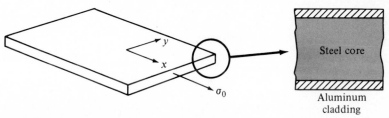

**Figure P7.8 to 7.10**

**7.8** The sheet is loaded in the $x$ direction with $\sigma_0$. At what value of $\sigma_0$ will the first yield occur? In which material?

**7.9** The unloaded sheet is heated. How large a temperature rise (or fall) is required for first yield?

**7.10** The temperature is taken beyond that determined in Prob. 7.9 and then returned to the starting temperature. What are the residual stresses (if any)?

**7.11** A high-strength machine element shown in the illustration below is subject to an oscillating force $F = \pm 1000$ lb at a speed of 3000 rpm. For alignment purposes, two very small holes have been drilled 1 in apart, as shown.

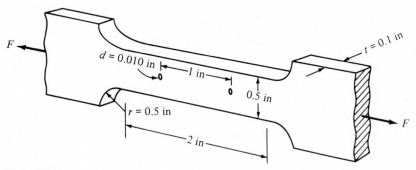

**Figure P7.11**

The material used is a high-strength bolting material whose fatigue strength is conservatively represented by:

$$\sigma_f = \sigma_{end} + \frac{4.5 \times 10^6}{\sqrt{N_f}} \quad \text{and} \quad \sigma_{end} = 50,000 \text{ psi}$$

where $N_f$ = the number of cycles to failure. Estimate the life of the part.

**7.12** Piping, in processing plants, is subject not only to the static pressures involved, but also to pressure oscillations or pulsations. The oscillations usually emanate from pumps, but can be amplified because of "resonance" in the piping itself (as in an organ pipe).

In a well-designed system, the oscillations may be less than 5 percent of the mean (static) value. With resonance, however, the oscillations can reach 15 to 20 percent of the mean, and are frequently very difficult to reduce.

Consider steel pipe of a 1-in inside diameter, having a minimum wall thickness at the root of the threads (for joining the pipe sections) of 0.1 in. If the pressure oscillations are ±20 percent of the mean, and we require a safety factor of 3, determine the maximum mean pressure allowable. Base the design on tangential stress:

$$K_f = 4.0$$

$$\sigma_{yp} = 70,000 \text{ psi}$$

$$\sigma_{end} = 40,000 \text{ psi}$$

$$\sigma_{ult} = 100,000 \text{ psi}$$

**7.13** As in Prob. 7.12 except for a minimum wall thickness of 0.25 in. (Although this is not a "thin-walled" tube, use the thin-wall approximation.)

**7.14** For various purposes, including the attenuation of pressure fluctuations in a piping system, "accumulators" are used. An accumulator is simply a tank, containing a volume of pressurized fluid under pressure, that acts like a "spring" in the system.

Determine the radius and wall thickness of a spherical accumulator to meet the following requirements (use the thin-wall approximations even though they are not conservative). If a single accumulator seems impractical, multiple units can be used, but will be more costly.

Pressure = 15,000 psi ± 20% oscillation
Volume = 500 in³
Safety factor = 3
$K_f = 3.0$ at fluid inlet fitting
$\sigma_{end} = 30,000$ psi
$\sigma_{ult} = 80,000$ psi
$\sigma_{yp} = 55,000$ psi

**7.15** As in Prob. 7.14 except pressure is 3000 psi ± 20% and volume required is 200 in³.

**7.16** As in Prob. 7.14 where the pressure is 3500 psi ± 15% oscillation, and the volume required is 1000 in³. Use SF = 1 for mean stress and SF = 2 for alternating stress.

**Probs. 7.17 to 7.21:** The bolts in any machine are very frequently subject to oscillating loads in addition to the mean tensile preload that is obtained by "torquing up" the bolt initially. Consider here a variety of situations involving bolts made from the same high-strength bolting material, AISI 4340, heat-treated to give:

$$\sigma_{yp} = 100 \text{ ksi}$$

$$\sigma_{ult} = 130 \text{ ksi}$$

$$\sigma_{end} = 50 \text{ ksi}$$

$$\sigma_{fat} = 50,000\left(1 + \frac{89}{\sqrt{N_f}}\right) \text{ psi}$$

where $\sigma_{fat}$ = fatigue strength at $N_f$ cycles

$K_f = 4.0$ at root of thread

(See Prob. 2.12 for bolt nomenclature.)

**7.17** $\frac{1}{4}''$-20 screw (minimum diameter 0.189 in) subject to $F_m = 500$ lb (mean load) and $F_a = \pm 100$ lb (alternating load). Determine the safety factor.

**7.18** $1''$-8 screw (0.847-in minimum diameter), $F_m = 8500$ lb, $F_a = 1500$ lb. Determine the safety factor.

**7.19** $1''$-8 screw (0.847-in minimum diameter), $F_m$ actually = 17,000 lb. The safety factor must be 2.0. Determine the maximum allowable alternating load $F_a$.

**7.20** $3.5''$-12 UN screw (3.400-in minimum diameter). Actual *prestress* equals 15,000 psi. Alternating *force* $F_a = 40,000$ lb. Determine the safety factor.

**7.21** $1.5''$-8 bolts are used in a situation where the load oscillations are due to thermal effects and are not frequent, but are severe. Estimate the number of cycles to failure given the following values:

Minimum diameter = 1.340 in
Mean stress =       15 ksi
Alternating stress =   10 ksi

# EIGHT

## BENDING OF BEAMS I

## 8.1 OVERVIEW

Beams are very common, very important, structural members. Quite possibly they are *the* most important structural members. Therefore, it is necessary to be able to deal with them in an effective way.

However, the total subject matter associated with "beams in bending" is both extensive and complex. The traditional treatment requires a very considerable expenditure of time and thought, and it is very easy to become lost in the various analytical manipulations.

In this text, we shall make a rather radical departure: We will present here the analysis of "simple" beams, subject to "simple" loading. In a later chapter we will introduce additional aspects to complete the presentation for those who desire it. There are two major incentives for this departure:

1. Many practicing engineers never deal with the more complex beam situations.
2. There are many problems of complex beam systems or structures that are very difficult to analyze by traditional methods. However, we can understand, through matrix structural analysis (FEM), the behavior of complex structures involving both bending and axial loading. The structural analysis programs that we will develop (Chap. 9) are based on "simple" elements, which are uniform beams with loading only at the ends.

Hence it seems appropriate to consider here an elementary analysis of beams. Then, in "Bending of Beams II" (Chap. 12), we will pick up the subject again and carry it further.

For much practical use, however, this chapter on elementary bending plus the development of the FEM for bending will, in fact, suffice.

There are a number of ways to approach the analysis of bending. Here, we will take what is often termed the "strength of materials" approach. Basically, we will first study the deformation of a beam when we apply a "bending moment." We will then deduce what the stresses must be if the beam is elastic; and then, through equilibrium, we will determine the magnitude of the bending moment required to give that deformation. Finally, we will determine how the overall beam deflection depends on the localized deformation of the beam at all points along its length.

We will consider the "buckling" of beams and will present a computer analysis that permits handling of much more complex situations than are normally considered. The analyses particular to "composite beams" will also be introduced.

Following this chapter, for those who are interested in the details, we will introduce matrix structural analysis including bending. This simply involves the incorporation of bending terms in the stiffness matrices.

## 8.2 DEFORMATION OF A BEAM IN PURE BENDING

We all know what a "beam" is: a relatively long member that can support loads perpendicular to its axis. It can also support applied moments that tend to bend it, and can support axial loads tending to make it longer or shorter. Later, after we consider the effects of twisting moments (torques) that tend to twist a beam about its axis (Chap. 10, "Torsion"), we will be able to handle a completely generalized set of loads.

For this chapter, we will consider *only* the effects of bending. You will recall that as long as the beam remains elastic, we can superpose the effects due to other loadings.

For this basic analysis, we want to be assured that if we push straight down, on a horizontal beam, it will move downward and will not be displaced to the side. That is, we want to load the beam along a *principal axis of compliance*.

Accordingly, we will restrict ourselves (for the time being) to bending of symmetrical beams. The beams must have at least one plane of symmetry, and the loading and deflections must lie in this plane. To clarify this restriction, Fig. 8.1 shows a number of different beam cross sections. Their characteristics are as follows:

(a) Full axial symmetry—can be loaded in any direction perpendicular to the $x$ axis
(b) Two planes of symmetry—can be loaded in $y$ or $z$ directions
(c) One plane of symmetry—can be loaded along the $y$ axis
(d) Same as (c)

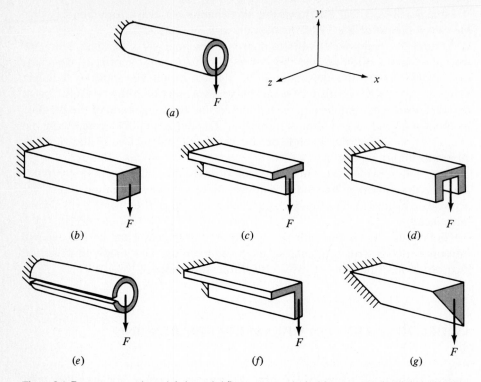

**Figure 8.1** Beam cross sections: (a) through (d) are symmetrical with respect to the plane of loading; (e) through (g) are not symmetrical relative to the plane of loading, and the analysis will not apply.

(e) One plane of symmetry, but not loaded in that plane—could be loaded along the z axis
(f) No plane of symmetry
(g) Same as (f)

For our derivations here, we will always let the beam axis be the x axis and the plane of loading and deflection the xy plane, as shown in Fig. 8.1. That is, bending will be about the z axis.

To ensure that the beam deflects only in the y direction, and not in the z direction too, not only must the section geometry be symmetrical with respect to the xy plane, but if the material properties vary (as in a composite beam made of two or more materials), the material property distribution must also be symmetrical about the y axis. (It is possible to have nonsymmetrical distributions of both area and properties in such a way that the effects of asymmetry just cancel, but we will not consider this case.)

While the above restrictions may seem severe and limiting, a large portion of actual beams do meet these requirements.

Now consider what happens physically when we subject a length of beam to "pure bending," i.e., subject it only to a moment which tends to "bend" the beam. For this analysis, a bending moment, $M_b$, is a moment vector along the $z$ axis. Figure 8.2 shows a beam with only $M_b = M_z$ applied. At every position $x$ along the beam (except at the very ends where the loads are being applied), the loading and load distribution are exactly the same as at every other location. A little contemplation should serve to convince you that if the beam is straight and uniform along its length, and if $M_b$ is constant along the length, then the beam *must* bend in a circular arc.

If you have a rubber eraser, bend it as a demonstration. You will see a thickening or widening at the top, as shown in Fig. 8.2b. As we bent the "beam," the top became shorter and the bottom became longer. The Poisson effect causes it to become thicker on top and thinner on the bottom.

Figure 8.3 is an exaggerated sketch of beam deformation when subject to pure bending ($M_b$). We see that the axis of the beam is bent into a circular arc centered at $O$ having a "radius of curvature" $\rho$, as shown. We also see that because of the Poisson effect there is a secondary curvature centered at $O'$, of radius $\rho'$. This secondary curvature, which is free to occur in narrow beams, is called "anticlastic curvature." Because the anticlastic curvature is due to the Poisson effect, it is not surprising that $\rho' = \rho/\nu$.

If we ignore for now the Poisson effect, every part of the beam is curved in circular arcs centered about an axis passing through $O$. If we look at the $xy$ plane (Fig. 8.4), we see the geometry that will apply to all planes parallel to that plane.

We have described the bending geometry in terms of the radius of curvature $\rho$. It is clear that $\rho$ starts at the center of curvature, $O$; but where does it end? At the top of the beam? At the bottom of the beam? At the center of the beam? Clearly, each level of the beam has a different radius of curvature, smaller at the top, larger at the bottom.

If we consider the physical deformation due to bending, we see that the top of the beam is shortened (compressed), while the bottom of the beam is stretched. There must be some location, between the top and the bottom, where the length remains constant. There is, and it does!

We call the surface that does not change length the "neutral surface," and

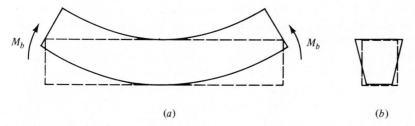

(a) (b)

**Figure 8.2** Rectangular beam in "pure bending": (*a*) Beam is bent into a circular arc; (*b*) beam becomes thicker on top, thinner at the bottom.

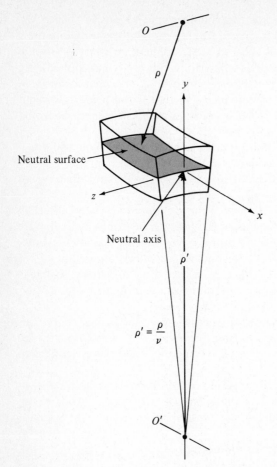

**Figure 8.3** Showing the details of beam deformation including the anticlastic radius of curvature $\rho' = \rho/\nu$.

the line in the $xy$ plane that does not change length the "neutral axis" (NA). Because the NA separates compression (above) from tension (below), we will measure $\rho$ from the center of curvature to the NA. Also, in order to discuss various levels (up and down) on the beam cross section, we will place the origin of the coordinate system (for the beam cross section) on the NA. In Fig. 8.4 we see that $y = 0$ is on the NA, and is positive upward. For our analyses, the beam cross sections will always be in the $yz$ plane.

We might note, as an aside, that the above comments regarding tension and compression are correct only when the beam is bent concave upward, as shown in Fig. 8.4. We will not go into the details for sign convention yet, but will merely stipulate that a positive bending moment is one that produces concave upward bending—i.e., the beam would "hold water." A negative $M_b$ will produce a concave downward shape; the center of curvature will be at $-\rho$; we will have tension on top, and compression at the bottom; it will "shed

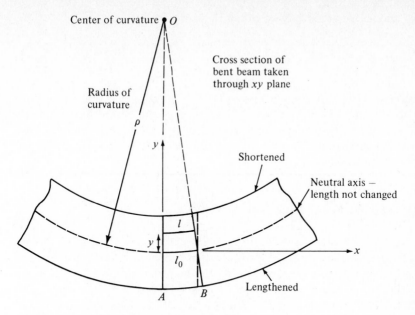

**Figure 8.4** Showing the geometry of pure bending.

water." As we proceed, we will see that this convention is completely consistent with our other sign conventions.

We have seen that for positive values of $y$, the beam material is compressed along its length, just like a strut in compression. For negative $y$, the material is stretched, just like a member in tension. The only difference is that in bending, the *strain* (compressive, tensile) varies with $y$ rather than being constant as in an axially loaded member. By studying the geometry of deformation, we can calculate the strain $\epsilon$ (in the $x$ direction) as a function of $y$.

In Fig. 8.4 we have shown two radial lines from the center of curvature $O$ to points $A$ and $B$. Along the neutral axis, the two lines are a distance $l_0$ apart. At a height $y$ above the neutral axis, they are a shorter distance $l$ apart. Clearly, before the beam was bent, the two distances were equal ($l = l_0$), i.e., the two radial lines were parallel ($\rho = \infty$).

Consider now the two similar triangles having apices at the center of curvature $O$ and the two lengths $l$ and $l_0$ as bases:

$$\frac{l_0}{\rho} = \frac{l}{\rho - y}$$

The compressive strain at height $y$ is just the change in length divided by the initial length of the line segment:

$$\epsilon = \frac{l - l_0}{l_0} = \frac{l}{l_0} - 1$$

Combining the above equations we obtain:

$$\boxed{\epsilon = -\frac{y}{\rho}}$$

(8.1)

Equation (8.1) is exceedingly important yet amazingly simple. It is fundamental to *all* analysis of bending. It tells us that the *strain* varies *linearly* above and below the neutral axis, *independent* of the material or materials and *independent* of whether the strain is elastic or not. The *only* requirements are those of symmetry discussed above.

## 8.3 STRESSES DUE TO PURE BENDING

Equation (8.1) gives us the strain in the $x$ direction when an appropriate beam is bent by a pure bending moment into an arc of radius $\rho$. In order to compute the corresponding stresses, we must specify the material (or materials), and we must specify whether the strains are purely elastic or not. For this chapter, we will study *only* elastic bending, i.e., bending in which the maximum strain must be less than that which would cause yielding. Given this stipulation, we can proceed to examine the stresses that must exist at any point on the beam cross section.

Given the strain in the $x$ direction, we can use the stress-strain relations discussed in Chap. 6 to determine the stresses. However, we know that the strain in the $x$ direction depends not only on the stress in the $x$ direction, but also on the stresses in the $y$ and $z$ directions. {*Note*: You will have no difficulty here if you have not yet studied Chap. 6, because for *narrow* beams, the stress-strain relation is the same as in a tensile test specimen [see Eq. (8.2)].}

Both the top and the bottom of the beam are stress-free, and there is no obvious mechanism to produce stresses in the $y$ direction. Hence it seems reasonable to assume that $\sigma_y = 0$ everywhere in the beam.

The sides of the beam are stress-free too, and if the beam is *narrow* (narrower than perhaps twice the height), we can assume that $\sigma_z = 0$ also.

We specified a narrow beam because we know that a "wide beam," such as a sheet of metal, plywood, or paper, will not develop anticlastic curvature when bent. Near the center of a "wide beam," the *stress* in the $z$ direction is not zero, but the *strain* is. "Wide beams" are generally called "plates," and are slightly stiffer than an equivalent "beam."

For the major developments of this text, we will limit ourselves to the analysis of beams where the assumption of zero stress in the $y$ and $z$ directions is appropriate. For these conditions:

$$\sigma_x = E\epsilon = -\frac{Ey}{\rho} \qquad \text{(elastic only)}$$

(8.2)

*Note*: The stresses due to the compression and extension in bending are

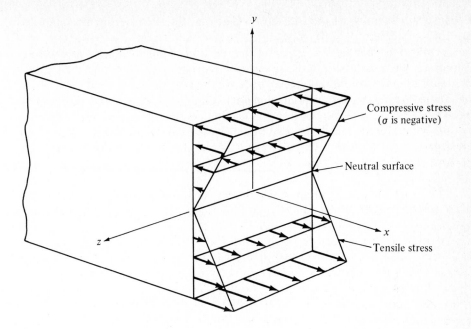

**Figure 8.5** *Elastic* stress distribution on the cross section of a beam in "pure bending."

frequently called the "bending stresses." For our geometry, they are normal stresses in the $x$ direction. When there is no ambiguity, we will omit the $x$ subscript.

Figure 8.5 shows how the elastic stresses would be distributed across a rectangular beam section. At this time, we could calculate the stresses ($\sigma$) *if* we knew the size of $\rho$, and *if* we knew where the neutral axis should be placed (where $y = 0$).

## 8.4 EQUILIBRIUM REQUIREMENTS

Our next step (which will lead us to a solution) is to ask what restrictions equilibrium places on the stress distribution shown in Fig. 8.5.

We are solving the case of "pure bending," i.e., if we isolate part of the beam, the *only* overall external loading is the bending moment $M_b$. The stress distribution shown in Fig. 8.5 must be such that it is equivalent to *only* $M_b$. We can say, then, that for the beam cross section, if we sum up the forces and moments:

$$\sum F_x = 0 \qquad \sum F_y = 0 \qquad \sum F_z = 0$$

$$\sum M_x = 0 \qquad \sum M_y = 0 \qquad \sum M_z = M_b$$

Because the stresses act only in the $x$ direction, the only pertinent equations are $\Sigma F_x = 0$, $\Sigma M_y = 0$, $\Sigma M_z = M_b$.

Equilibrium conditions apply to forces and moments, not to stresses. Therefore, we must multiply the stresses by areas to get forces, and then apply the equilibrium relationships. Here, because the stresses *vary*, we must look at the forces acting on differential elements. Figure 8.6 shows the differential force $(dF)$ due to a bending stress $(+\sigma)$ acting on the differential area $(dA)$.

*Note*: We know that a positive bending moment, at a positive value of $y$, must give negative stress. We will write the equilibrium relations in terms of positive stresses (Fig. 8.6), and let the equation for stress [Eq. (8.2)] introduce the negative sign.

The equilibrium conditions require that $\Sigma F_x = 0$ (no axial force acting on the beam); the sum of all $dF$ must be zero. Let the symbol $\int_A$ indicate integration over the area $A$:

$$\int_A dF = 0$$

But
$$dF = \sigma \, dA$$

and
$$\sigma = -\frac{Ey}{\rho}$$

Thus
$$\int_A -\frac{Ey \, dA}{\rho} = 0$$

If the beam is made of one material, $E$ is constant; $\rho$ is a constant. Therefore:

$$\int_A y \, dA = 0 \tag{8.3}$$

Equation (8.3) tells us that if a beam is made of one material (constant $E$), and there is no net axial force $F_x$, then $\int_A y \, dA$ must be zero.

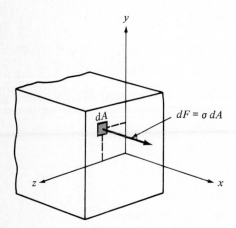

**Figure 8.6** Showing the force $dF$ acting on the differential area $dA$ due to bending stress $\sigma$.

You may recognize this integral; it is the definition for the centroid of an area. Therefore, if the beam is elastic, if $E$ is constant, and if $F_x = 0$, then the *neutral axis* is at the *centroid* of the area.

For $\Sigma M_y = 0$, and following the above use of $dF$:

$$dM_y = z \, dF$$

$$\int_A z \, dF = 0 = -\frac{E}{\rho} \int_A yz \, dA$$

Again, if $E$ is constant,

$$\int_A yz \, dA = 0$$

This integral, which is automatically zero for our symmetrical beams, is called the "product of inertia" of an area with respect to the centroid. It can be written:

$$I_{yz} = \int_A yz \, dA = \text{product of inertia} \tag{8.4}$$

We will have more to say about $I_{yz}$ toward the end of this chapter.

For $\Sigma M_z = M_b$, and again following the same procedure:

$$\int_A y \, dF = -M_b$$

But

$$dF = -\frac{Ey}{\rho} \, dA$$

Thus

$$\frac{E}{\rho} \int_A y^2 \, dA = M_b$$

(the $+dF$ in Fig. 8.6 is due to a $-M_b$). Again the integral is given a name, the "moment of inertia" of an area with respect to its centroid. We will label it $I_{yy}$:

$$I_{yy} = \int_A y^2 \, dA = \text{moment of inertia} \tag{8.5}$$

(By analogy we could also write $I_{zz} = z^2 \, dA$.) We now have:

$$M_b = \frac{EI_{yy}}{\rho} \tag{8.6}$$

Finally, combining Eqs. (8.2) and (8.6), we can solve for the bending stress:

$$\boxed{\sigma = -\frac{M_b y}{I_{yy}}} \tag{8.7}$$

*Note*: Various texts and handbooks may use a different notation for the properties of a beam area [Eqs. (8.4), (8.5), and (8.6)]. The system used here has two major advantages:

(1) The subscript is just the same as the variable to be integrated

$$yy = y^2 \qquad zz = z^2 \qquad \text{and} \qquad yz = yz$$

(2) We will see that the area properties are tensor quantities that can effectively be handled by the Mohr's circle method. The subscripting used here is completely *consistent* with that used for stress, strain, and compliance.

A common subscript system is based on the fact that bending due to $M_z$ produces bending about the $z$ axis, and the quantity that we call $I_{yy}$ is called either $I_z$ or $I_{zz}$. When using other resources, take care!

**Example 8.1** A rectangular beam (see the illustration below), 1 in wide by 2 in high, is subject to bending moment of 10,000 in·lb. Determine the maximum bending stress ($\sigma$), and the radius of curvature ($\rho$).

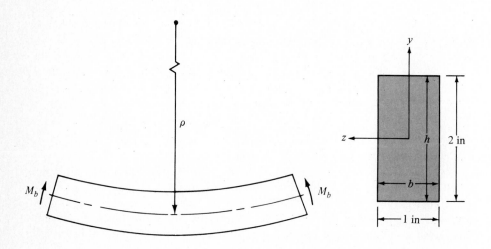

If we know where the centroid of the area is, and if we know $I_{yy}$, we can use Eqs. (8.6) and (8.7) to solve this problem. For a rectangle, the centroid is simply at the center.

The moment of inertia $I_{yy}$ [Eq. (8.5)] tells us how the beam area is distributed in the vertical ($y$) direction. Area that is at a large $y$ contributes more to $I_{yy}$ than area near the neutral axis.

To calculate $I_{yy}$ for a rectangular area, using Eq. (8.5), we can let $dA$ be a horizontal slice of height $dy$, as shown in the illustration below. Then

$$I_{yy} = \int_A y^2 \, dA = b \int_{-h/2}^{h/2} y^2 \, dy = \frac{bh^3}{12} \qquad \text{(for a rectangle)} \qquad (8.8)$$

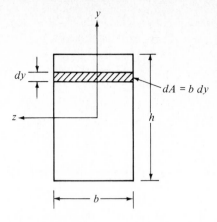

For this case,

$$I_{yy} = \frac{1(2)^3}{12} = \frac{8}{12} = 0.667 \text{ in}^4$$

Solving for the stress $\sigma$, it will be maximum where $y$ is maximum (at the top or bottom of the beam):

$$\sigma = -\frac{M_b y}{I_{yy}} = -\frac{10,000(1)}{0.667} = -15,000 \text{ psi}$$

(The minus sign applies at the positive $y$.)
    From Eq. (8.6),

$$\rho = \frac{EI_{yy}}{M_b}$$

$$= \frac{30 \times 10^6(0.667)}{10,000} = 2000 \text{ in} \qquad \text{***}$$

## 8.5 SHEAR FORCE AND BENDING MOMENT

In the treatment above, we studied the behavior of a (symmetrical) beam when subject only to a bending moment (pure bending). Sometimes this is the case. More generally, however, the beam area must support forces in the $y$ direction, forces which are tangential to the cross section. Because they act tangentially to the area rather than normal to it, these forces are called "shear forces" (from the shearing type of action they produce).

    Consider the "cantilever beam" (supported only at one end) shown in Fig. 8.7. The external force $F$, applied at $x = L$, tends to bend the beam in a concave upward fashion; i.e., it produces a positive bending moment. Obviously $M_b$ is largest at the wall, and zero at $x = L$. To find out what

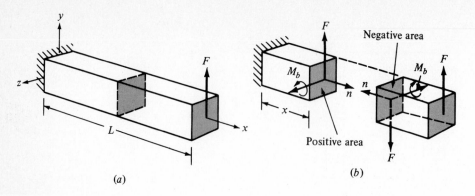

**Figure 8.7** Shear forces and bending moments: (*a*) cantilever beam subject to external load $F$; (*b*) force and moment carried by the cross section at $x$.

happens at intermediate positions, we must cut the beam at some position $x$ and isolate either portion of the beam. We can then solve for the forces acting on the cross section in question. By cutting the beam, the forces and moments that had been *internal* now become *external*.

Figure 8.7*b* shows the beam sectioned at $x$. We see that the section must carry a shear force equal to $F$, and a bending moment $M_b = F(L - x)$. There seems to be a bit of a problem, however! If we look at the left-hand part of the beam, we see that $F$ is in the $+y$ direction, and $M_b$ is in the $+z$ direction. However, if we look at the right-hand portion of the beam, the directions are reversed (as they must be, being equal and opposite, action and reaction). What sign should we give to $F$ and $M_b$? Plus or minus?

This problem of ambiguous signs becomes resolved when we realize that, when considering shear forces and bending moments, we must be concerned not only with the direction of the force or moment vector, but also with the "direction" of the area over which it acts (precisely as we were when considering stresses acting on an area).

It is general practice to define the *direction* of an area in terms of the "outwardly directed normal vector" $n$. Looking again at Fig. 8.7*b*, on the left-hand portion the outwardly directed normal vector is in the $+x$ direction, and that face can be considered to be a positive area; the *same* area on the right-hand portion has a negative $n$ and can be considered to be a negative area.

To determine the sign of a shear force or bending moment, multiply the sign of the force or moment vector by that of the area normal vector. Hence, the positively directed $F$ and $M_b$ acting on the left-hand portion (+ area) are defined as positive ($++ = +$). The negatively directed $F$ and $M_b$ on the right-hand section are also positive because they act on a negative area ($-- = +$).

To distinguish (when necessary) the sign of a force associated with a particular face from the sign of a force alone, we will use a double-subscripting system to identify both the face and the force. Figure 8.8 shows the complete set of force and moment components that can act on a given face.

The first subscript refers to the face, the second to the direction of the force or moment vector. We read $F_{xy}$ as: force $F$ acting on the $x$ face in the $y$ direction. A positive $F_{xy}$ can be either a positively directed $F_y$ acting on a positive $x$ face, or a negative $F_y$ acting on a negative $x$ face.

Clearly, the subscript convention and the sign convention for force and moment acting on an area is completely consistent with that for stress acting on an area.

To simplify discussion, and to associate the various forces and moments with the actions they produce, we name all of the components shown in Fig. 8.8:

$F_{xx}$ = "axial force," which changes the beam length.

$F_{xy}, F_{xz}$ = "shear forces," which tend to shear one portion relative to the other. For some reason, shear forces have historically been labeled $V$; $F_{xy} = V_y$; $F_{xz} = V_z$

$M_{xx}$ = "twisting moment" $M_t$, which twists the beam about its axis.

$M_{xy}, M_{xz}$ = "bending moments" $M_b$ about the $y$ and $z$ axes.

When we deal with a beam where the bending moment is *not* constant along the length, we must search to locate the position $x$ where the bending stress $\sigma$ is a maximum. If the beam is made of a brittle material, which may have different limiting stresses in tension and compression, we must search for both the largest tensile stress and the largest compressive stress. The maximum stress point will be the "weakest link" and will determine the maximum load that can be applied.

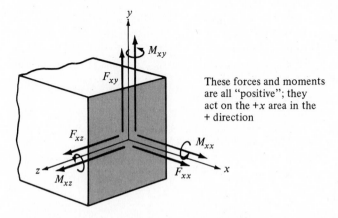

These forces and moments are all "positive"; they act on the $+x$ area in the $+$ direction

**Figure 8.8** Defining the double-subscript convention for forces and moments acting on an area.

To find the maximum stress points, it is generally most useful to plot the bending moment $M_b$ as a function of position along the beam ($x$). The beam strength *can* also be limited by the shear force $V$, so we generally plot both $V$ and $M_b$ as functions of $x$.

Figure 8.9 shows four loading diagrams for the cantilever beam illustrated in Fig. 8.7. The sketch of the beam and the free-body diagram should be clear to you. To draw the shear force and bending moment diagrams, we merely section the beam at any position $x$, solve for $V$ and $M_b$ (being careful about the sign convention), and plot the values.

In Chap. 12, we will concern ourselves with shear and bending moment diagrams for beams with complex loading. Here, where we are going to limit ourselves to simple end loading, there is a limited number of possibilities, as shown in Fig. 8.10. We see that for *end* loading, the *maximum* bending moment occurs only at an end of the beam, never in the central portion.

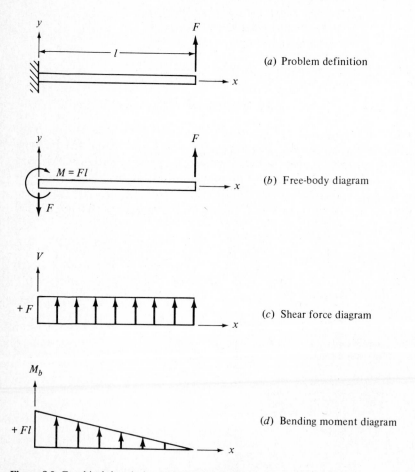

(*a*) Problem definition

(*b*) Free-body diagram

(*c*) Shear force diagram

(*d*) Bending moment diagram

**Figure 8.9** Graphical description of loading conditions on a cantilever beam.

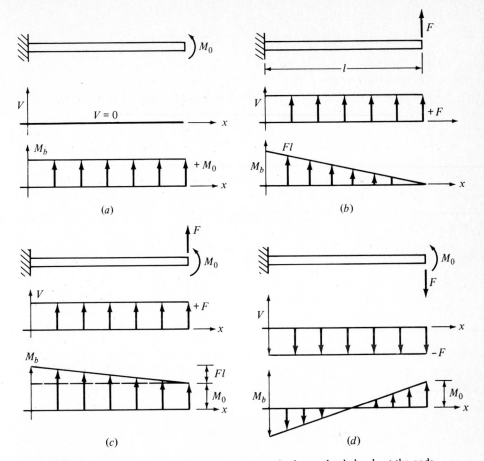

**Figure 8.10** Shear force and bending moment diagrams for beams loaded only at the ends.

**Example 8.2** A 1-in-diameter steel rod (see the illustration below) is used as a cantilever beam. It is loaded at the right-hand end with a force $F$ and a moment $M_0$, where $M_0 = 15F$ in·lb. A tensile test of the same material showed a yield stress $\sigma_{yp} = 60,000$ psi. How large can $F$ be before the rod yields because of the bending stresses?

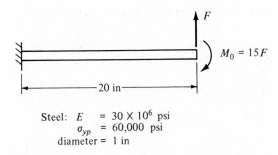

Steel: $E = 30 \times 10^6$ psi
$\sigma_{yp} = 60,000$ psi
diameter = 1 in

Plan the attack: Find $I_{yy}$ using Eq. (8.5)

Find $M_b$ at yield [Eq. (8.7)]

Find $F$ to give $M_b$ (loading diagrams)

First determine $I = I_{yy}$ (when there is no possibility for confusion, we can omit the subscripts and use only $I$).

The centroid is at the center. From the diagram below, we can determine both $y$ and $dA$ in terms of $r$ and $\theta$:

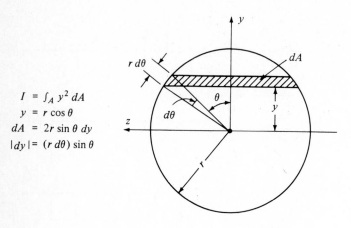

$$I = \int_A y^2 \, dA$$
$$y = r \cos \theta$$
$$dA = 2r \sin \theta \, dy$$
$$|dy| = (r \, d\theta) \sin \theta$$

Thus

$$I = \int_0^\pi 2r^4 \sin^2 \theta \cos^2 \theta \, d\theta = \frac{\pi r^4}{4}$$

$$I_{yy} = \frac{\pi r^4}{4} \quad \text{(for circular sections)} \tag{8.9}$$

In our case,

$$I = \frac{\pi 0.5^4}{4} = 0.0491 \text{ in}^4$$

The bending moment $M_b$ that will give a tensile (or compressive) stress of 60,000 psi is:

$$M_b = -\frac{\sigma_{yp} I}{y} \quad \text{[from Eq. (8.7)]}$$

$$\doteq -\frac{60{,}000(0.0491)}{0.5} = -5892 \text{ in·lb}$$

*Note*: The minus sign on $M_b$ above indicates that it requires a negative $M_b$ to produce a $+\sigma$ at a $+y$. Because we are interested in either $+\sigma$ or $-\sigma$, we can limit the *magnitude* of $M_b$ to 5892 in·lb.

Finally, plot the shear force and bending moment diagrams as shown on page 255.

The maximum moment is at $x = 20$, and equals $-15F$, which cannot exceed 5892 in·lb. Clearly the maximum load is 393 lb.

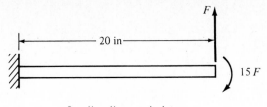

Loading diagram, isolate
and solve for support
forces

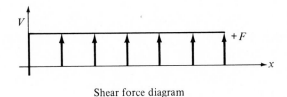

Shear force diagram

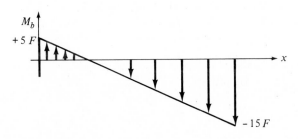

Bending moment diagram                                    ***

## 8.6 BEAM DEFLECTION

Recalling (or looking back at) Fig. 8.4 and Eq. (8.6), we saw that a bending
moment $M_b$ will bend a straight beam to an arc with radius of curvature $\rho$. For
elastic bending:

$$\frac{1}{\rho} = \frac{M_b}{EI} = \text{curvature} \tag{8.10}$$

We can reasonably interpret Eq. (8.10) to mean that the curvature $(1/\rho)$ is
directly proportional to $M_b$ at *all* points along the beam, and that the
proportionality constant is $1/EI$. If we know the curvature at *all* points along
the beam, it just requires some mathematical manipulation (or a numerical
method) to determine the deflection at any point.

Figure 8.11 shows (in a grossly exaggerated fashion) the deflection of a bent
cantilever beam. We can describe the shape of the beam in terms of either the

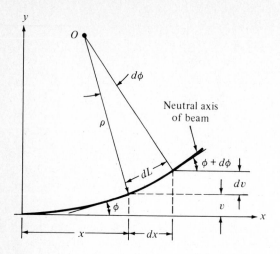

**Figure 8.11** Geometry for calculating beam deflections (grossly exaggerated).

slope ($\phi$), or the displacement ($v$), as functions of position ($x$) along the length of the beam:

$$v(x) = \text{the deflection (displacement in the } +y \text{ direction)}$$
$$\phi(x) = \text{the slope of the beam where } \tan\phi = dv/dx$$

As we move outward along the beam, we can compute the incremental deflection ($dv$) associated with an increment of beam length ($dL$), and then sum up the $dv$'s (integrate).

From Fig. 8.11, the increment of beam length ($dL$) is:

$$dL = \rho\, d\phi \qquad \text{or} \qquad \frac{1}{\rho} = \frac{d\phi}{dL}$$

If $\phi$ is small (less than 5° to 10°), it is a good approximation to write:

$$dx \cong dL \qquad \text{and} \qquad \phi \cong \frac{dv}{dx}$$

Incorporating these small-angle approximations we have:

$$\frac{1}{\rho} = \frac{d\phi}{dx} \qquad \frac{d\phi}{dx} = \frac{d^2v}{dx^2}$$

Thus, for small $\phi$:

$$\boxed{\frac{d^2v}{dx^2} = \frac{d\phi}{dx} = \frac{M}{EI}}$$

(8.11)

If $\phi$ is not small, we must use the proper relationship between $\rho$, $v$, and $x$.

Most elementary calculus texts show:

$$\text{Curvature} = \frac{1}{\rho} = \frac{d^2v/dx^2}{[1+(dv/dx)^2]^{3/2}} \tag{8.12}$$

Use of the nonlinear Eq. (8.12) is very difficult in a closed-form analytic solution. However, it is very easy to use in computerized numerical solutions. Looking again at Eq. (8.11), we see that:

$$\phi = \int \frac{M_b}{EI} \, dx + C_1$$

$$\tag{8.13}$$

$$v = \int \phi \, dx + C_2$$

So we merely have to integrate $M_b/EI$ with respect to $x$ to obtain $\phi$, and then integrate $\phi$ with respect to $x$ to get $v$. The two constants of integration $C_1$ and $C_2$ must be determined from the displacement and rotational boundary conditions at the ends of the beam.

**Example 8.3** Determine the slope and deflection of a cantilever beam loaded at the end as shown in the accompanying illustration.

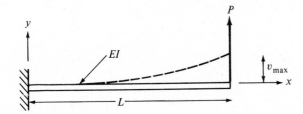

To use Eqs. (8.13), we can note that $EI$ is constant along the beam and can be taken outside of the integral. We have already seen in Fig. 8.10$b$ that the bending moment is linear with $x$ and is given by $M_b = +P(L-x)$; therefore:

$$\phi = \frac{1}{EI} \int P(L-x) \, dx + C_1$$

$$= \frac{P}{EI}\left(Lx - \frac{x^2}{2}\right) + C_1$$

$$v = \int \phi \, dx = \frac{P}{EI}\left(\frac{Lx^2}{2} - \frac{x^3}{6}\right) + C_1x + C_2$$

In almost every situation, $C_1$ equals the slope at the origin ($x = 0$), and $C_2$ the deflection at the origin. In this case, with a "built-in" support, both $C_1$ and $C_2$ are zero; that is:

$$\text{At } x = 0, \ \phi = 0 \qquad \text{thus } C_1 = 0$$

$$\text{At } x = 0, \ v = 0 \qquad \text{thus } C_2 = 0$$

The equations for slope and deflection are:

$$\phi = \frac{P}{EI}\left(Lx - \frac{x^2}{2}\right)$$

$$v = \frac{P}{EI}\left(\frac{Lx^2}{2} - \frac{x^3}{6}\right)$$

Both $\phi$ and $v$ are maximum at $x = L$, thus:

$$\phi_{\max} = \frac{PL^2}{2EI} \qquad v_{\max} = \delta = \frac{PL^3}{3EI}$$

The above example is exceedingly simple; the problem becomes considerably more difficult when loads are applied at intermediate positions along the beam. Then we cannot write a single, continuous equation for $M_b$ as a function of $x$, and we must (in general) divide the beam into segments such that we can write a continuous equation for each segment and then integrate those equa-

**Table 8.1 Beam deflection relations for end-loaded beams**

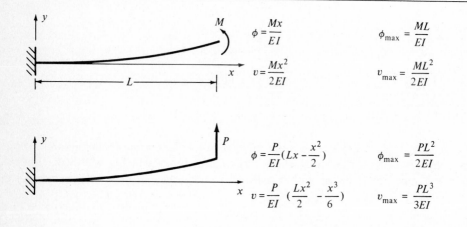

$$\phi = \frac{Mx}{EI} \qquad \phi_{\max} = \frac{ML}{EI}$$

$$v = \frac{Mx^2}{2EI} \qquad v_{\max} = \frac{ML^2}{2EI}$$

$$\phi = \frac{P}{EI}\left(Lx - \frac{x^2}{2}\right) \qquad \phi_{\max} = \frac{PL^2}{2EI}$$

$$v = \frac{P}{EI}\left(\frac{Lx^2}{2} - \frac{x^3}{6}\right) \qquad v_{\max} = \frac{PL^3}{3EI}$$

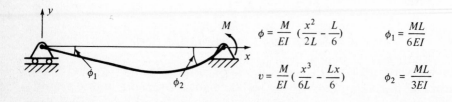

$$\phi = \frac{M}{EI}\left(\frac{x^2}{2L} - \frac{L}{6}\right) \qquad \phi_1 = \frac{ML}{6EI}$$

$$v = \frac{M}{EI}\left(\frac{x^3}{6L} - \frac{Lx}{6}\right) \qquad \phi_2 = \frac{ML}{3EI}$$

tions, matching their boundary conditions. We will consider this problem in Chap. 12.

For the development of computerized structural analysis, the solutions for simple end loading will suffice, and will give us programs which can then be used for beams with intermediate loads.

For those who enjoy elegant methods, we will develop (Chap. 12) a set of "singularity functions" that will permit us to integrate discontinuous functions as though they were continuous.

Table 8.1 gives the slope and deflection relations for a simple beam with force and moment end loadings. Remember that if we have a combination of such loads, we can use superposition to find the result.

**Example 8.4** For the beam shown in the accompanying illustration, using superposition and Table 8.1, find the deflection at the end ($x = L$).

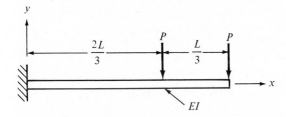

First we will find the deflection at $x = L$ due to one load, and then that due to the other load. We can then add (superpose) the deflections to obtain the result.

1. Deflection due to the end load:

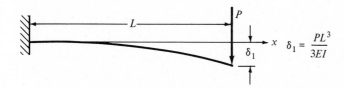

$$\delta_1 = \frac{PL^3}{3EI}$$

2. Deflection due to the intermediate load:

$$\delta_2 = \delta_3 + \delta_4$$

$$\delta_3 = \frac{"PL^3"}{3EI} \qquad \text{where } "L" = \tfrac{2}{3}L$$

Therefore
$$\delta_3 = \frac{P(2L/3)^3}{3EI}$$

For $\delta_4$, note that the beam is straight from $x = \frac{2}{3}L$ to $L$ (no moment); therefore, $\delta_4 = \phi_4 L/3$. But

$$\phi_4 = \frac{``PL^2"}{2EI} \qquad \text{where } ``L" = \frac{2}{3}L$$

Thus
$$\delta_4 = \frac{P(2L/3)^2}{2EI} \frac{L}{3} = \frac{4PL^3}{54EI}$$

Finally,
$$\delta_2 = \frac{PL^3}{EI}\left(\frac{8}{81} + \frac{4}{54}\right) = \frac{14PL^3}{81EI}$$

and
$$\delta = \frac{PL^3}{EI}\left(\frac{1}{3} + \frac{14}{81}\right) = \frac{41PL^3}{81EI} \qquad\qquad ***$$

In addition to the superposition method used above, we could have solved the problem by a method that has been called "juxtaposition" (placing two parts together). In juxtaposition, we would cut the beam at $x = 2L/3$ and isolate the two parts. As shown in the illustration below, we could solve for the end deflection of the left-hand part and use that position as the origin for the right-hand part. See if *you* can obtain the same answer arrived at above by this method.

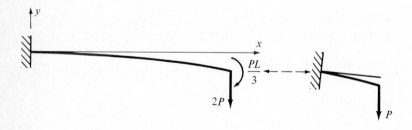

## 8.7 PROPERTIES OF BEAM CROSS SECTIONS

If we looked back to Chap. 3, where we considered struts under axial loading of tension or compression, we would see that the only important characteristic of the strut cross section was its area $A$. The product $EA$ carried all the information necessary to calculate the stiffness of a given length.

In this chapter, we have seen that when we consider bending, it is not just the area that is important, but the moment of inertia $I_{yy}$, which tells us how the area is distributed in the $y$ direction. Again, the product $EI$ carries all the information necessary to compute the bending stiffness of a given length of beam when it is bent in the $xy$ plane. So far, we have considered only rectangular and circular cross sections. Frequently other sections are used, and we should be able to handle them as well.

You should recall that when we studied the equilibrium requirements for elastic bending stresses (Sec. 8.4), we obtained three geometric requirements:

1. The neutral axis is at the centroid of the area.
2. The product of inertia ($I_{yz}$) must be zero.
3. The moment of inertia ($I_{yy}$) controls bending stresses and deflections (when bent in the $xy$ plane).

In this section, we will address these three requirements for beams having more complex cross-sectional shapes.

### 8.7.1 Centroid

Figure 8.12 shows a generalized area $A$ where the centroidal position is not known. We want to determine the centroidal coordinates $\bar{y}_1$ and $\bar{z}_1$ relative to an arbitrary set of $y_1z_1$ axes.

We can compute the $\int_A y_1 \, dA$ relative to the arbitrary axes by noting that $y_1 = \bar{y}_1 + y$:

$$\int_A y_1 \, dA = \int_A (\bar{y}_1 + y) \, dA = \int_A \bar{y}_1 \, dA + \int_A y \, dA$$

But $\qquad \int_A \bar{y}_1 \, dA = \bar{y}_1 \int_A dA = \bar{y}_1 A \qquad$ and $\qquad \int_A y \, dA = 0 \qquad$ (centroid)

Thus $\qquad \bar{y}_1 = \dfrac{\displaystyle\int_A y_1 \, dA}{A} \qquad$ and $\qquad \bar{y}_1 A = \int_A y_1 \, dA \qquad$ (8.14)

We find $\bar{z}_1$ in just the same way.

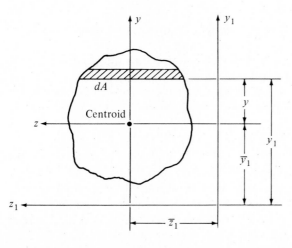

**Figure 8.12** Calculation of centroidal position $\bar{y}_1, \bar{z}_1$.

## 8.7.2 Moment of Inertia

In a development similar to that used to find the centroid position, we can determine the moment of inertia of an area about an arbitrary set of axes ($I_{y_1y_1}$) in terms of the moment of inertia about the centroidal axes ($I_{yy}$). $I_{y_1y_1}$ is given by:

$$I_{y_1y_1} = \int_A y_1^2 \, dA = \int_A (\bar{y}_1 + y)^2 \, dA = \int_A \bar{y}_1^2 \, dA + 2 \int_A \bar{y}_1 y \, dA + \int_A y^2 \, dA$$

But $\qquad \int_A \bar{y}_1^2 \, dA = \bar{y}_1^2 A \qquad$ and $\qquad 2\bar{y}_1 \int_A y \, dA = 0 \qquad$ (centroid)

Thus, letting $I_{yy}$ represent the moment of inertia about the centroidal axis,

$$I_{y_1y_1} = I_{yy} + \bar{y}_1^2 A \tag{8.15}$$

Equation (8.15) is usually called the "parallel axis theorem" because it allows us to calculate the moment of inertia about axes parallel to the centroidal axes; if we know the moment of inertia about the centroid (and we do for simple shapes), we can calculate it about other parallel axes. This will enable us, for instance, to determine the moment of inertia for a T beam based on the "T" being made of two rectangles (see Example 8.5).

## 8.7.3 Product of Inertia

For completeness, we will consider the third property of an area, the product of inertia $I_{yz}$:

$$I_{yz} = \int_A yz \, dA \qquad \text{(relative to centroidal axes } yz\text{)} \tag{8.16}$$

In a manner similar to that for the moment of inertia, we can develop an equivalent parallel axis theorem:

$$I_{y_1z_1} = I_{yz} + \bar{y}_1 \bar{z}_1 A \tag{8.17}$$

The product of inertia ($I_{yz}$) can be thought of as telling us how symmetrical a beam section is; for any truly symmetrical beam, $I_{yz} = 0$.

However, $I_{yz}$ can also be zero for nonsymmetrical beams, and when it is, the beam bends (when subject to pure bending) as though it were symmetrical.

## 8.7.4 Mohr's Circle

For any beam cross section we can compute three properties of the area:

$$I_{yy} \qquad I_{zz} \qquad I_{yz}$$

where the origin ($y = z = 0$) is at the centroid. If we were to compute these area properties relative to a set of $y'z'$ axes which are rotated relative to the original axes, we would find that the resulting equations are completely analogous to those for compliance, stress, and strain.

Accordingly, the area properties form a tensor quantity which can be treated by using the Mohr's circle construction. We find that there are *principal axes* of inertia where the product of inertia is zero. For any beam cross section, then, whether it is symmetrical or not, there exists an orthogonal set of principal axes I and II such that the product of inertia is zero.

Early in this chapter, we stipulated that the beam analysis was valid only if the beam was loaded and bent in a plane of symmetry. Now we can relax that requirement for the case of pure bending by stipulating that the beam be bent about principal axes of inertia.

This entire analysis applies only to *elastic* deformation, and we know that we can use superposition for elastic problems. Therefore, if we load an asymmetrical beam along a nonprincipal axis, we can resolve the load into components along the two principal directions, solve for the deflections in the two principal directions, and superpose the results.

We will see in Chap. 12 that if we load certain thin-walled, asymmetrical sections with a shear force, they may twist as well as bend. To avoid twisting, the load must be offset from the centroid so as to pass through the "shear center." For instance, with any angle section, the load should pass through the intersection of the two flat parts rather than through the centroid.

**Example 8.5** A T beam has the cross section shown in the accompanying illustration. Determine the location of the centroid ($\bar{y}_1$) and the moment of inertia ($I_{yy}$).

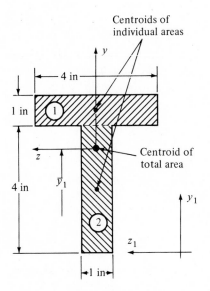

First, we note that the area can be divided into two rectangular areas, 1 and 2, as shown by the dashed line. Being rectangles, we know where their individual centroids are $(\bar{y}_1)_1$, $(\bar{y}_1)_2$, and we know how to calculate the moments of inertia of the individual areas about their centroids $(I_{yy})_1$ and $(I_{yy})_2$.

Measuring $\bar{y}_1$ from the bottom of the section, and using Eqs. (8.14), we can write:

$$\bar{y}_1 A = \int_A y_1 \, dA = \int_{A_1} y_1 \, dA + \int_{A_2} y_1 \, dA$$

$$= (\bar{y}_1)_1 A_1 + (\bar{y}_1)_2 A_2$$

Thus
$$\bar{y}_1 = \frac{\Sigma_i A_i (\bar{y}_1)_i}{A}$$

Setting up a method for $i = 2$ areas, we have:

| $i$ | $A_i$ | $(\bar{y}_1)_i$ | $A_i(\bar{y}_1)_i$ |
|---|---|---|---|
| 1 | 4 | 4.5 | 18 |
| 2 | 4 | 2.0 | 8 |
| | 8 | | 26 |

$$\bar{y}_1 = \frac{26}{8} = 3.25 \text{ in}$$

Now to compute the moment of inertia: We will calculate $(I_{yy})_i$ for each area about its own centroidal axis ($bh^3/12$ for a rectangle), and add the parallel axis terms to get the total moment of inertia:

$$I_{yy} = \sum_i (I_{yy})_i + \sum_i A_i(\bar{y})_i^2$$

Note that we now use the $yz$ axes that pass through the overall centroid.

Setting up a procedure for $i = 2$ areas:

| $i$ | $(I_{yy})_i$ | $A_i$ | $\bar{y}_i$ | $A_i\bar{y}_i^2$ |
|---|---|---|---|---|
| 1 | 4/12 | 4 | +1.25 | 6.25 |
| 2 | 64/12 | 4 | −1.25 | 6.25 |
| | 5.667 | | | 12.50 |

$$I_{yy} = 5.667 + 12.5 = 18.167 \text{ in}^4$$

As long as the area can be divided into rectangular or circular areas, we can use the above technique. For complex shapes we can always resort to numerical integration.

***

**Example 8.6** For the asymmetrical angle section shown in the accompanying illustration, determine the magnitudes of the principal moments of inertia and the principal axis directions.

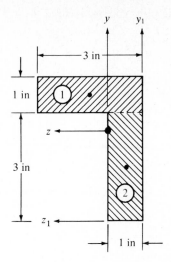

Divide the area in two as shown, and using the tabular methods shown above, calculate first the centroid position, then $I_{yy}$, $I_{zz}$, and $I_{yz}$. Finally, apply the Mohr's circle technique to find the principal axes and moments of inertia.

To determine the centroid location:

| $i$ | $A_i$ | $(\bar{y}_1)_i$ | $(\bar{z}_1)_i$ | $A_i(\bar{y}_1)_i$ | $A_i(\bar{z}_1)_i$ |
|---|---|---|---|---|---|
| 1 | 3 | 3.5 | 1.5 | 10.5 | 4.5 |
| 2 | 3 | 1.5 | 0.5 | 4.5 | 1.5 |
|  | 6 |  |  | 15.0 | 6.0 |

$$\bar{y}_1 = \frac{15}{6} = 2.5 \text{ in} \qquad \bar{z}_1 = \frac{6}{6} = 1 \text{ in}$$

For the moments and products of inertia:

| $i$ | $(I_{yy})_i$ | $(I_{zz})_i$ | $A_i$ | $(\bar{y})_i$ | $(\bar{z})_i$ | $(A\bar{y}^2)_i$ | $(A\bar{z}^2)_i$ | $(A\bar{y}\bar{z})_i$ |
|---|---|---|---|---|---|---|---|---|
| 1 | 0.25 | 2.25 | 3 | 1.0 | 0.5 | 3.0 | 0.75 | +1.5 |
| 2 | 2.25 | 0.25 | 3 | −1.0 | −0.5 | 3.0 | 0.75 | +1.5 |
|  | 2.5 | 2.5 |  |  |  | 6 | 1.5 | +3.0 |

$$I_{yy} = 2.5 + 6 = 8.5 \text{ in}^4$$

$$I_{zz} = 2.5 + 1.5 = 4 \text{ in}^4$$

$$I_{yz} = +3.0 \text{ in}^4$$

We can now plot these values on a Mohr's circle diagram just as we did for compliance, stress, or strain. We plot $I_{yy}$ and $I_{zz}$ along the horizontal axis, and $I_{yz}$ vertically. Our sign convention (for the vertical axis) was originally established relative to the $x$ axis; here we have no $x$ axis in the plane of the section. The axis that is equivalent to the $x$ axis (if our axes were labeled $xy$

instead of $yz$), is the $y$ axis. Accordingly, we will plot $y$ downward if $I_{yz}$ is positive, and upward if it is negative.

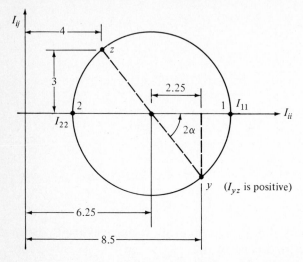

In the Mohr's circle diagram above, the point labeled $y$ (plotted at $+I_{yy}$ and $I_{yz}$ down) refers to bending in the $xy$ plane. Point $z$ refers to bending in the $xz$ plane. We see that the two principal axes, given by points 1 and 2, refer to bending in the $x1$ plane and the $x2$ plane, respectively. We can compute the magnitude of $I_{11}$, $I_{22}$, and their orientation as shown below:

$$\text{Center} = \frac{8.5 + 4}{2} = 6.25 \text{ in}^4$$

$$\text{Radius} = \sqrt{2.25^2 + 3^2} = 3.75 \text{ in}^4$$

$$\text{Maximum moment of inertia} = 10.0 \text{ in}^4$$

$$\text{Minimum moment of inertia} = 2.25 \text{ in}^4$$

$$\text{Inclination of principal axes} = 26.57°$$

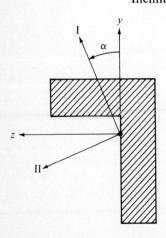

Looking at the diagram above, we can see intuitively that the I axis is the stiffest axis.                                                                                      ***

## 8.8 SHEAR FLOW AND SHEAR STRESS

In the previous sections, the effects of the shear force were essentially ignored. For many beam situations the effect of shear is negligible. Here we will examine these effects and determine when they must be considered.

In order to put this analysis into context, consider the cantilever beam shown in Fig. 8.13. The beam cross section is $bh$ and the load $F$ (downward) is applied at the origin. Also shown are the shear force ($V$) and bending moment ($M_b$) diagrams. The shear force is positive with a constant value of $+F$. The bending moment is negative, and becomes increasingly negative with increasing $x$. Because the bending moment (and the resulting bending stresses) *varies* with $x$, we will study the problem by isolating a differential element of the beam of length $dx$.

Figure 8.14 shows two representations of the isolated element: In ($a$) the shear forces and bending moments are shown; in ($b$), the stress distributions (bending stress $\sigma_x$) due to the bending moments are shown. Clearly the stresses are greater on the right than on the left. When we satisfy moment equilibrium for the element, we find that:

$$\frac{dM_b}{dx} = -V \qquad (8.18)$$

Figure 8.15 shows two representations of the "slice" of the "element." In Fig. 8.15$a$, which is a side view, it is clear that for equilibrium in the horizontal direction, there *must* be a force ($dF$) acting to the left as shown. Where can this force come from?

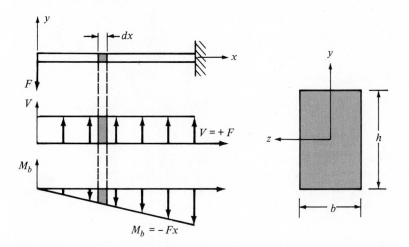

**Figure 8.13** Cantilever beam for shear stress analysis.

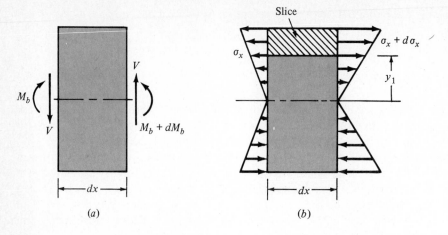

**Figure 8.14** Isolation of beam element of length $dx$.

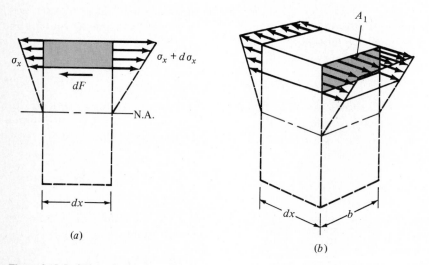

**Figure 8.15** Isolation of a portion of the beam element of length $dx$.

The only possibility for a force in the $x$ direction (other than on the end faces which have already been considered) is a shear force acting *on* the top slice *by* the lower portion of the element. The force $dF$ must be due to the stress $d\sigma_x$ because the forces due to $\sigma_x$ just cancel. Referring to Fig. 8.15b, the force $dF$ must be due to the integral of $d\sigma_x$ over the area $A_1$, where $A_1$ is the cross-sectional area *above* the height $y_1$. To put this into equation form:

$$dF = \int_{A_1} d\sigma_x \, dA_1$$

$$\frac{dF}{dx} = \int_{A_1} \frac{d\sigma_x}{dx} \, dA_1$$

But $\qquad\qquad \sigma_x = -\dfrac{M_b y}{I_{yy}} \quad$ and $\quad \dfrac{d\sigma_x}{dx} = -\dfrac{y}{I}\dfrac{dM_b}{dx}$

We know that $\qquad\qquad\qquad \dfrac{M_b}{dx} = -V$

Therefore $\qquad\qquad\qquad \dfrac{dF}{dx} = \dfrac{V}{I} \int_{A_1} y \, dA_1 \qquad\qquad\qquad$ (8.19)

The quantity $dF/dx$ is the shear force per unit length, in a horizontal direction, required for equilibrium of the slice defined by $A_1$. For reasons having to do with fluid analogies that were used some years ago, the quantity $dF/dx$ is called the "shear flow," and labeled $q$.

Note: The standard terminology in this field uses $q$ to represent force per length, both shear force and normal force. We will now confuse the issue by introducing $Q$, a traditional notation associated with the computation of shear stresses in beams.

The area integral in Eq. (8.19) is normally given the symbol $Q$, which is the integral of $(y \, dA_1)$ over the area of interest, $A_1$. Thus

$$q = \frac{VQ}{I} \qquad\qquad\qquad (8.20)$$

where $\qquad\qquad\qquad q = \dfrac{dF}{dx}$

$$Q = \int_A y \, dA_1$$

To summarize:

$q$ is the horizontal force per unit length required to hold a portion of the beam in equilibrium. The horizontal force must come from the remainder of the beam.

The quantity $Q$ is the integral of $y \, dA_1$ over the cross-sectional area to be held in equilibrium. If the shapes are simple, it is useful to recall that the integral of $y \, dA_1$ equals the area $A_1$ times the distance $\bar{y}_1$ up to its centroid.

The moment of inertia $I$ is that for the entire beam cross section, not just the area $A_1$.

To return to the problem of a rectangular beam:

$$Q = \int_{A_1} y \, dA_1 = \bar{y}_1 A_1$$

where $A_1 = b\left(\dfrac{h}{2} - y_1\right)$

$$\bar{y}_1 = \frac{h/2 + y_1}{2}$$

So
$$Q = \frac{b(h^2/4 - y_1^2)}{2}$$

$$I = \frac{bh^3}{12}$$

Thus
$$q = \frac{3V}{2h}\left[1 - 4\left(\frac{y_1}{h}\right)^2\right]$$

For this solid, rectangular beam section, if we divide the shear force per unit length, $q$, by the width $b$, we obtain the shear force per unit area, or shear stress $\tau$:

$$\tau = \frac{q}{b} = \frac{VQ}{Ib}$$

$$= \frac{3V}{2bh}\left[1 - 4\left(\frac{y_1}{h}\right)^2\right] \tag{8.21}$$

The shear stress we have calculated is the stress on the bottom of the slice determined by the area $A_1$ above height $y_1$. This is $\tau_{yx}$. Because $\tau_{xy} = \tau_{yx}$, the same shear stress must act at $y = y_1$ on the vertical $x$ face as shown in Fig. 8.16.

Thus Eq. (8.21) holds for the shear stress on the beam cross section. The quantity $V/bh$ is simply the "average" shear stress acting on the cross section; the shear force divided by the section area. Equation (8.21) shows that (for a rectangular cross section) the shear stress is distributed in a parabolic fashion, that the maximum shear stress is just three-halves of the average, and that the maximum shear stress occurs at the neutral axis ($y = 0$).

For solid sections (such as a rectangle), the maximum shear stress occurs at

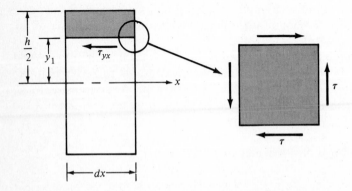

**Figure 8.16** The shear stress on the $x$ face at $y = y_1$ equals that on the $y$ face.

the neutral axis, and is usually relatively small if the beam is long. This is why we can often ignore the shear stresses relative to bending stresses.

**Example 8.7** Consider the cantilever beam shown in Fig. 8.13 where:

$$b = 2 \text{ in} \qquad L = 60 \text{ in}$$

$$h = 4 \text{ in} \qquad F = 1000 \text{ lb}$$

Determine the maximum shear stresses due to the bending moment and due to the shear force.

The maximum shear stress due to bending is just half the normal stress $\sigma_x$ at the top or bottom of the beam:

$$\sigma_x = \frac{M_b y}{I}$$

where $M_b = 60,000 \text{ in} \cdot \text{lb}$  (at $x = L$)
  $y = 2 \text{ in}$
  $I = \dfrac{2(4)^3}{12} = 10.67 \text{ in}^4$

Thus $\qquad\qquad \sigma_x = 11,246 \text{ psi}$

$$\tau_{max} = 5623 \text{ psi} \qquad \text{(due to the bending moment)}$$

To compute the shear stress due to the shear force:

$$\tau = \frac{VQ}{Ib}$$

where $V = 1000 \text{ lb}$
  $Q = 2(2)(1) = 4 \text{ in}^3$
  $I = 10.67 \text{ in}^4$
  $b = 2 \text{ in}$

So $\qquad\qquad \tau = 187.5 \text{ psi} \qquad \text{(due to the shear force)}$

(Because this is a rectangular section, the shear stress is just three-halves the mean stress of 1000/8 psi.)

The shear stress due to bending is 30 times that due to the shear force. For the two to be equal, the beam would have to be shortened to $L = 2 \text{ in}$ (our bending equations will hold for such a short beam only if we apply the loads in a distributed manner that matches the computed stress distribution). This should provide a qualitative feeling for the importance of the shear force in solid rectangular beams—not very important!  \*\*\*

From a qualitative point of view, we see that the bending moment is carried primarily by the material away from the neutral axis, while the shear force is carried primarily by the material near the neutral axis. Also, for solid rectangular beams, the stresses due to the shear force are relatively small.

Accordingly, for conservation of material and weight, various beams have been designed that more effectively utilize the material near the neutral axis. Typical of these are the familiar I beams, T beams, channels, box beams, and a wide variety of composite beams using weaker, lighter, or cheaper materials near the neutral axis.

An I beam is made with "flanges" top and bottom joined by the vertical "web." At first glance it might seem that the web could be thinned down until the maximum web shear stress (due to $V$) just equaled the flange shear stress (due to $M_b$).

However, for very thin webs, we might find that the web buckled before it yielded (if you draw the Mohr's stress circle for the web at the neutral axis, you will find that the shear stress due to $V$ produces tensile and compressive stresses at 45° to the beam axes—the compressive stress causes buckling).

Also, if the flanges are made too wide, the flange under compression will buckle. Accordingly, a set of standardized beam sections has been developed which has small probability of buckling under normal use.

In order to understand the shear-stress distribution in an I beam, consider the following example of a "built-up" beam.

**Example 8.8** An I beam is made from three 1-in-thick steel plates welded together with fillet welds, as shown in the accompanying illustration. The beam is subject to:

$$V = 10,000 \text{ lb} \qquad \text{(downward)}$$

$$M_b = 10^6 \text{ in} \cdot \text{lb}$$

Determine (a) the shear stress distribution due to $V$, (b) the maximum normal stress due to $M_b$, and (c) the force that must be carried by the weld.

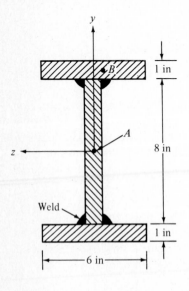

First calculate the moment of inertia ($I$) about the $z$ axis (using the parallel-axis theorems given in Sec. 8.8):

$$I = 2\left[\frac{6(1)^3}{12} + 6(4.5)^2\right] + \frac{1(8)^3}{12} = 286.67 \text{ in}^4$$

To answer part (a), the shear flow $q = VQ/I$, and will be maximum at the neutral axis where $Q$ is greatest. The $Q$ at the neutral axis is simply the sum of $Q$ for one flange plus $Q$ for the upper half of the web:

$$Q = \int_{A_1} y \, dA_1 = \sum_i \bar{y}_1 A_i = 4.5(6)(1) + 2(4)(1) = 35 \text{ in}^3$$

and

$$q = \frac{10,000(35)}{286.67} = 1221 \text{ lb/in}$$

At the neutral axis, the shear flow of 1221 lb/in acts on the 1-in web thickness; thus the shear stress $\tau$ is:

$$\tau = \frac{q}{1} = 1221 \text{ psi} \qquad \text{(at the NA)}$$

To determine $\tau$ elsewhere, we must find the $q$ based on the proper $Q$ and divide $q$ by the appropriate width of material ($b$) that carries the shear flow $q$. In the I-beam flanges, we can determine $Q$ and $q$ for an outboard segment of the flange as shown below.

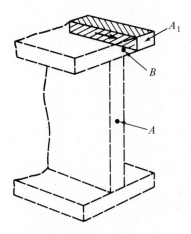
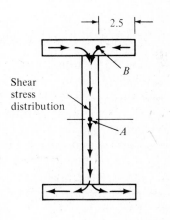

Shear stress distribution

The area of interest, $A_1$, varies from zero for points at the outer edge of the flange to a maximum where the flange joins the web. The $Q$ for any $A_1$ on the flange is just the area $A_1$ times the vertical distance ($\bar{y}_1$) to the centroid of $A_1$ (4.5 in in this case). Thus, for the flanges, $Q$ varies linearly from zero at the outer edges to a maximum where the web is welded to the flange.

If we were to consider point $B$, 2.5 in in from the edge of the flange, we would find:

$$Q = \bar{y}_1 A_1 = 4.5(2.5)(1) = 11.25 \text{ in}^3$$

$$q = \frac{VQ}{I} = \frac{10,000(11.25)}{286.67} = 392 \text{ lb/in}$$

Because the flange thickness at $B$ is 1 in, the shear stress $\tau = 392/1 = 392$ psi.

Note that in the flange, the shear stress is horizontal rather than vertical as in the web ($\tau_{xz}$).

The distribution of shear stress over the cross section, then, is as shown above; in the top flange, horizontal shear stress varying linearly from zero at the edges to a maximum near the web. For the bottom flange, the directions are reversed.

In the web, the stresses vary with $Q$, and have a maximum at the centroid. Because the flange and web thicknesses were equal and uniform, the maximum shear stress was at the neutral axis.

We note that when the shear stresses are depicted as arrows on the cross section, there is a definite "flow" of the shear from the top to the bottom. Shear stress and shear-flow patterns will always exhibit a "smooth flow" across the beam section. We always know the direction of the vertical shear stresses (same direction as the shear force), so the concept of a "smooth flow" helps to determine the sense of the shear stresses in the flanges.

To answer part ($b$), the maximum bending stress is computed from the usual beam equation:

$$\sigma_x = -\frac{M_b y}{I} = \frac{10^6(5)}{286.67} = 17,442 \text{ psi}$$

*Note*: The equation for bending stresses does have the minus sign as shown above. However, when we are only interested in magnitude we may avoid confusion by ignoring it.

To answer part ($c$), the weld metal bonds the three pieces together so that they act as a single beam instead of three separate beams. The primary requirement is to ensure that the flanges do not "slide" axially relative to the web.

If we consider the entire upper (or lower) flange to constitute the area $A_1$, we can calculate the shear flow $q = VQ/I$. This $q$ is the horizontal shear force, per unit length of the flange, that *must* be applied for the three pieces to act together as a single beam.

Because there are two weld lines for each flange, each weld must carry $q/2$. For this case, considering the shear flow $q$ at the flange-web juncture,

$$Q = \bar{y}_1 A_1 = 4.5(6)(1) = 27 \text{ in}^3$$

$$q = \frac{VQ}{I} = \frac{10,000(27)}{286.67} = 942 \text{ lb/in}$$

Thus each inch of each weld must be able to support a shear force of $942/2 = 471$ lb.                                                                           ***

Using the methods outlined in Example 8.8 we can compute the joining requirements for any built-up beam. The beam can be welded, glued, nailed, bolted, riveted, etc. If we know the capability of the joint material to carry shear, we can calculate the maximum allowable shear force $V$.

## 8.9 COMPOSITE BEAMS

Frequently, as we try to optimize the use of materials, we design composite beams made from two or more materials. The design rationale is quite straightforward. Utilize stiff, strong, heavy, or expensive material far away from the neutral axis at places where its effect will be greatest. Utilize weaker, lighter or less expensive material in the central portions of the beam where its effect will be less.

At one extreme is a steel-reinforced concrete beam, where weight is not a major concern, but strength and cost are. At the other extreme is a "sandwich" structure used in an aircraft made with graphite-fiber-reinforced plastic on the outside and a plastic foam core. Here, stiffness and weight are of the essence but cost is not.

In this section, we will consider three somewhat different analyses for three different structural members. In all cases we will assume symmetry about the plane of loading:

Symmetrical sections—such as the sandwich structure mentioned above.
Asymmetrical sections—not symmetrical about the axis of bending (the $z$ axis).
Reinforced concrete beams—concrete is assumed to have zero strength in tension.

### 8.9.1 Symmetrical Sections

Figure 8.17 shows a *sandwich* structure composite beam having outer layers of material 1 bonded to a core of material 2. The analysis which follows can easily be extended to three or more materials.

Any composite beam, when bent, will have a radius of curvature which is measured from the neutral axis. For the symmetrical section, we know where the NA is—right at the center. The bending strain, due to the bending moment, is proportional to the distance from the NA, as in all of our beam analyses. Thus the bending strain is continuous as we cross the interface between the two materials. However, because the elastic moduli are generally different, the bending stresses will be discontinuous across the interface as shown in Fig. 8.17b.

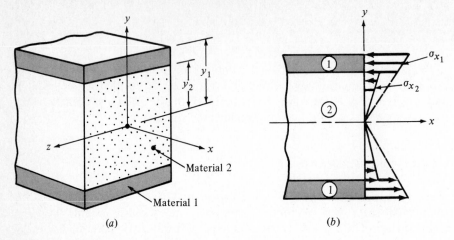

**Figure 8.17** Sandwich-type composite beam.

For this case, we can analyze the composite beam in terms of two separate beams, 1 and 2, having equal radii of curvature. It is necessary, of course, that the bonding between the layers is strong enough to assure that they do have a common center of curvature. This is primarily a shear requirement as we have seen.

We can write:

$$\frac{1}{\rho_1} = \frac{1}{\rho_2} \qquad \left(\frac{M_b}{EI}\right)_1 = \left(\frac{M_b}{EI}\right)_2 \qquad (8.22)$$

Letting

$$r = \frac{(EI)_2}{(EI)_1}$$

and

$$y_1 = \text{extreme } y \text{ for material } 1$$

$$y_2 = \text{extreme } y \text{ for material } 2$$

$$\frac{M_2}{M_1} = r \qquad M_b = M_1 + M_2 = M_1(1 + r) = M_2\left(1 + \frac{1}{r}\right)$$

The maximum bending stresses in the two materials are:

$$\sigma_{x_1} = \frac{M_1 y_1}{I_1} = \frac{M_b y_1}{I_1(1 + r)}$$

$$\sigma_{x_2} = \frac{M_2 y_2}{I_2} = \frac{M_b y_2}{I_2(1 + 1/r)} \qquad (8.23)$$

We can develop an approximate solution, which is both simple and conservative, based on the assumption that *all* of the bending moment is carried by the outer layers of material 1, that is, that $r = 0$.

If we further assume that the outer layers (of thickness $t$) are thin relative to the beam height $h$, then the distributed load can be replaced by two point forces ($F_x$) a distance $h$ apart, as shown in Fig. 8.18.

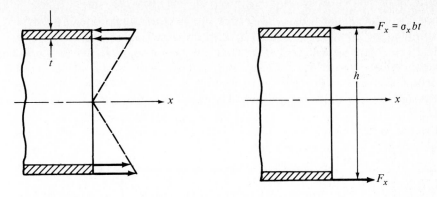

**Figure 8.18** Approximate solution for a sandwich beam.

The relationships for force, moment, and stress are:

$$F_x = \sigma_x bt \qquad M_b = F_x h \qquad \sigma_x = -\frac{M_b}{bth} \qquad (8.24)$$

If we now consider the effect of shear in this simplified model, where all of the bending moment is carried by the outer layers, we conclude that the core must simply supply sufficient shear force to hold the outer layers in (horizontal) equilibrium. In a development parallel to the earlier shear analysis, we find:

$$q = \frac{dF_x}{dx} = bt \frac{d\sigma_x}{dx}$$

From Eqs. (8.24):

$$\frac{d\sigma_x}{dx} = -\frac{1}{bth} \frac{dM_b}{dx}$$

but

$$\frac{dM_b}{dx} = -V$$

Thus, $q = V/h$, and the shear stress in the core is constant everywhere and is equal to:

$$\tau = \frac{V}{bh} = \text{shear force/beam area} \qquad (8.25)$$

To calculate the bending deflection of a composite beam, we use the usual procedures. For the approximate case we use the approximate moment of inertia (for the outer layers only):

$$I_{yy} = \frac{bth^2}{2} \qquad (8.26)$$

For most beams we can ignore the additional deflection due to the shear stresses. For a sandwich beam with thin outer layers, we must take care. In essence, the entire shear force $V$ is carried by the core, thus producing a shear stress in the core equal to $V$ divided by the core area [Eq. (8.25)].

The core material usually has a relatively low shear modulus $G$, so that the shear deflections can be significant. When the shear stress is constant across the section (as is the case here), it is very easy to calculate the shear deflection. The shear deflection, per unit length, is equal to the shear strain.

**Example 8.9** A cantilever sandwich beam of the type discussed above has the following dimensions and properties:

$$b = 1 \text{ in} \qquad E_1 = 20 \times 10^6 \text{ psi}$$

$$h = 2 \text{ in} \qquad G_2 = 0.1 \times 10^6 \text{ psi}$$

$$t = 0.05 \text{ in} \qquad \sigma_{1\,max} = 50{,}000 \text{ psi}$$

$$L = 20 \text{ in} \qquad \tau_{2\,max} = 1000 \text{ psi}$$

If a load of 100 lb is applied to the free end, how large is the safety factor, and how large is the deflection?

From Eqs. (8.24), the maximum bending stress is:

$$\sigma_x = \frac{20(100)}{1(0.05)(2)} = 20{,}000 \text{ psi}$$

And from Eq. (8.25), the shear stress in the core is:

$$\tau = \frac{100}{2} = 50 \text{ psi}$$

So the safety factor is $50{,}000/20{,}000 = 2.5$ (not $1000/50 = 20$).

From Eq. (8.26), the moment of inertia (for the outer layers only) is $0.1 \text{ in}^4$. The *bending deflection* is given by the usual cantilever equation:

$$\delta = \frac{FL^3}{3EI} = \frac{100(20^3)}{3(20 \times 10^6)(0.1)} = 0.133 \text{ in}$$

The *shear deflection* (for constant shear stress) is:

$$\delta_s = \gamma L = \frac{\tau}{G} L = \frac{50(20)}{100{,}000} = 0.01 \text{ in}$$

The *total deflection*, being the sum of the shear and bending deflections, equals $0.143 \text{ in}$. Although the shear deflection is relatively small, as the beam becomes shorter it will become relatively larger. ***

### 8.9.2 Asymmetrical Sections

When the cross section of a composite beam is not symmetrical about the $z$ axis (when loaded in the $y$ direction), we do not know the location of the neutral axis and the problem becomes more difficult.

We have previously learned how to analyze completely an asymmetrical beam *made of one material*: how to find the centroid, the moment of inertia, bending stresses, shear stresses, etc. Our approach to this current problem will be to replace the actual composite beam with an "equivalent beam" made of one material.

Figure 8.19a shows a composite beam having a layer of high-modulus material 1 attached to a large section of material 2. In Fig. 8.19b is shown an equivalent beam made from material 1. The two beams will be considered equivalent if the distributions of strain and force, as functions of y, are identical.

Thus, at any height y, the strain must be the same in both beams, and the forces carried by slices of thickness dy must be the same. These conditions will be met if at any height y the layer of low-modulus material is replaced by a narrower layer of high-modulus material. For a given strain, the force in the narrow slice will equal the force in the wider slice if the widths are inversely proportional to the elastic moduli. The proper ratio of the widths is:

$$\frac{b_2}{b_1} = n = \frac{E_2}{E_1} \tag{8.27}$$

Thus, the rectangular, composite beam is replaced with a T beam made from material 1. Given the T beam, we know how to find the centroid, the moment of inertia, stresses, deflections, etc.

When calculating stresses, we must remember that the actual beam is wider (in the replaced areas) so that the actual stresses are n times those calculated for the equivalent beam. (Clearly we could have replaced the high-modulus material with wider layers of low-modulus material.)

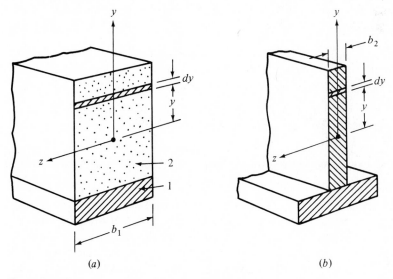

**Figure 8.19** Showing a composite beam replaced by an "equivalent" beam of one material.

**Example 8.10** In order to stiffen an $8 \times 10$ in wooden beam in a home, a 0.25-in layer of steel is bolted onto the bottom as shown in the accompanying illustration. How much stiffer will the resulting beam be?

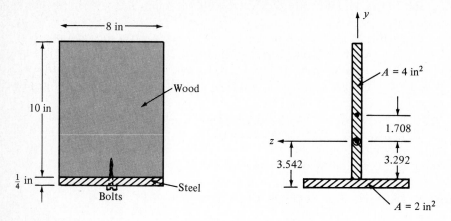

The elastic moduli are:

$$\text{Wood } E = 1.5 \times 10^6 \text{ psi}$$

$$\text{Steel } E = 30 \times 10^6 \text{ psi}$$

The modulus ratio $n$ equals 20, so the low-modulus wood can be replaced by a steel beam $\frac{8}{20} = 0.4$ in wide. This is now a standard T-beam problem.

The neutral axis of the T beam is found (using standard procedures) to be 3.54 in above the bottom of the beam; the moment of inertia about the centroidal axis is 68.37 in⁴.

The moment of inertia for the original $8 \times 10$ in beam is 666.67 in⁴.

The ratio of the stiffness of the T beam to the original wooden beam equals the ratio of $EI$ for the two:

$$\frac{(EI)_{\text{new}}}{(EI)_{\text{old}}} = \frac{68.37(30)}{666.7(1.5)} = 2.05$$

The new beam is twice as stiff *if* the bolting is sufficiently strong. How would you determine the bolting requirements? \*\*\*

### 8.9.3 Reinforced Concrete Beams

Because concrete has poor tensile properties, steel-reinforcing bars are used to carry the tensile load, while the concrete carries the compressive load associated with a bending moment. It is generally assumed that the concrete has zero strength in tension, so that the entire tensile load is carried by the steel.

Figure 8.20a shows the beam section; the re-bars are at a distance $d$ from the top, while the neutral axis is at some unknown distance $kd$.

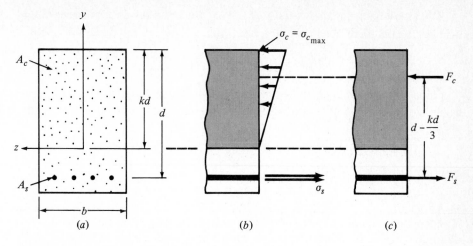

**Figure 8.20** Analysis of a reinforced concrete beam.

The effective area of the concrete $(A_c)$ is the area *above* the NA. We assume that the area below, which would be in tension, carries no axial load—does it carry a shear load?

The area of the steel $(A_s)$, at an unknown distance below the NA, carries the entire tensile load.

The strain will vary linearly from the NA, and the stresses will equal strain times the respective moduli. In Fig. 8.20*b* and *c*, the stress distributions and the resultant forces are shown.

Because there must be zero net horizontal force, the compressive force in the concrete $(F_c)$ must equal the tensile force in the steel $(F_s)$. The neutral axis location $(k)$ is determined from this equality (as it was for normal beams):

$$\sum F_x = 0$$

Therefore

$$F_s = F_c \quad \text{(magnitude only)}$$

$$F_s = \sigma_s A_s \qquad F_c = \frac{\sigma_c b k d}{2}$$

$$\sigma_s = \epsilon_s E_s = \frac{-y_s}{\rho} E_s = \frac{d(1-k)}{\rho} E_s$$

$$\sigma_c = \epsilon_c E_c = \frac{-y_c}{\rho} E_c = -\frac{kd}{\rho} E_c$$

Letting $m = E_c b d / E_s A_s$ and combining the above equations, we can solve the resulting quadratic equation for $k$:

$$k = \frac{-1 + \sqrt{1 + 2m}}{m}$$

The bending moment is equal to $F_s$ or $F_c$ times the distance between them, as shown in Fig. 8.20c:

$$M_b = F_c d\left(1 - \frac{k}{3}\right) = F_s d\left(1 - \frac{k}{3}\right)$$

Once $k$ is determined, the problem can be treated based on the equations above, or an equivalent beam can be constructed made entirely of steel and analyzed accordingly.

## 8.10  BEAM BUCKLING

Back in Chap. 3 we noted that the compressive-load-carrying ability of long slender members could be limited by buckling. Buckling can be considered to be a catastrophic bending instability due to excessive compressive axial loads acting on the beam. The beam may be very strong in compression, but it may not be *stable*. That is, when we have axial compression, if the beam deviates slightly from perfect straightness, or if the loading is not perfectly axial, then the load causes bending moments along the beam. The bending moments, in turn, produce deflections tending to increase the moments.

If the load is small enough, the beam is stable; but if the load exceeds a critical value ($F_{crit}$) the beam will undergo large, usually catastrophic, deformation. Because a buckling instability is a phenomenon of elastic bending, the critical load is determined by $EI$ and $L$ rather than $\sigma_{yp}$.

Figure 8.21 shows an originally straight beam that is being bent upward by the horizontal compressive load $F$. When $F$ was applied, it was exactly along the beam axis (the $x$ axis). A small external perturbation caused the beam to deflect and to assume the position shown. It would appear that the beam is in the process of buckling. Let us determine the deflection of the beam due to $F$.

The process of analysis should be clear: We can isolate a section of the beam, determine the bending moment, and then calculate the deflection. In Fig. 8.21b a beam element of length $dx$ is isolated. The only loading is due to the axial force $F$ and the moments due to $F$. Moment equilibrium gives:

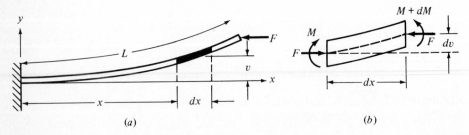

**Figure 8.21** Determination of "buckling" load: (a) Beam bent by axial load $F$; (b) isolation of beam element.

For $\Sigma M_z = 0$: $\qquad\qquad dM + F\,dv = 0$

$\qquad\qquad$ or $\qquad \dfrac{dM}{dx} + F\dfrac{dv}{dx} = 0$

We cannot calculate the bending moment directly because it depends on the beam deflection (which in turn depends on the bending moments). What equilibrium provides is a relationship that controls the variation of moment along the beam ($dM/dx$). Using the moment-curvature relation of Eq. (8.11) (for small deformations) we can write:

$$\frac{dM}{dx} = \frac{d}{dx}\left(EI\,\frac{d^2v}{dx^2}\right)$$

For constant $EI$ we have:

$$EI\,\frac{d^3v}{dx^3} + F\frac{dv}{dx} = 0$$

or

$$\frac{d^3v}{dx^2} + A\frac{dv}{dx} = 0$$

where $A = \dfrac{F}{EI}$

If you will recall your calculus, you will remember that certain functions are proportional to their second derivatives—namely, $\sin\theta$, $\cos\theta$, and $e^{j\theta}$. Figure 8.22 shows that for this particular case, where the beam is built in at $x = 0$, a cosine function might appropriately represent the shape of the buckled beam.

If we assume that the beam deflection is given by:

$$v = B\left(1 - \cos\frac{\pi x}{2L}\right)$$

where $B$ = a constant, then the derivatives are:

$$\frac{dv}{dx} = B\,\frac{\pi}{2L}\sin\frac{\pi x}{2L}$$

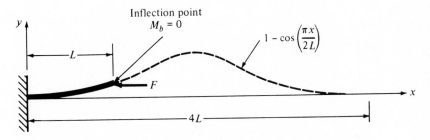

**Figure 8.22** Showing that the buckled beam is shaped like a cosine function.

$$\frac{d^2v}{dx^2} = B\left(\frac{\pi}{2L}\right)^2 \cos\frac{\pi x}{2L}$$

$$\frac{d^3v}{dx^3} = -B\left(\frac{\pi}{2L}\right)^3 \sin\frac{\pi x}{2L}$$

Inserting the proper derivatives into the differential equation, we obtain:

$$-B\left(\frac{\pi}{2L}\right)^3 \sin\frac{\pi x}{2L} + \frac{F}{EI} B \frac{\pi}{2L} \sin\frac{\pi x}{2L} = 0$$

The quantity
$$B \frac{\pi}{2L} \sin\frac{\pi x}{2L}$$

drops out, leaving:
$$-\left(\frac{\pi}{2L}\right)^2 + \frac{F}{EI} = 0$$

Thus we have the critical value for $F$:

$$F_{\text{crit}} = EI\left(\frac{\pi}{2L}\right)^2 = \frac{\pi^2}{4}\frac{EI}{L^2} \tag{8.28}$$

That is, a solution exists only when $F = F_{\text{crit}}$. For smaller loads, the beam will not remain bent, but will return to the stable, straight position. For greater loads, the beam will already have buckled.

Equation (8.28) gives the buckling load for a simple cantilever beam. Obviously $F_{\text{crit}}$ will be different if the end conditions are different. Figure 8.23

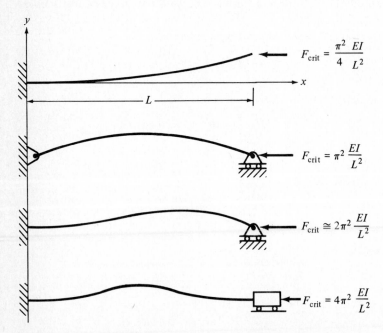

**Figure 8.23** Critical buckling loads for various conditions of end constraint.

gives the buckling load for various end conditions. We see a ratio of 16:1 as we go from a completely free right-hand end to one which can only move along the beam axis. End conditions are seldom as rigid as we would like, so prudence dictates a rather cautious and conservative application of the buckling relationships.

The results for the second and fourth beams shown in Fig. 8.23 can be obtained by superposition (or juxtaposition) from the result of the first beam. Can you see how this is done?

The buckling conditions shown in Fig. 8.23 are for the simplest situation. Frequently we have other loads acting on the beam, loading eccentricities, nonuniform beams, etc. In Sec. 8.11 we will show how easy it is to solve such complex problems numerically.

## 8.11 COMPUTER EXAMPLE

The computer is ideally suited to solving a wide variety of beam problems. The programs can range from just a few lines to something quite complex. We will show one example here and a number of others in Chap. 13.

### 8.11.1 The Effect of Axial Loads on Beam Deflections

Figure 8.24 shows a cantilever beam subject to three end loads: a bending force $P$, an end moment $M_0$, and an axial load $F$. A positive $F$ is a tensile load; a negative $F$ would tend to buckle the beam. As we have done previously, we divide the beam into $N$ elements. We write the bending moment equation for the $i$th element, calculate the resulting change in slope, and finally determine the deflection.

The only difficulty is that we must know the deflection at the end $(D_0)$ in order to compute the moment. We can achieve a solution in an iterative fashion by first assuming a value for $D_0$ (based on an approximate calculation), then computing the end deflection $v$. We then choose a new $D_0$, nearer to $v$, and recompute. We continue until the calculated value $v$ is essentially equal to the assumed value $D_0$.

Program Listing 8.1 solves the problem. The REMarks should be sufficient

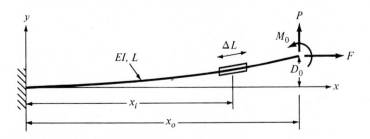

**Figure 8.24** Cantilever beam subject to both axial and bending loads.

```
10 REM        BENDING + AXIAL LOADING
20 REM
30 L=1        \REM  Beam length
40 X0=L       \REM Horizontal projection of L
50 D=.1       \REM  Diameter
60 E=30*10^6 \REM  Modulus
70 E1=3.14159*E*D^4/64              \REM  EI
80 F0=2.4674*E1/(L*L)              \REM  Fcrit buckle
90 N=100      \REM  Number of elements
100 L1=L/N    \REM  Delta-L
110 P=1       \REM  Normal loading (Fy)
120 F=F0      \REM  Axial load (+ is tension)
130 M0=P*L    \REM  End moment
140 D2=P*L^3/(3*E1) + M0*L^2/(2*E1)  \REM  Defl when F=0
150 D0=D2*(1-F*L^2/(2*E1))          \REM  Assumed initial defl.
160 D1=D0     \REM  Assumed intitial deflection
170 P1=0      \REM Phi
180 V =0      \REM Defl. v
190 X =0      \REM Location x
200 REM
210 FOR I=1 TO N
220  M=P*(X0-X) +M0 -F*(D0-V)       \REM  Moment at x
230  P2=M*L1/E1                     \REM  Delta-phi
240  P1=P1+P2                       \REM  Phi
250  V1=L1*SIN(P1)                  \REM  Delta-v
260  X1=L1*COS(P1)                  \REM  Delta-x
270  V=V+V1  \  X=X+X1              \REM  Deflection v at x
280 NEXT
290 X0=X
300 REM  Iterate if deflection is not within 1% of the assumed deflection
310 IF ABS(ABS(V/D0)-1)<.01 THEN 330
320 D0=(D0+V)/2 \ GOTO 170
330 !%10F2, "F = ",F," P = ",P," Mo = ",M0," Fo = ",F0
340 !
350 !%10F6, "Xo = ",X0," v = ",V," v/Vo = ",V/D2
```

**Program Listing 8.1**

for understanding of the logic. The output for a run is also in Program Output 8.1.

Note that in line 80 we calculated F0, the critical buckling load for the beam; and in line 140 we compute D2, the value of the end deflection if the axial force had been zero.

F =    363.35   P =     1.00   Mo =     1.00   Fo =    363.35

Xo =   .999995   v =   .002812   v/Vo =   .496892
READY

**Program Output 8.1**

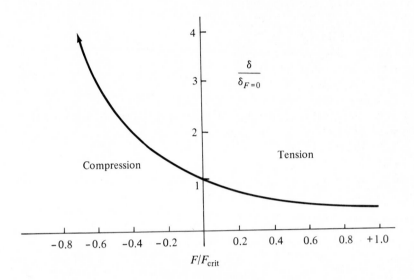

**Figure 8.25** Effect of axial loading on beam deflection.

In the print statement on line 350 we print the ratio of the actual deflection $v$ to D2. If the program is run for a variety of values for $F$, $P$, and $M_0$, and we plot $v$/D2 vs. the ratio $F/F_{crit}$, we find that the results can be plotted as shown in Fig. 8.25. The curve shown is not a "line" but rather a fairly narrow "band," hardly wider than the line.

## 8.12 PROBLEMS

**8.1** To improve the physical properties, wire is "cold-drawn" (nominally at room temperature) through a "draw-die." In essence, the die is a few percent (5 to 10 percent) smaller in diameter than the original wire, so that the wire comes out smaller and longer. The plastic deformation involved can increase the yield stress considerably.

   We wish to wind the wire onto a drum of radius $R$ such that the wire is bent elastically around the drum rather than plastically (see the accompanying illustration). If we ignore the tension in the wire, how large must $R$ be?

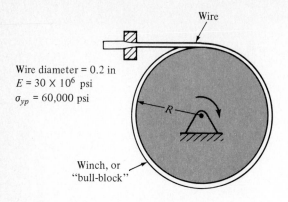

Wire diameter = 0.2 in
$E = 30 \times 10^6$ psi
$\sigma_{yp}$ = 60,000 psi

Winch, or
"bull-block"

**Figure P8.1**

In Probs. 8.2 to 8.7, draw and label the shear-force and bending moment diagrams for the following cases. Be careful about the signs.

**8.2**

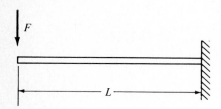

**Figure P8.2**

**8.3**

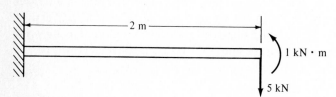

**Figure P8.3**

**8.4**

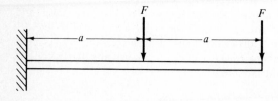

**Figure P8.4**

**8.5**

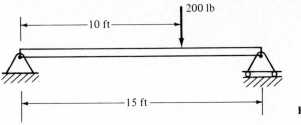

Figure P8.5

**8.6**

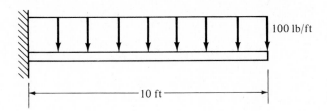

Figure P8.6

**8.7**

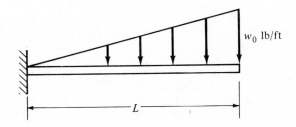

Figure P8.7

**8.8** A steel tube is loaded as shown in the accompanying illustration. Determine the maximum bending stress.

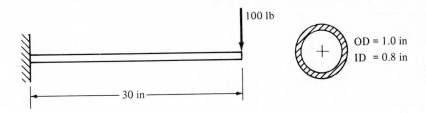

Figure P8.8

**8.9** If $E = 30 \times 10^6$ psi, how far down does the load deflect in Prob. 8.8?

**8.10** A 20-ft steel mast shown in the accompanying illustration must withstand a modest force $F$ at the top. The mast is a steel tube with OD $= 2.0$ in, ID $= 1.5$ in, $E = 30 \times 10^6$ psi and $\sigma_{yp} = 80,000$ psi. How far will the top deflect prior to any yielding?

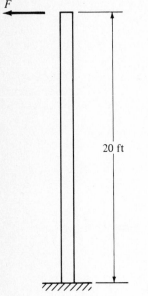

F

20 ft

**Figure P8.10**

**8.11** Using integration techniques, determine the slope and deflection at the free end of the beam shown in the accompanying illustration.

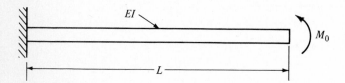

$EI$

$M_0$

$L$

**Figure P8.11**

**8.12** The cantilever beam shown below is supported at the right-hand end. When $P = 0$ there is no force on the support. How large is the support force when $P$ is applied? (*Hint*: Use superposition.)

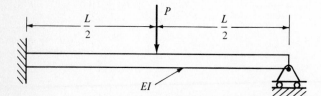

$\dfrac{L}{2}$

$P$

$\dfrac{L}{2}$

$EI$

**Figure P8.12**

**8.13** A beam of stiffness $EI$ and length $L$ shown in the accompanying illustration is built in at both ends. When the central load $P$ is applied, how far does it move down?

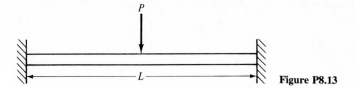

**Figure P8.13**

**8.14** The beam shown below was originally built in at both ends as in Prob. 8.13. Then, over time, the right-hand end "settled" downward a distance $\delta$ as shown. Where, and how large, is the maximum bending moment in the beam? (*Hint*: Use symmetry and/or superposition.)

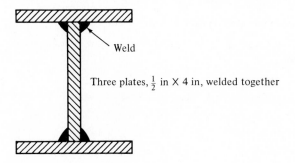

**Figure P8.14**

**8.15–8.19** For the beam cross sections illustrated in Probs. 8.15 to 8.19, determine the location of the centroid and the area moment of inertia $I_{yy}$.

**8.15**

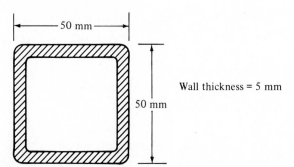

Weld

Three plates, $\frac{1}{2}$ in $\times$ 4 in, welded together

**Figure P8.15**

**8.16**

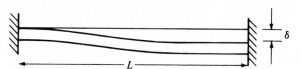

50 mm

50 mm

Wall thickness = 5 mm

**Figure P8.16**

**8.17**

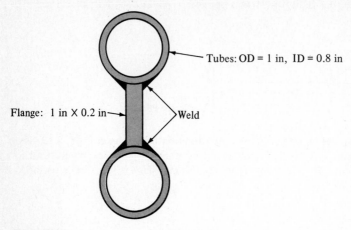

Tubes: OD = 1 in, ID = 0.8 in

Flange: 1 in × 0.2 in

Weld

**Figure P8.17**

**8.18**

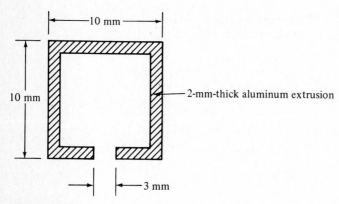

10 mm

10 mm

2-mm-thick aluminum extrusion

3 mm

**Figure P8.18**

**8.19**

Three 2-in × 8-in wooden beams, nailed together

**Figure P8.19**

**8.20–8.23** For the beam cross sections shown in Probs. 8.20 to 8.23, locate the centroid and then determine the principal moments of inertia. Show the principal axes, properly oriented, in a sketch.

**8.20**

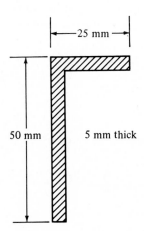

Three pieces, $\frac{1}{2}$ in × 2 in, welded together

**Figure P8.20**

**8.21**

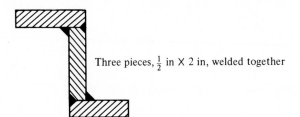

**Figure P8.21**

**8.22**

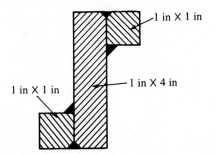

**Figure P8.22**

**8.23**

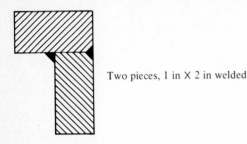

Two pieces, 1 in × 2 in welded

**Figure P8.23**

**8.24** A T beam is loaded as shown in the accompanying illustration. Determine the location and magnitude of:
  (a) The maximum tensile stress
  (b) The maximum compressive stress

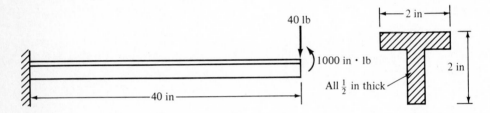

**Figure P8.24**

**8.25** If the T beam shown in Prob. 8.24 is made from aluminum with $E = 10 \times 10^6$, what will be the slope and deflection of the free end?

**8.26** A 1-in by 2-in beam, 10 in long, is built in at a 45° angle as shown below. A vertical load of 1000 lb is applied. Assuming no yielding, what is the deflection of point $A$?

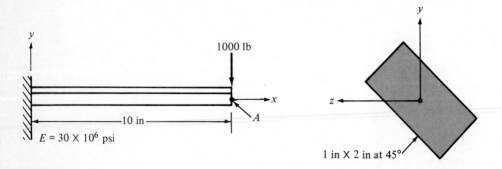

**Figure P8.26**

**8.27–8.29** For Probs. 8.27 to 8.29 write a computer program to determine their solutions.
**8.27** Prob. 8.9.

**8.28** Prob. 8.10 using all the same information except that the tube tapers from a 2-in diameter at the base to 1-in diameter at the top. Thickness is uniform.

**8.29** A vertical steel rod, 1.60 in in diameter and 25 ft long, is set into a concrete base. A 180-lb man climbs to the top and leans outward. If he can lean outward to such an extent that his center of gravity is 20 in from the rod, how much will the rod deflect? How large is the bending stress at the base? Would dynamic sway affect your answer? Would you climb the rod?

**8.30** Two 1-in² pieces of wood are glued together to form a beam, as shown in the accompanying illustration. How strong in shear must the glue be?

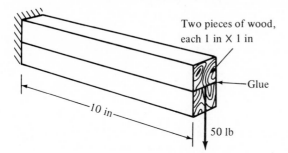

Two pieces of wood, each 1 in X 1 in

Glue

10 in

50 lb

**Figure P8.30**

**8.31** A square box beam is made by nailing together four pieces of wood as shown in the accompanying illustration. The beam can be used with the wide pieces on the top and bottom (*a*), or on the sides (*b*). Which way (if any) is preferable? Why?

    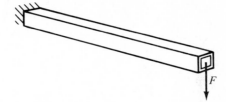

(*a*)          (*b*)

*F*

**Figure P8.31**

**8.32** A steel box beam is made by welding four plates together as shown below. If the vertical shear force in the beam (*V*) equals 10,000 lb, how strong in shear (in/lb) must the welds be?

Welds

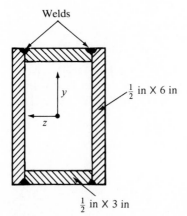

$\frac{1}{2}$ in X 6 in

*y*

*z*

$\frac{1}{2}$ in X 3 in          **Figure P8.32**

**8.33** A T beam is fabricated from two steel plates, as shown in the accompanying illustration. If the shear force $V = 4000$ lb, how strong must the welds be (in lb/in)?

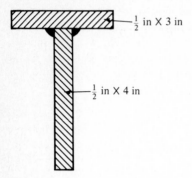

$-\frac{1}{2}$ in X 3 in

$-\frac{1}{2}$ in X 4 in

**Figure P8.33**

**8.34** An extruded aluminum beam section has the dimensions shown in the drawing below. The maximum allowable shear stress is 50 MPa. How large is the maximum allowable shear force and the maximum allowable bending moment?

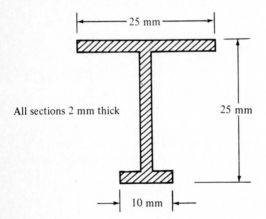

25 mm

25 mm

All sections 2 mm thick

10 mm

**Figure P8.34**

**8.35** A composite beam (see the drawing below) is made from a 1-in² piece of wood ($E = 1.5 \times 10^6$ psi) sandwiched between two thin ($t = 0.05$ in) sheets of aluminum ($E = 10 \times 10^6$ psi). The 10-in long cantilever beam supports a load $P = 50$ lb. Determine the following:

    (a) The required shear strength of the glue

    (b) The maximum bending stress in the aluminum

    (c) The deflection at the load $P$

State your assumptions.

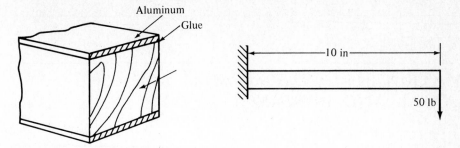

**Figure P8.35**

# NINE

## MATRIX STRUCTURAL ANALYSIS (AXIAL AND BENDING)

### 9.1 OVERVIEW

As you will recall, the FEM developments described in Chap. 4 allowed us to solve complex two- and three-dimensional structural problems as long as all members were two-force members, pinned (or having ball and socket joints) at their ends; only axial loading was permitted.

In Chap. 8, we considered simple uniform beams in bending—beams that were loaded only at their ends, either by forces normal to the beam axis or by moments perpendicular to the axis. Now we can extend the FEM analysis and permit both axial loading and bending of the elements. We will consider only two-dimensional systems such that both axial and bending deformations are in the $xy$ plane.

In Chap. 4 we organized the FEM into four distinct steps:

1. Defining the problem (geometry, degrees of freedom, loading, misfit, and temperature)
2. Establishing the equilibrium equations through the displacement method and stiffness matrices
3. Solving for the nodal displacements using the gaussian elimination technique
4. Determining element forces from the displacements and the local stiffness matrices

In this chapter, we will proceed through the same four steps, but will be able to do so very quickly because the overall problem organization does not change. We will merely add the appropriate terms to account for bending.

## 9.2 DEFINING THE STRUCTURE

Structure definition is only slightly changed from that given in Chap. 4. We still must number the elements and nodes; we still must give the global *xy* coordinates of each node; and we still must state which elements lie between which nodes. Now however, because our elements can have bending stiffness *EI*, and bending "slopes" are permitted, we must provide for a rotational dof at each node.

Figure 9.1 shows a generalized element which can now have *EI* as well as *EA* to describe its stiffness, and which can have a rotational dof at each end. Now when we specify the support boundary constraints (in terms of active or inactive dof), we must stipulate whether or not rotation ($\alpha$) is permitted.

We must be very careful here to incorporate a system that will properly differentiate between two-force members and those that can carry bending loads. If we think back, we realize that even in the pin-jointed trusses, the elements did rotate at the nodes. However, they rotated on frictionless pin joints, so there was no moment involved.

Now we have the possibility of having both bending members and two-force members joined at a node. When is the rotational dof "active," and how do we distinguish a bending member (which will "feel" a moment if its end is rotated) from a two-force member (which will not)?

Depending on how the program is structured, various conventions could be used. To be completely general, we would have to provide a separate rotational dof for each element joined at a node, i.e., so that if they were pinned, each could have a different angle of rotation even though the displacements were identical.

For simplicity the programs considered here (and you can alter them quite easily if you desire) will permit only 1 rotational dof at each node. This

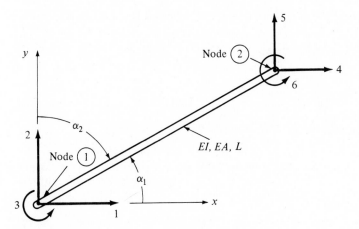

**Figure 9.1** Geometry for calculating the local stiffness matrix [*k*]. The 6 dof; are numbered 1 to 6, as shown.

means that at internal nodes, all bending members must rotate together (welded joints). For two-force members, if we set $EI = 0$, they will feel no moment and their ends will act as pinned.

In order to avoid singularities due to zero stiffness at an active dof, we will use the following conventions:

1. A rotational dof can be active *only* if at least one of the elements joined at the node is a bending member with $EI > 0$.
2. A rotational dof must be inactive if ($a$) no rotation is permitted, or ($b$) only two-force members are joined at the node.
3. For every two-force member, we will declare $EI = 0$. In this way, there will be no end moment even if an end is at an active dof.
4. Bending members ($EI > 0$) *cannot* be pinned together within the structure. The joint is "rigid" (as though welded). They *can* be pinned at support points.

As in Chap. 4, we will consider an example as we develop the FEM. Figure 9.2 shows a simple two-element structure with one bending member and one two-force member. Be sure you understand thoroughly why the 3 rotational dof are treated as they are.

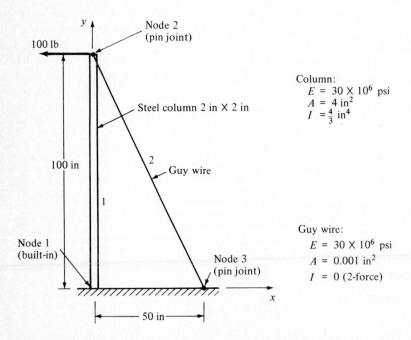

Figure 9.2 Example having a bending element (1), and a two-force element (2).

Node 1: all dof inactive; built-in support (rule 2)

Node 2: all dof active; free to move in $x$, $y$, and $\alpha$ directions, *and* element 1 is a bending member (rule 1)

Node 3: all dof inactive; although element 2 can rotate, it is a two-force member (rule 2)

Whenever we have an active dof, we must state the externally applied load in the direction of that dof (either zero or some finite value). When a rotational dof is active, we must specify the externally applied moment (zero or finite). A positive applied moment will be in the direction of a positive rotation (counterclockwise when looking down the $z$ axis).

The format we will use for problem definition together with the appropriate values for the example shown in Fig. 9.2 is shown in Table 9.1.

You might wonder why we have set $EI = 0$ for element 2. This is in accordance with rule 3 above. Basically, we must have an active rotational dof at node 2 because the steel column is a bending element, and in fact, there will be a positive rotation at node 2 when the column bends to the left. *If* we gave a nonzero value for the $EI$ of the guy wire, the rotation at the node would apply a moment to the wire. Obviously, this is not correct. We could use alternative conventions, but for simplicity we set $EI = 0$.

Even if a two-force member is very stiff, we will call

$$EI = 0 \qquad \text{(all two-force members)}$$

Given the input shown, the FEM will proceed to solve the problem. For simply *using* the FEM program, you have sufficient information. In the remainder of this chapter, we will show how the pertinent equations must be modified to account for bending.

**Table 9.1**

| | | | | | | | | | |
|---|---|---|---|---|---|---|---|---|---|
| | | | | Number of *elements* = 2 | | | | | |
| | | | | Number of *nodes* = 3 | | | | | |

| | dof | | | Forces and moments | | | | Coordinates | |
|---|---|---|---|---|---|---|---|---|---|
| Node | $x$ | $y$ | $z$ | $x$ | $y$ | $a$ | | $x$ | $y$ |
| 1 | 0 | 0 | 0 | 0 | 0 | 0 | | 0 | 0 |
| 2 | 1 | 1 | 1 | −100 | 0 | 0 | | 0 | 100 |
| 3 | 0 | 0 | 0 | 0 | 0 | 0 | | 50 | 0 |

| Element No. | Located between nodes | | $EA \times 10^{-6}$, lb | $EI \times 10^{-6}$, lb·in$^2$ | $\delta_0$, in | Temp. strain × 10$^6$ |
|---|---|---|---|---|---|---|
| 1 | 1 | 2 | 120 | 40 | 0 | 0 |
| 2 | 2 | 3 | 0.03 | 0 | 0 | 0 |

## 9.3 ESTABLISHING THE STIFFNESS MATRIX

The local stiffness matrix $[k_1]$ will be a $6 \times 6$ matrix because we have 6 dof for each element. As before, the coefficients of $[k_1]$, $k_{ij}$, represent the force (moment) in the direction of the $i$th dof due to a unit displacement (rotation) in the direction of the $j$th dof. The tension-compression terms due to extension or compression of the beam will be just as before. Now, however, we must add the forces and moments associated with bending of the beam.

Figure 9.3$a$ shows an element subject to a unit displacement along dof No. 1. Note that now (as opposed to the case of two-force elements), the *slope* of the beam must not change (unless we are applying a unit rotation in the direction of dof No. 3 or No. 6). Figure 9.3$b$ shows the forces and moments that must exist due to bending only; the axial forces have been omitted for clarity.

In Fig. 9.4 we have reoriented the element shown in Fig. 9.3 so that it looks more like the beam bending examples of Chap. 8 (see Prob. 8.14).

We must determine $F$ and $M$ when the deflection is as shown in Fig. 9.4. There are many ways to do this; perhaps the most simple is to note that the beam is symmetrical about its midpoint. Because of symmetry, there must be a point of inflection, and therefore zero curvature, at the midpoint. The problem can be reduced to a simple beam of length $L/2$, deflected a distance $\delta/2$ by the force $F$ as shown in Fig. 9.4$b$. Because there is no curvature at $L/2$, there can be no moment there either. Now we can use the simple cantilever beam equation from Table 8.1:

$$\text{``}\delta\text{''} = \frac{\text{``}PL^3\text{''}}{3EI}$$

where $\text{``}\delta\text{''} = \dfrac{\delta}{2} = \dfrac{\sin \alpha}{2}$

$\text{``}L\text{''} = \dfrac{L}{2}$

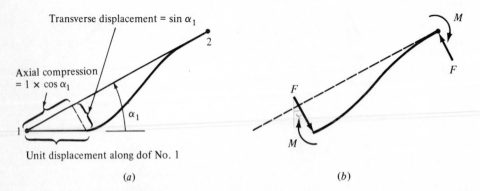

Unit displacement along dof No. 1

$(a)$

$(b)$

**Figure 9.3** Unit displacement along dof No. 1: ($a$) displacement and rotation magnitudes; ($b$) forces and moments due only to bending.

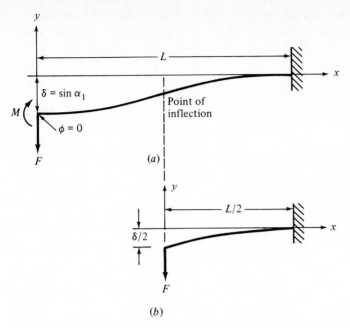

**Figure 9.4** Solve for $F$ and $M$ required to deflect the beam $\delta = \sin \alpha_1$ while the slope is zero. In (b) the solution is simplified through symmetry.

This gives:
$$F = \frac{12EI}{L^3} \sin \alpha_1$$

$$M = \frac{6EI}{L^2} \sin \alpha_1$$

The force $F$ has components along both dof No. 1 and dof No. 2. The stiffness matrix terms associated with a unit displacement along dof No. 1 are finally given by:

$$k_{11} = F \sin \alpha_1 = +\frac{12EI}{L^3} \sin^2 \alpha_1 = +\frac{12EI}{L^3} \cos^2 \alpha_2$$

$$k_{21} = F \cos \alpha_1 = -\frac{12EI}{L^3} \sin \alpha_1 \cos \alpha_1 = -\frac{12EI}{L^3} \cos \alpha_1 \cos \alpha_2$$

$$k_{31} = -\frac{6EI}{L^2} \sin \alpha_1 = -\frac{6EI}{L^2} \cos \alpha_2$$

At the other end of the beam, the moment is identical, but the forces are of opposite sign:

$$k_{14} = -k_{11}$$

$$k_{15} = -k_{12}$$

$$k_{16} = k_{13}$$

## Table 9.2 Local stiffness matrix terms (two-dimensional, axial, and bending)

|  |  | 1 | 2 | 3 | 4 | 5 | 6 |
|---|---|---|---|---|---|---|---|
|  | 1 | $B_1 + T_1$ | $-B_3 + T_3$ | $-B_4$ | $-B_1 - T_1$ | $B_3 - T_3$ | $-B_4$ |
|  | 2 |  | $B_2 + T_2$ | $B_5$ | $B_3 - T_3$ | $-B_2 - T_2$ | $B_5$ |
| $[k_1] =$ | 3 |  |  | $4B_6$ | $B_4$ | $-B_5$ | $2B_6$ |
|  | 4 | symmetrical |  |  | $B_1 + T_1$ | $-B_3 + T_3$ | $B_4$ |
|  | 5 |  |  |  |  | $B_2 + T_2$ | $-B_5$ |
|  | 6 |  |  |  |  |  | $4B_6$ |

where $B_1 = \dfrac{12EI}{L^3} \sin^2 \alpha_1$   $\qquad B_6 = \dfrac{EI}{L}$

$B_2 = \dfrac{12EI}{L^3} \cos^2 \alpha_1$   $\qquad T_1 = \dfrac{AE}{L} \cos^2 \alpha_1$

$B_3 = \dfrac{12EI}{L^3} \sin \alpha_1 \cos \alpha_1$   $\qquad T_2 = \dfrac{AE}{L} \sin^2 \alpha_1$

$B_4 = \dfrac{6EI}{L^2} \sin \alpha_1$   $\qquad T_3 = \dfrac{AE}{L} \sin \alpha_1 \cos \alpha_1$

$B_5 = \dfrac{6EI}{L^2} \cos \alpha_1$

Using similar methods, we can solve for the rest of the stiffness matrix terms. All the terms are shown in Table 9.2.

Thus, for each element, knowing $L$, $\alpha_1$, $EI$, and $EA$, we can evaluate each term in the local matrix, add the appropriate terms to the global matrix, and store sufficient data to quickly reconstruct the local matrix when we want to determine the element forces.

The local stiffness matrices for the example shown in Fig. 9.2 are given below:

$$[k_1]_1 = \begin{bmatrix} 480 & 0 & -24{,}000 & -480 & 0 & -24{,}000 \\ & 1.2 \times 10^6 & 0 & 0 & -1.2 \times 10^6 & 0 \\ & & 1.6 \times 10^6 & 24{,}000 & 0 & 0.8 \times 10^6 \\ & & & 480 & 0 & 24{,}000 \\ & & & 0 & 1.2 \times 10^6 & 0 \\ & & & 24{,}000 & 0 & 1.6 \times 10^6 \end{bmatrix}$$

$$[k_1]_2 = \begin{bmatrix} 53.7 & -107.3 & 0 & -53.7 & 107.3 & 0 \\ & 214.7 & 0 & 107.3 & -214.7 & 0 \\ & & 0 & 0 & 0 & 0 \\ & & & 53.7 & -107.3 & 0 \\ & & & -107.3 & 214.7 & 0 \\ & & & 0 & 0 & 0 \end{bmatrix}$$

The only active dof are the three at node 2, so the global stiffness matrix $[k]$ will be only $3 \times 3$ in size. As in Chap. 4, we label the active dof from 1 to $M$ (1, 2, 3), and assign a specific dof to each number to identify its position in $[k]$. In this case, we can use:

$$1 = x \text{ dof at node 2}$$

$$2 = y \text{ dof at node 2}$$

$$3 = \alpha \text{ dof at node 2}$$

These dof correspond to those within the dashed lines on the local stiffness matrices above. Adding these together, we have the global matrix $[k]$:

$$[k] = \begin{bmatrix} 533.7 & -107.3 & 24,000 \\ -107.3 & 1,200,215 & 0 \\ 24,000 & 0 & 1.6 \times 10^6 \end{bmatrix} \quad \{F\} = \begin{Bmatrix} -100 \\ 0 \\ 0 \end{Bmatrix}$$

For the load vector $\{F\}$, we have only the single external load of $-100$ lb in the $-x$ direction as shown above. We can now write the displacement-based equilibrium equations:

$$[k]\{u\} = \{F\}$$

## 9.4 SOLVING THE EQUATIONS

Given the equilibrium equation above, we solve for the displacements $\{u\}$, and then combine these with the local stiffness matrices to determine the element forces.

This particular set of equations is very easily solved. The displacements are given below, and the element forces are shown in Fig. 9.5. If you are not clear regarding any of these steps, work out the individual steps for yourself. That is the best way to clarify matters.

$$u_1 = -0.576 \text{ in}$$

$$u_2 = -0.000051 \text{ in}$$

$$u_3 = +0.00864 \text{ rad}$$

It should be clear that the only significant difference between this FEM procedure and that in Chap. 4 is the inclusion of the bending terms in the local stiffness matrices.

**Example 9.1** To demonstrate the FEM developed above, the example shown in Fig. 9.2 has been run for four different "loadings"; the first set is the same loading worked out in this chapter, while the others show different loads and misfits.

The results are shown in Program Output 9.1 just as they appear on the

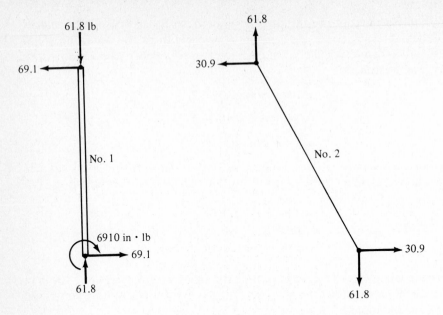

**Figure 9.5** Forces and moments acting *on* the two elements in the example.

computer screen. The computational time is about 5 s for each set of loads (printing the result takes much longer). The program used is given in the next section (Program Listing 9.1).

```
STRUCTURAL ANALYSIS

2-DIMENSIONAL STRUCTURES
    Axial loading    (E*A)
    Bending          (E*I)


*****************************************************************


        Input data to define the problem


            Number of ELEMENTS = 2
            Number of NODES    = 3


Input data for each NODE:
    For Degrees of Freedom, 'active'=1,  'non-active'=0
```

**Program Output 9.1**

| | DOF | Applied Forces | | Moment | Coordinates | | |
|---|---|---|---|---|---|---|---|
| Node | X Y A | X | Y | A | X | Y | |
| 1 | 0 0 0 | 0 | 0 | 0 | 0 | 0 | OK? (CR/N) |
| 2 | 1 1 1 | -100 | 0 | 0 | 0 | 100 | OK? (CR/N) |
| 3 | 0 0 0 | 0 | 0 | 0 | 50 | 0 | OK? (CR/N) |

Input data to define ELEMENT location, stiffness, and initial conditions
    Note multipliers for E*A, E*I, and Temp-Strain
        For 'pin-joint', negate (-) node number

| Element No. | Located Between Nodes | E*A *(10^-6) (lbs.) (N) | E*I *(10^-6) (lb-in^2) (N-m^2) | Delta- Zero (in.) (m) | Temp. Strain *(10^6) | |
|---|---|---|---|---|---|---|
| 1 | 1  2 | 120 | 40 | 0 | 0 | OK? (CR/N) |
| 2 | 2  3 | .03 | 0 | 0 | 0 | OK? (CR/N) |

There are  3 degrees of freedom
Wait for computation

Nodal Displacements and Rotations

| Node | X | Y | Alpha |
|---|---|---|---|
| 1 | .00000 | .00000 | .00000 |
| 2 | -.57585 | -.00005 | .00864 |
| 3 | .00000 | .00000 | .00000 |

| Element | Node | Forces Fx | Fy | Moments Mz |
|---|---|---|---|---|
| 1 | 1 | 69.1 | 61.8 | -6910.2 |
| 1 | 2 | -69.1 | -61.8 | .0 |
| 2 | 2 | -30.9 | 61.8 | .0 |
| 2 | 3 | 30.9 | -61.8 | .0 |

**Program Output 9.1** (continued)

Run other loadings for this structure? (CR/N) Y

```
               Applied Forces and Moments
       Node       X        Y        M

        2       -100      1000      0    OK?  (CR/N)
```

```
     Element      Delta-Zero     T-Strain #10^6

        1             0             0    OK? (CR)
        2             0             0    OK? (CR)
```

                    Wait for computation

```
        Nodal Displacements and Rotations
    Node        X          Y         Alpha

     1       .00000     .00000     .00000
     2      -.57534     .00078     .00863
     3       .00000     .00000     .00000
```

```
                             Forces           Moments
    Element     Node      Fx        Fy          Mz

       1          1      69.0     -938.1      -6904.0
       1          2     -69.0      938.1          .0
       2          2     -31.0       61.9          .0
       2          3      31.0      -61.9          .0
```

Run other loadings for this structure? (CR/N) Y

```
               Applied Forces and Moments
       Node       X        Y        M

        2         0        0        0    OK?  (CR/N)
```

**Program Output 9.1** (continued)

| Element | Delta-Zero | T-Strain #10^6 |
|---------|------------|----------------|
| 1 | 0 | 100   OK? (CR) |
| 2 | 0 | 0   OK? (CR) |

Wait for computation

### Nodal Displacements and Rotations

| Node | X | Y | Alpha |
|------|-------|-------|--------|
| 1 | .00000 | .00000 | .00000 |
| 2 | .00618 | .01000 | -.00009 |
| 3 | .00000 | .00000 | .00000 |

| Element | Node | Forces Fx | Fy | Moments Mz |
|---------|------|-----|------|------|
| 1 | 1 | -.7 | 1.5 | 74.2 |
| 1 | 2 | .7 | -1.5 | .0 |
| 2 | 2 | -.7 | 1.5 | .0 |
| 2 | 3 | .7 | -1.5 | .0 |

Run other loadings for this structure? (CR/N) Y

### Applied Forces and Moments

| Node | X | Y | M |
|------|---|---|---|
| 2 | 0 | 0 | 0   OK?  (CR/N) |

| Element | Delta-Zero | T-Strain #10^6 |
|---------|------------|----------------|
| 1 | 0 | 0   OK? (CR) |
| 2 | -.2 | 0   OK? (CR) |

Wait for computation

**Program Output 9.1** (continued)

Nodal Displacements and Rotations

| Node | X | Y | Alpha |
|------|------|------|------|
| 1 | .00000 | .00000 | .00000 |
| 2 | .13818 | -.00003 | -.00207 |
| 3 | .00000 | .00000 | .00000 |

| Element | Node | Forces Fx | Fy | Moments Mz |
|---------|------|------|------|------|
| 1 | 1 | -16.6 | 33.2 | 1658.2 |
| 1 | 2 | 16.6 | -33.2 | .0 |
| 2 | 2 | -16.6 | 33.2 | .0 |
| 2 | 3 | 16.6 | -33.2 | .0 |

**Program Output** (continued)

## 9.5 STRUCTURAL ANALYSIS PROGRAM LISTING

```
20  !"                    STRUCTURAL ANALYSIS"
40  !
50  !"               2-DIMENSIONAL STRUCTURES"
60  !"                 Axial loading   (E*A)"
70  !"                 Bending         (E*I)" \ !
90  REM
100 !" ***********************************************************************"\!\!
110 REM
120 REM
130 REM    1.  DEFINE THE STRUCTURE
140 REM        a. Number of elements and nodes
150 REM        b. Input element data
160 REM        c. Input node data
170 REM        d. Input element data
180 REM        e. Determine and number the active degrees of freedom (1,2,--M)
190 REM        f. Dimension arrays
200 REM        g. Establish the applied force (moment) vector {f}
210                  GOSUB 790
220 REM
230 REM
```

**Program Listing 9.1**

```
240 REM     2.  ESTABLISH THE GLOBAL STIFFNESS MATRIX  (k)
250 REM         For each element,
260 REM         a. Determine element constants and store in array
270 REM         b. Establish local stiffness matrix (k1)
280 REM         c. Put appropriate terms into global matrix (k)
290                        GOSUB 1750
300 REM
310 REM
320 REM     3.  ESTABLISH THE INITIAL FORCE VECTOR  {f0}
330 REM         For each element,
340 REM         a. Calculate initial forces due to temperature or misfit
350 REM            and put into vector {f0} and final element force
360 REM            array E1
370 REM         b. Subtract initial forces from the applied forces
380 REM            {f}={f}-{f0}
390                        GOSUB 2280
400 REM
410 REM
420 REM     4.  SOLVE FOR DISPLACEMENTS {u} BY GAUSSIAN ELIMINATION
430 REM            (k)*{u}={f}
440 REM         a. Triangularize stiffness matrix (k)
450 REM         b. Calculate displacements {u}
460 REM         c. Put displacements into nodal array N1
470                        GOSUB 2540
480 REM
490 REM
500 REM     5.  DETERMINE FORCES ON EACH ELEMENT  (at both nodes)
510 REM         For each element\
520 REM         a. Determine nodes N1 AND N2
530 REM         b. Establish local displacement vector {u1}
540 REM         c. Establish local stiffness matrix (k1)
550 REM         d. Calculate forces due to displacements {f1}
560 REM            {f1}=(k1)*{u1}
570 REM         e. Add to initial forces in array E1
580                        GOSUB 2930
590 REM
600 REM
610 REM     6.  PRINT OUT RESULTS
620 REM         a. Nodal motion
630 REM         b. Forces on each element at both ends
640                        GOSUB 3290
650 REM
660 REM
```

**Program Listing 9.1** (continued)

```
670 REM    7.  FURTHER CALCULATIONS
680 REM        a. Other loadings for this structure?
690 REM        b. Alter Structural Data?
700                     GOSUB 3570
710 REM
720 REM
730 REM ***************************************************************************
740 REM
750 REM    1.  DEFINE THE STRUCTURE
760 REM
770 REM        **** a. Number of elements and nodes ****
780 REM
790 IF R1>0 THEN 1400              \REM  R1=0 for first pass
800 !"              Input data to define the problem"\!\!
810 INPUT"               Number of ELEMENTS = ",I9    \REM I=ELEMENTS
820 INPUT"               Number of NODES   = ",J9    \REM J=NODES
830 !\!
840 REM        **** b. Dimension arrays ****
850 REM
860 DIM N(J9,8)    \REM NODAL INPUT DATA
870 DIM N1(J9,3)   \REM NODAL DISPLACEMENTS FOR OUTPUT
880 DIM E(I9,8)    \REM ELEMENT INPUT DATA
890 DIM E1(I9,8)   \REM ELEMENT FORCES FOR OUTPUT
900 DIM K0(I9,12)  \REM ELEMENT CONSTANTS
910 DIM M(6)       \REM GLOBAL DOF NUMBERS FOR LOCAL ELEMENT DOF
920 DIM K1(6,6)    \REM LOCAL ELEMENT STIFFNESS MATRIX
930 DIM F1(6)      \REM LOCAL FORCE VECTOR
940 DIM U1(6)      \REM LOCAL DISPLACEMENT VECTOR
950 R1=0           \REM FIRST RUN FOR THIS STRUCTURE
960 REM
970 REM        **** c. Input nodal data ****
980 REM
990 !"Input data for each NODE: "
1000 !"    For Degrees of Freedom, 'active'=1,  'non-active'=0"
1010 !\!
1020 !"         DOF        Applied Forces    Moment      Coordinates" \ !
1030 !"Node    X Y A        X         Y         A         X       Y"\!
1040 IF R1>0 THEN 1060
1050 FOR J=1 TO J9
1060 !" ",J,
1070 INPUT1"     ",N(J,1) \ INPUT1" ",N(J,2) \ INPUT1" ",N(J,3)
1080 !TAB(23), \ INPUT1" ",N(J,4) \ !TAB(33), \ INPUT1" ",N(J,5)
1090 !TAB(42), \ INPUT1" ",N(J,6) \ !TAB(53), \ INPUT1" ",N(J,7)
```

**Program Listing 9.1** (continued)

```
1100 !TAB(61), \ INPUT1" ",N(J,8) \ INPUT"   OK? (CR/N)",Z$
1110 IF Z$="N" THEN 1060
1120 IF R1>0 THEN RETURN
1130 NEXT \ !\!
1140 REM
1150 REM         **** d. Input element data
1160 REM
1170 !"Input data to define ELEMENT location, stiffness, and initial conditions"
1180 !"      Note multipliers for E*A, E*I, and Temp-Strain"
1190 !"           For 'pin-joint', negate (-) node number"
1200 !\!
1210 !"Element   Located     E*A      E*I      Delta-     Temp."
1220 !"  No.    Between    *(10^-6)  *(10^-6)  Zero      Strain"
1230 !"         Nodes      (lbs.)   (lb-in^2)  (in.)    *(10^6)"
1240 !"                    (N)      (N-m^2)    (m)"\!
1250 IF R1>0 THEN 1270
1260 FOR I=1 TO I9
1270 !"   ",I,TAB(11),\INPUT1"",E(I,1)\!TAB(14),\INPUT1"",E(I,2)
1280 !TAB(20), \ INPUT1"  ",E(I,3)
1290 !TAB(30), \ INPUT1"  ",E(I,4)
1300 !TAB(40), \ INPUT1"  ",E(I,5)
1310 !TAB(50), \ INPUT1"  ",E(I,6)
1320 E(I,3)=E(I,3)*10^6 \ E(I,4)=E(I,4)*10^6 \ E(I,6)=E(I,6)*10^-6
1330 INPUT"   OK? (CR/N)",Z$
1340 IF Z$="N" THEN 1270
1350 IF E(I,1)<0 THEN E(I,7) = 1 ELSE E(I,7)=0 \ E(I,1)=ABS(E(I,1))
1360 IF E(I,2)<0 THEN E(I,8) = 2 ELSE E(I,8)=0 \ E(I,2)=ABS(E(I,2))
1370 IF R1>0 THEN RETURN
1380 NEXT \!\!\!
1390 REM
1400 REM         **** e. Determine and number active degrees of freedom ****
1410 REM
1420 M=0
1430 FOR J=1 TO J9
1440 FOR K=1 TO 3
1450 IF N(J,K)=0 THEN 1470
1460 M=M+1 \ N(J,K)=M
1470 NEXT
1480 NEXT
1490 !"           There are ",M," degrees of freedom"
1500 !"              Wait for computation"\!\!
1510 REM
1520 REM         **** f. Dimension arrays ****
```

**Program Listing 9.1** (continued)

```
1530 REM
1540 IF R1>0 THEN 1590
1545 M=M+3              \REM SPACE FOR 3 MORE DOF
1550 DIM K(M,M)         \REM GLOBAL STIFFNESS MATRIX
1560 DIM F(M)           \REM APPLIED FORCE VECTOR
1570 DIM F0(M)          \REM INITIAL FORCE VECTOR
1580 DIM U(M)           \REM DISPLACEMENT VECTOR
1585 M=M-3              \REM ACTUAL NUMBER OF DOF
1590 REM
1600 REM          **** g. Establish the applied force vector {f} ****
1610 REM
1620 FOR J=1 TO J9 \ FOR K=1 TO 3
1630 IF N(J,K)=0 THEN 1660
1640 M=N(J,K)
1650 F(M)=N(J,K+3)
1660 NEXT
1670 NEXT
1680 RETURN
1690 REM
1700 REM *************************************************************************
1710 REM
1720 REM    2.  ESTABLISH THE GLOBAL STIFFNESS MATRIX {k}
1730 REM
1740 REM          **** a. Determine and store element constants ****
1750 REM
1760 REM
1770 FOR I=1 TO I9
1780 N1=E(I,1) \ N2=E(I,2)                \REM N1 IS NODE AT LOCAL ORIGIN
1790 REM
1800 FOR K=1 TO 3 \ M(K)=N(N1,K) \ M(K+3)=N(N2,K) \ NEXT
1810 X=N(N2, 7)-N(N1, 7)
1820 Y=N(N2, 8)-N(N1, 8)
1830 L=SQRT(X*X+Y*Y)
1840 IF X<>0 THEN 1870
1850 C=0 \ IF Y<0 THEN S=-1 ELSE S=1
1860 GOTO 1920
1870 A1=ATN(Y/X)
1880 C=COS(A1)
1890 S=SIN(A1)
1900 IF X<0 THEN C=-C
1910 IF X<0 THEN S=-S
1920 L2=L*L
1930 S1=E(I,3)/L \ S2=E(I,4)/L            \REM EA/L    EI/L
```

**Program Listing 9.1** (continued)

```
1940 T1=S1#C#C \ T2=S1#S#S \ T3=S1#S#C
1950 B0=12#S2/L2
1960 B1=B0#S#S
1970 B2=B0#C#C
1980 B3=B0#S#C
1990 B4=6#S2#S/L
2000 B5=6#S2#C/L
2010 REM
2020 REM         #### b. File coef. for local stiffness matrix ####
2030 REM
2040 K0(I,7)=E(I,7)+E(I,8)
2050 K0(I,8)=L \ K0(I,9)=C \ K0(I,10)=S \ K0(I,11)=S1 \ K0(I,12)=S2/L
2060 IF K0(I,7))0 THEN 2100
2070 K0(I,1)=B1+T1 \ K0(I,2)=B2+T2 \ K0(I,3)=-B3+T3
2080 K0(I,4)=-B4    \ K0(I,5)=B5    \ K0(I,6)=4#S2
2090 GOTO 2160
2100 IF K0(I,7)=3 THEN 2140
2110 K0(I,1)=B1/4+T1 \ K0(I,2)=B2/4+T2 \ K0(I,3)=-B3/4+T3
2120 K0(I,4)=-B4/2    \ K0(I,5)=B5/2    \ K0(I,6)=3#S2
2130 GOTO 2160
2140 K0(I,1)=T1 \ K0(I,2)=T2 \ K0(I,3)=T3
2150 K0(I,4)=0  \ K0(I,5)=0  \ K0(I,6)=0
2160 GOSUB 4120
2170 REM
2180 REM         #### c. Put appropriate terms into global matrix (k) ####
2190 REM
2200 FOR U=1 TO 6 \ FOR V=1 TO 6
2210 K(M(U),M(V))=K(M(U),M(V)) + K1(U,V)
2220 NEXT \ NEXT \ NEXT
2230 RETURN
2240 REM
2250 REM #################################################################
2260 REM
2270 REM    3.  ESTABLISH THE INITIAL FORCE VECTOR {f0}
2280 REM
2290 REM         #### a. Calculate initial forces, put in {f0} and E1 ####
2300 REM
2310 FOR I=1 TO I9
2320 N1=E(I,1) \ N2=E(I,2)                 \REM N1 IS NODE AT LOCAL ORIGIN
2330 FOR K=1 TO 3 \ M(K)=N(N1,K) \ M(K+3)=N(N2,K) \ NEXT
2340 F=-K0(I,11)#(E(I,5)+E(I,6)#K0(I,8))
2350 IF F=0 THEN 2410
2360 C=K0(I,9) \ S=K0(I,10)
```

**Program Listing 9.1** (continued)

```
2370 FO(M(1))=FO(M(1))-F#C  \ E1(I,3)=-F#C
2380 FO(M(2))=FO(M(2))-F#S  \ E1(I,4)=-F#S
2390 FO(M(4))=FO(M(4))+F#C  \ E1(I,6)=F#C
2400 FO(M(5))=FO(M(5))+F#S  \ E1(I,7)=F#S
2410 NEXT
2420 REM
2430 REM           #### b. Subtract initial forces {f}={f}-{f0} ####
2440 REM
2450 FOR K=1 TO M \ F(K)=F(K)-FO(K) \ NEXT
2460 RETURN
2470 REM
2480 REM ####################################################################
2490 REM
2500 REM    4.  SOLVE FOR DISPLACEMENTS (u) BY GAUSSIAN ELIMINATION
2510 REM
2520 REM           #### a. Triangularize global matrix (k) ####
2530 REM
2540 FOR I=1 TO M
2550 IF K(I,I)=0 THEN K(I,I)=.000001      \REM  Avoid divide by zero
2560  FOR J=I+1 TO M
2570   Z=-K(J,I)/K(I,I)
2580    FOR K=I TO M
2590    K(J,K)=K(J,K)+Z#K(I,K)
2600    NEXT
2610    F(J)=F(J)+Z#F(I)
2620  NEXT
2630 NEXT
2640 REM
2650 REM           #### b. Calculate displacement vector {u} ####
2660 REM
2670 FOR I=M TO 1 STEP-1
2680 Z=0
2690 FOR J=I+1 TO M
2700  Z=Z+U(J)#K(I,J)
2710 NEXT
2720 U(I)=(F(I)-Z)/K(I,I)
2730 NEXT
2740 REM
2750 REM           c. Put displacements in nodal array N1 ####
2760 REM
2770 M1=1
2780 FOR J=1 TO J9 \ FOR K=1 TO 3
2790 IF N(J,K)<>M1 THEN 2820
```

**Program Listing 9.1** (continued)

```
2800 N1(J,K)=U(M1)
2810 M1=M1+1
2820 NEXT \ NEXT
2830 RETURN
2840 REM
2850 REM ###################################################################
2860 REM
2870 REM
2880 REM    5.   DETERMINE FORCES ON EACH ELEMENT
2890 REM
2900 REM
2910 REM          #### a. Determine nodes N1 and N2 ####
2920 REM
2930 FOR I=1 TO I9
2940 N1=E(I,1) \ N2=E(I,2)
2950 E1(I,1)=N1 \ E1(I,2)=N2
2960 REM
2970 REM          #### b. Establish local displacement vector {u1} ####
2980 REM
2990 FOR K=1 TO 3 \ M1=N(N1,K) \ U1(K)=U(M1) \ NEXT
3000 FOR K=1 TO 3 \ M1=N(N2,K) \ U1(K+3)=U(M1) \ NEXT
3010 REM
3020 REM          #### c. Establish local stiffness matrix (k1) ####
3030 REM
3040 GOSUB 4120
3050 REM
3060 REM          #### d. Calculate local forces {f1} due to {u1} ####
3070 REM          #### e. Add {f1} to initial forces in E1 ####
3080 REM
3090 FOR U=1 TO 6
3100 F1(U)=0
3110 FOR V=1 TO 6
3120 F1(U)=F1(U)+K1(U,V)#U1(V)
3130 NEXT
3140 E1(I,U+2)=E1(I,U+2)+F1(U)
3150 NEXT
3160 NEXT
3170 RETURN
3180 REM
3190 REM
3200 REM ###################################################################
3210 REM
```

**Program Listing 9.1** (continued)

```
3220 REM     6.  PRINT OUT THE RESULTS
3230 REM
3240 REM         **** a. Nodal displacements ****
3250 REM         **** b. Forces at both ends of each element ****
3260 REM
3270 Z3=0        \REM    Line counter for printing on screen
3280 REM
3290 !"          Nodal Displacements and Rotations"
3300 !"Node           X        Y        Alpha" \!
3310 FOR J=1 TO J9
3320 ERRSET 3340,Z8,Z8                    \REM  Free format if numbers too large
3330 !" ",J, TAB(11), %8F5,N1(J,1), TAB(22),N1(J,2),TAB(33),N1(J,3) \GOTO 3350
3340 !" ",J, TAB(11), %#,  N1(J,1), TAB(22),N1(J,2),TAB(33),N1(J,3)," (error)"
3350 Z3=Z3+1 \ IF Z3<15 THEN 3370
3360 Z3=0 \ INPUT"CR to cont",Z$
3370 NEXT
3380 ERRSET
3390 !\!\!
3400 Z3=0
3410 !"                        Forces        Moments"
3420 !"Element      Node     Fx      Fy      Mz"
3430 !
3440 FOR I=1 TO I9
3450 !%4I,I, %13I, E1(I,1), %10F1, E1(I,3), E1(I,4), E1(I,5)
3460 !%4I,I, %13I, E1(I,2), %10F1, E1(I,6), E1(I,7), E1(I,8)
3470 Z3=Z3+1 \ IF Z3<15 THEN 3490
3480 Z3=0 \ INPUT "CR to cont",Z$
3490 NEXT
3500 !\!\!\RETURN
3510 REM
3520 REM
3530 REM *******************************************************************
3540 REM
3550 REM     7.  FURTHER CALCULATIONS
3560 REM
3570 INPUT"Further calculations for this problem?  (CR/N) ",Z$ \!
3580 IF Z$="N" THEN STOP
3590 REM
3600 R1=R1+1                      \REM  Not First Run
3610 FOR U=1 TO M                 \REM  Zero Arrays
3620 F(U)=0 \ F0(U)=0 \ FOR V=1 TO M \ K(U,V)=0 \ NEXT
3630 NEXT
3640 FOR I=1 TO I9 \ FOR K=3 TO 8 \ E1(I,K)=0 \ NEXT \ NEXT
```

**Program Listing 9.1** (continued)

```
3645 FOR J=1 TO J9 \ FOR K=1 TO 3 \ N1(J,K)=0 \ NEXT \ NEXT
3650 REM
3660 REM          #### a. Other loadings for this structure? ####
3670 REM
3680 !\ INPUT "Run other loadings for this structure? (CR/N) ",Z$ \ !
3690 IF Z$="N" THEN 3980
3700 !
3710 !"              Applied Forces and Moments  "
3720 !"   Node         X         Y         M"
3730 !
3740 FOR J=1 TO J9
3750 FOR K=1 TO 3 \ IF N(J,K)<>0 THEN EXIT 3770 \ NEXT
3760 GOTO 3840
3770 !"    ",J,
3780 INPUT1 "              ",N(J,4) \ INPUT1 "         ", N(J,5)
3790 INPUT1"        ",N(J,6) \ INPUT"  OK?  (CR/N) ",Z$ \!
3800 IF Z$="N" THEN 3770
3810 FOR K=1 TO 3 \ M1=N(J,K)
3820 F(M1)=N(J,K+3)
3830 NEXT
3840 NEXT
3850 !\!
3860 !"  Element      Delta-Zero    T-Strain #10^6 "\!
3870 FOR I=1 TO I9
3880 !%7I,I, \ INPUT1"            ",E(I,5)
3890 INPUT1"          ",E(I,6) \ INPUT"   OK? (CR)",Z$
3900 IF Z$="N" THEN 3880
3910 E(I,6)=E(I,6)#10^-6
3920 NEXT
3930 !\!"            Wait for computation"\!\!
3940 GOTO 290
3950 REM
3960 REM
3970 REM          #### b. Alter Structural Data? ####
3980 REM
3990 INPUT"Alter Nodal Data?  (CR/N) ",Z$\!
4000 IF Z$="N" THEN 4040
4010 INPUT"  Alter Data for Node Number ",J \!\GOSUB 1020
4020 !\GOTO 3990
4030 REM
4040 INPUT"Alter Element Data?  (CR/N) ",Z$\!
4050 IF Z$="N" THEN 4080
4060 INPUT"  Alter Data for Element Number ",I \!\GOSUB 1210
```

**Program Listing 9.1** (continued)

```
4070 !\GOTO 4040
4080 !\!"                    Wait for computation"\!\!
4090 GOTO 210
4100 REM
4110 REM *******************************************************************
4120 REM
4130 REM *** LOCAL STIFFNESS MATRIX SUBROUTINE ***
4140 REM
4150 FOR U=1 TO 6 \ FOR V=1 TO 6 \ K1(U,V)=0 \ NEXT \ NEXT
4160 B=K0(I,1)\K1(1,1)=B\K1(1,4)=-B\K1(4,4)=B
4170 B=K0(I,2)\K1(2,2)=B\K1(2,5)=-B\K1(5,5)=B
4180 B=K0(I,3)\K1(1,2)=B\K1(1,5)=-B\K1(4,5)=B\K1(2,4)=-B
4190 B=K0(I,4)\K1(1,3)=B\K1(1,6)=B \K1(3,4)=-B\K1(4,6)=-B
4200 B=K0(I,5)\K1(2,3)=B\K1(2,6)=B \K1(3,5)=-B\K1(5,6)=-B
4210 B=K0(I,6)\K1(3,3)=B\K1(6,6)=B \K1(3,6)=B/2
4220 FOR U=1 TO 6 \ FOR V=1 TO U \ K1(U,V)=K1(V,U) \ NEXT\NEXT
4230 IF K0(I,7)=0 THEN RETURN
4240 IF K0(I,7)=3 THEN RETURN
4250 IF K0(I,7)=1 THEN 4270
4260 FOR V=1TO6\K1(V,6)=0\NEXT\RETURN
4270 FOR V=1 TO 6\K1(3,V)=0 \ K1(V,3)=0 \ NEXT \ RETURN
```

**Program Listing 9.1** (continued)

## 9.6 PROBLEMS

Referring to the local stiffness matrix $[k_1]$ given in Sec. 9.3, derive the stiffness coefficients indicated in Probs. 9.1 to 9.5.

**9.1** Derive $k_{21}, k_{22}, \ldots, k_{26}$.

**9.2** Derive $k_{31}, k_{32}, \ldots, k_{36}$.

**9.3** Derive $k_{41}, k_{42}, \ldots, k_{46}$.

**9.4** Derive $k_{51}, k_{52}, \ldots, k_{56}$.

**9.5** Derive $k_{61}, k_{62}, \ldots, k_{66}$.

**9.6–9.11** A steel bar (see the accompanying illustrations), 25 mm by 50 mm and 0.6 m long, is

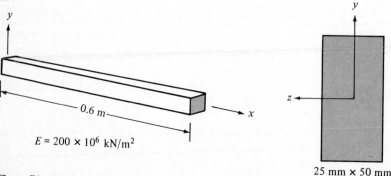

$E = 200 \times 10^6 \ kN/m^2$

25 mm × 50 mm

**Figure P9.6 to 9.11**

supported and loaded in various ways. Determine the FEM input coding and solve for all forces and deflections using the structural analysis program. Also, determine the maximum bending stress for each case.

**9.6** (*Hint*: Place an extra node at the first load. Check with Example 8.4.)

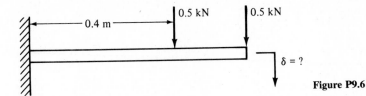

0.5 kN    0.5 kN

0.4 m

$\delta = ?$

**Figure P9.6**

**9.7**

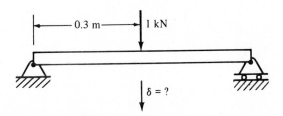

0.3 m    1 kN

$\delta = ?$

**Figure P9.7**

**9.8**

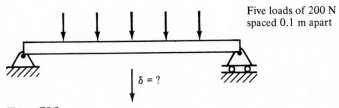

Five loads of 200 N
spaced 0.1 m apart

$\delta = ?$

**Figure P9.8**

**9.9**

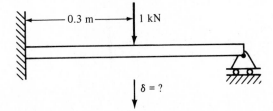

0.3 m    1 kN

$\delta = ?$

**Figure P9.9**

**9.10**

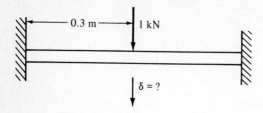

Figure P9.10

**9.11**

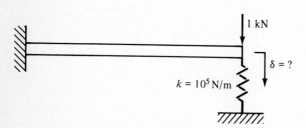

Figure P9.11

**9.12–9.18** A frame, as shown below, is made from three pieces of 1-in² steel rigidly welded together at the corners. We will support the frame in various ways and apply the load $F = 100$ lb, as shown. Determine the deflection $\delta$ for each case.

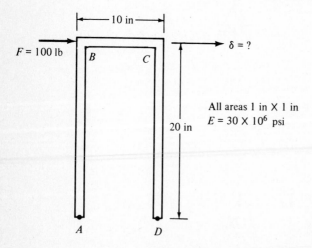

Figure P9.12 to 9.18

**9.12**

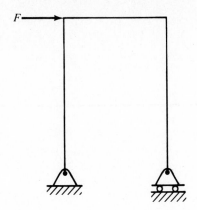

**Figure P9.12**

**9.13**

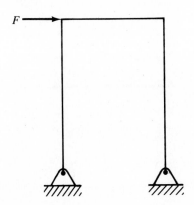

**Figure P9.13**

**9.14**

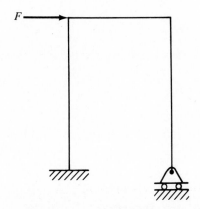

**Figure P9.14**

**9.15**

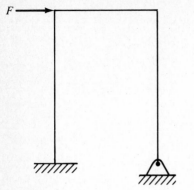

Figure P9.15

**9.16**

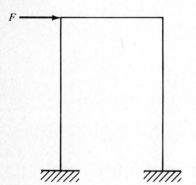

Figure P9.16

**9.17**

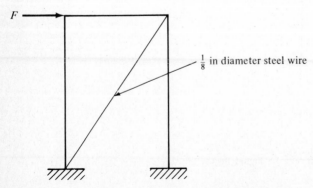

$\frac{1}{8}$ in diameter steel wire

Figure P9.17

**9.18**

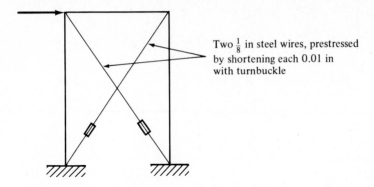

Two $\frac{1}{8}$ in steel wires, prestressed by shortening each 0.01 in with turnbuckle

**Figure P9.18**

**9.19** A support structure is made from 2-in-diameter steel pipe and pipe fittings and is embedded in a concrete floor, as shown in the accompanying illustration. How large, and where, is the maximum bending stress:

(a) When a 500-lb weight is hung from the center of the top member?

(b) When the right-hand vertical member is heated 200°F?

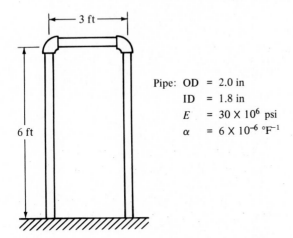

Pipe:  OD  = 2.0 in
          ID  = 1.8 in
          $E$  = 30 × 10⁶ psi
          $\alpha$  = 6 × 10⁻⁶ °F⁻¹

**Figure P9.19**

**9.20** A large steel frame is made from lengths of 6-in pipe welded together at the joints, as shown in the drawing (Fig. P9.20, p. 326). The supports at $A$ and $B$ are embedded in concrete foundations in the ground. Over the years, point $B$ moves relative to $A$. It moves 2 in to the right and 1 in downward. Where, and how large, are the maximum bending stresses?

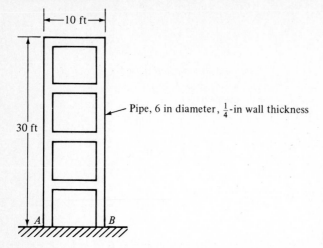

Pipe, 6 in diameter, $\frac{1}{4}$-in wall thickness

**Figure P9.20**

**9.21** Castings are being designed for part of a machine-tool structure. Two designers are arguing about which of the two designs shown in the accompanying illustration will deflect less under the load *F*. Would you please settle this dispute.

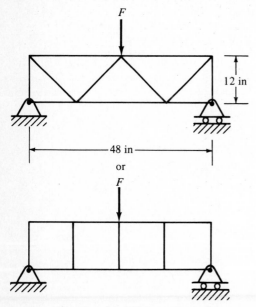

All sections 2-in × 2-in cast iron          **Figure P9.21**

**9.22** A wooden roof truss is fabricated from three long "two-by-sixes" (1-3, 3-5, and 1-5), and three short bracing members. All joints are fastened through the use of "gussets"; side plates nailed to the truss members as shown in the accompanying illustration. We wish to calculate the deflection of point 6 due to the three 5000-lb loads, but we do not know whether to consider the joints rigid (very good gussets) or pinned (rather poor gussets). Check to see how much difference it actually makes. (Let $E = 2 \times 10^6$ psi.)

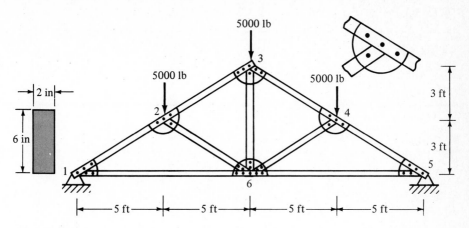

**Figure P9.22**

**9.23–9.24** Solve Probs. 9.23 and 9.24 using the FEM programs and making whatever assumptions are required.

**9.23** Solve Prob. 2.14.

**9.24** Solve Prob. 2.15.

**9.25** The "octagonal ring" shown in the accompanying illustration is 0.5 in wide perpendicular to the drawing and is made of steel. Determine the vertical deflection due to a force $F = 200$ lb. Draw the bending moment diagram for one of the 45° sections.

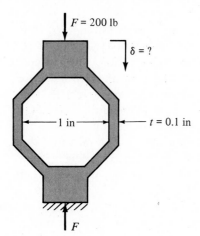

**Figure P9.25**

# TEN

## TORSION

### 10.1 OVERVIEW

We have previously considered long slender members under axial loading (struts in a truss), and under bending loads (beams). We now want to complete the possible loading on long members by considering the effects of "twisting moments," a subject generally called "torsion."

Torsion arises in many practical applications:

Power transmission, from one point to another, is often through a "shaft" (generally circular) which rotates while subject to twisting moments at each end. If the only load is the torque, it is called "pure torsion." This is typical of the shaft between a turbine and a generator, or the "torque tube" between transmission and differential on a rear-wheel-drive vehicle.

A power-transmission shaft is often subject to axial loads and/or bending as well as torque. This would be the case in a ship's propeller shaft, a "line shaft" having numerous gears or pulleys attached, or a screwdriver shaft. Here we have "combined stresses" due to the combined loading.

Static structural members are very often subject to torsion, usually in combination with axial and/or bending loads. Many curved members are subject to varying torsion and bending along their lengths: a sign post, a telephone pole, or an aircraft fuselage can all be subject to combined loading.

Springs, both coil springs and "torsion bar" suspension springs, are subject mainly to pure torsion.

Most of the torsion elements mentioned above are circular, either solid cylinders or hollow "tubes." Accordingly, in this chapter, we shall concentrate primarily on circular sections. Coincidentally, circular sections in torsion are far easier to analyze than any other sections. We will briefly consider rectangular and other sections, but in less detail. We will also consider the "residual stresses" that arise after a shaft has been twisted into the plastic regime.

The overall approach to torsion will be exactly the same as that we used for bending. First we will look at the deformation and deduce what the strains must be; then we will apply the stress-strain relations to determine the stresses (for elastic deformation); and finally, we will use equilibrium to relate the total twisting moment to the deformation and to the stresses.

## 10.2 TORSIONAL DEFORMATION OF CIRCULAR SECTIONS

Consider a solid circular cylinder made of isotropic material and subject to pure twisting as shown in Fig. 10.1. The entire body is symmetrical about its axis and also about any diameter. Let us focus our attention on an initially plane cross section and on an initially straight diameter in that plane.

In Fig. 10.1*b* we have sectioned the cylinder through the plane in question. During twisting, the surfaces *A* and *B* may have become "warped" in an axial direction, i.e., the plane may have "humped up" or "dished down." However, if we rotate the upper part 180° about the *xx* axis, surface *B* must appear exactly the same as surface *A* because the cylinder and its loading are symmetrical about *xx*.

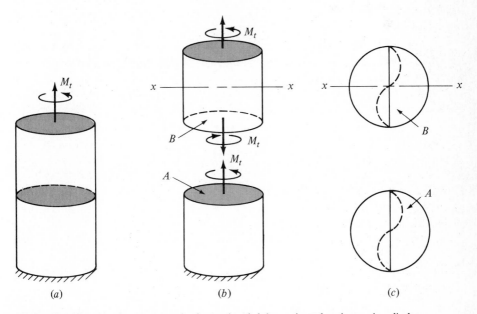

(*a*)                    (*b*)                   (*c*)

**Figure 10.1** Showing the symmetry in the torsional deformation of an isotropic cylinder.

The only way that surfaces $A$ and $B$ can fit together and also be identical is if they are flat. That is, plane sections must remain plane.

By similar reasoning, looking now at Fig. 10.1$c$, we find that any diametral line on the cross section must remain a straight, diametral line.

We conclude, then, that an initially plane cross section remains plane after twisting, and that straight lines in that plane must remain straight. We have not stipulated elastic behavior or plastic behavior; we have only used geometry and symmetry.

Because the cross sections remain plane and only rotate, we might consider torsional deformation to be similar to the rotational sliding that occurs between coins in a stack of silver dollars or chips in a stack of poker chips, as shown in Fig. 10.2. In this instance, the proper rotation is achieved without any change in either diameter or height of the stack. Although the analogy is poor, it suggests the *assumption* that there is actually no change in height or diameter when we twist a circular shaft.

Accordingly, we will assume that:

$$\text{Axial strain} = \epsilon_z = 0$$

$$\text{Radial strain} = \epsilon_r = 0$$

(10.1)

(An exact solution, according to the theory of elasticity, shows these assumptions to be correct.)

Figure 10.3 shows a portion of the deformed cylinder of height $L$ and radius $r$. The top of the cylinder has been twisted an angle $\phi$ relative to the bottom. The vertical line $AB$ and the radial line $OB$ (before twisting) have moved to $AB'$ and $OB'$ after twisting. The vertical line has become a helix at the angle $\gamma$.

If we look at the initially square element near point $A$, we see that it has become a parallelogram at the angle $\gamma$. Clearly, the element has *sheared* with a shear strain $\gamma$ (for small angles).

Now let us calculate the shear strain $\gamma$: Point $B$ has moved to $B'$, a distance $r\phi$; thus (for small angles):

$$\gamma \cong \frac{BB'}{L} = \frac{r\phi}{L}$$

(10.2)

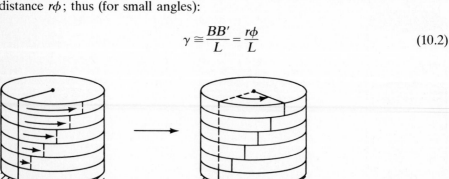

**Figure 10.2** Torsional deformation is likened to rotary sliding of a stack of silver dollars.

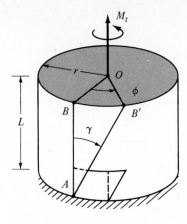

**Figure 10.3** Showing the geometry of torsional strain.

Equation (10.2) holds for interior parts of the cylinder as well. At smaller values of $r$, the distance $BB'$ is smaller, but is still equal to $\phi r$. So the strain $\gamma$ must vary *linearly* from a maximum at the outer diameter to zero at the axis. (Do you recognize the very close parallel to the deformation of a beam in bending?)

Figure 10.4 shows sections through the $\theta$ plane and the $z$ plane. We would expect the right angles shown to remain right angles such that there is zero shear strain in these planes.

Finally, then, in the torsion of a solid, isotropic cylinder, the only nonzero strain (in the $r\theta z$ coordinate system) is $\gamma_{z\theta}$ which is shown in Fig. 10.3 and given by Eq. (10.2).

We have not stipulated either elastic or plastic deformation; the strain relation holds for either. We can now even relax the isotropy requirement a bit: The material properties, if not uniform, must vary symmetrically about the axis and about any diameter. Clearly a hollow cylinder, or two concentric cylinders, meets these requirements.

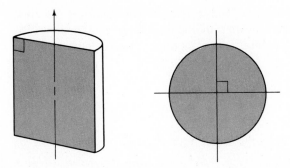

**Figure 10.4** Demonstrating the lack of shear strain on diametral and axial sections.

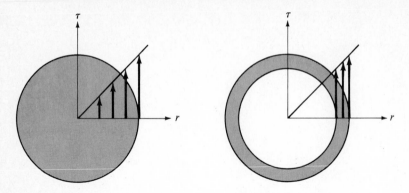

**Figure 10.5** In elastic torsion, the shear stress varies linearly with the radius.

## 10.3 TORSIONAL STRESSES FOR CIRCULAR SECTIONS (ELASTIC)

Knowing what the strains must be when we twist a cylinder of length $L$ through an angle $\phi$ (one end relative to the other), we can now calculate the appropriate stresses. At any point in the cylinder the only strain is $\gamma_{z\theta}$. For an isotropic material, within the elastic region, our stress-strain relations (Chap. 6) show that there can be only one stress component $\tau_{z\theta}$:

$$\tau_{z\theta} = \gamma_{z\theta}G = \frac{r\phi G}{L} \tag{10.3}$$

Because the strain varies linearly with the radius within a cylinder, the stress must do so also (if $G$ is constant). Figure 10.5 shows that the linear behavior holds for hollow as well as solid cylinders. What could be simpler?

## 10.4 EQUILIBRIUM REQUIREMENTS FOR CIRCULAR SECTIONS

If we section the cylinder, we know that the stresses, distributed over the area, must be equivalent to the applied twisting moment $M_t$. Figure 10.6 shows the total cross section with a few elementary areas identified. Each element of area $dA$, when multiplied by the appropriate shear stress $\tau$, carries a differential force $dF$:

$$dF = \tau \, dA = \frac{r\phi G}{L} \, dA$$

Each $dF$, when multiplied by the appropriate moment arm $r$, will provide an

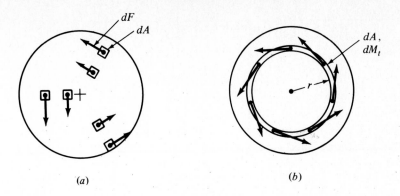

**Figure 10.6** Showing the distribution of forces due to twisting of a solid cylinder.

increment of twisting moment $dM_t$. If we sum these over the total area, we obtain the total moment:

$$dM_t = r \, dF = \frac{r^2 \phi G}{L} \, dA$$

$$M_t = \frac{\phi G}{L} \int_A r^2 \, dA \tag{10.4}$$

The integral in Eq. (10.4) is a property of the area. Because the same integral appears when computing the mass moment of inertia of a cylinder about its axis, the integral is called the "polar moment of inertia," and labeled $I_p$ in a manner similar to the area moments of inertia for beam cross sections:

$$\int_A r^2 \, dA = I_p = \text{area polar moment of inertia} \tag{10.5}$$

(In accordance with our subscript convention, $I_p = I_{rr}$.)

If we combine Eqs. (10.4) and (10.5), we can solve for the angle of twist $\phi$:

$$\boxed{\phi = \frac{M_t L}{G I_p}} \tag{10.6}$$

Combining Eqs. (10.3) and (10.6), we get the value of the shear stress at any point on the cross section:

$$\boxed{\tau = \frac{M_t r}{I_p}} \tag{10.7}$$

In order to use the above relations, we first must compute $I_p$, the polar moment of inertia for the area. This is most easily done by letting $dA$ be an annular area of radius $r$ and width $dr$, such that:

$$\int_A r^2 \, dA = \int_0^r r^2 \, (2\pi r \, dr) = 2\pi \int_0^r r^3 \, dr$$

$$I_p = \frac{\pi r^4}{2} \tag{10.8}$$

(Because $r^2 = y^2 + z^2$, $I_{rr} = I_{yy} + I_{zz}$.)

Equation (10.8) refers to a solid cylinder of radius $r$; if we were to compute $I_p$ for a hollow cylinder of outer radius $r_o$ and inner radius $r_i$, we would find that we simply must subtract the $I_p$ for the hole from that for the total area; that is,

$$I_p = \frac{\pi}{2} (r_o^4 - r_i^4) \tag{10.9}$$

**Example 10.1** A steel shaft, 3 cm in diameter and 1 m long, is subject to a pure twisting moment of 1 kilonewton-meter (kN·m). The steel has a yield stress in tension ($\sigma_{yp}$) of 80,000 psi, and a shear modulus $G$ of $12 \times 10^6$ psi. (a) How large is the angle of twist $\phi$? (b) How large is the maximum shear stress? (c) How large is the safety factor (SF) (relative to yielding)?

First, change all quantities to a common set of units:

$$r = 0.015 \text{ m} \qquad = 0.59 \text{ in}$$

$$L = 1 \text{ m} \qquad = 39.37 \text{ in}$$

$$M_t = 1000 \text{ N·m} \qquad = 8851 \text{ in·lb}$$

$$\sigma_{yp} = 5.52 \times 10^8 \text{ Pa} = 80,000 \text{ psi}$$

$$G = 82.74 \times 10^9 \text{ Pa} = 12 \times 10^6 \text{ psi}$$

The polar moment of inertia $I_p$ [Eq. (10.8)] is:

$$I_p = \frac{\pi r^4}{2} = 7.95 \times 10^{-8} \text{ m}^4 = 0.190 \text{ in}^4$$

The angle of twist [Eq. (10.6)] is:

$$\phi = \frac{M_t L}{G I_p} = 0.152 \text{ rad} = 8.71°$$

The maximum shear stress [Eq. (10.7)] occurs at the outer radius and is:

$$\tau = \frac{M_t r}{I_p} = 1.89 \times 10^8 \text{ Pa} = 27,485 \text{ psi}$$

To calculate the safety factor relative to yield we must (a) select a yield criterion and (b) compare the maximum applied stresses (or loads) with the maximum permitted before yield.

The Mohr's stress circle for pure shear is shown in the accompanying illustration. Because the maximum shear stress is the applied stress calculated above, we might as well use the maximum shear-stress yield criterion.

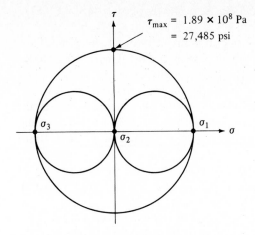

$$\tau_{max} = 1.89 \times 10^8 \text{ Pa}$$
$$= 27,485 \text{ psi}$$

The maximum allowable shear stress prior to yield is simply half of the tensile yield stress. Therefore,

$$\text{SF} = \frac{2.76}{1.89} = 1.46 \qquad \frac{40,000}{27,485} = 1.46 \qquad\qquad ***$$

## 10.5 YIELD AND POSTYIELD BEHAVIOR

As we saw in Example 10.1, yielding in pure torsion occurs when the shear stress at the outer radius reaches a critical value. However, because the shear strain varies linearly from the center outward, the entire central core is still elastic when the outer layer yields. Figure 10.7 shows the distribution of shear stress with radius for various amounts of twist of a circular cylinder. The

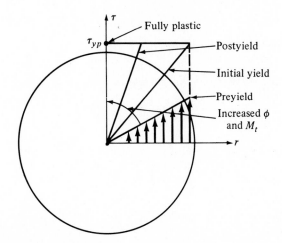

**Figure 10.7** Showing the linear stress distribution prior to yield and the saturated postyield behavior (for an ideal elastic-plastic material).

particular curves shown are for an "ideal elastic perfectly plastic material" (Fig. 6.1).

Prior to yield, because the strain variation is linear and the shear modulus $G$ is constant, the shear stress varies linearly with radius. This remains true until initial yield occurs at the outer radius of the cylinder.

As we continue to twist the shaft, the yield shear stress (and yield shear strain) is reached at smaller and smaller radii. Thus the elastic core gets smaller and smaller as the plastic annulus becomes larger and larger. Eventually, for all practical purposes, the entire section becomes plastic, a condition referred to as "fully plastic."

This postyield behavior is quite different from that of a tensile specimen where the entire section yields at the same time. We will see later that the behavior in bending is very similar to that in torsion.

It is very easy to calculate the "fully plastic twisting moment" $(M_{t_{fp}})$, because the shear stress is constant and equal to the yield shear stress at all points. Referring back to Fig. 10.6b, we can write:

$$dM_{t_{fp}} = \tau_{yp} r \, dA = \tau_{yp} r (2\pi r \, dr)$$

$$M_{t_{fp}} = \tau_{yp} \frac{2\pi r^3}{3} \tag{10.10}$$

We can easily calculate the twisting moment required to initiate yield $(M_{t_{yp}})$ from Eqs. (10.7) and (10.8):

$$M_{t_{yp}} = \tau_{yp} \frac{\pi r^3}{2} \tag{10.11}$$

Thus, the fully plastic twisting moment for a solid cylinder is just four-thirds of that required for yield. The variation of twisting moment with angle of twist is shown in Fig. 10.8. The moment approaches the maximum value

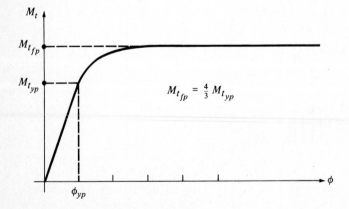

**Figure 10.8** Showing the postyield behavior of a solid cylinder in torsion.

rapidly, reaching 99 percent of $M_{t_{yp}}$ when the angle is only three times the yield angle.

It should be clear that the increase in moment as we twist beyond yield is due to the large elastic core. If we twist a thin-walled cylinder, the elastic core is very small, and there is little increase in moment beyond that required for initial yield.

## 10.5.1 Residual Stresses

Frequently (perhaps always), when we purchase a piece of metal, it is not "stress-free"; there are "residual stresses" locked up in the material. It is not possible to determine by simple inspection whether the residual stresses are small or large. When we load the material, the residual stresses are superposed on those due to loading, and the resulting conditions can be worse (or better) than we would have thought.

Sometimes it is possible to introduce residual stresses purposely to increase the effective strength of the part. We saw earlier that fatigue strength could be very much affected by application of compressive (good) or tensile (bad) surface residual stresses.

It is also possible to increase (or decrease) the apparent yield strength of a part through residual stress control. This is effective in torsion, bending, and for pressure vessels. The study of residual stresses can become quite complex, and perhaps should be considered beyond the scope of an elementary text. However the effects can be quite dramatic so we will discuss them briefly. Consider here the effect in torsion.

Suppose we twisted a solid cylinder until it was fully plastic and then simply "let go." What would be the state of stress in the shaft?

Whenever we "let go" of a loaded member (made of one material), it undergoes whatever *elastic* deformations are necessary to be consistent with zero net applied load. Even though the strains during loading were plastic, any recovery, upon release of the load, is elastic. Because the recovery strains and stresses are elastic, it is permissible to superpose the elastic recovery stresses on the stresses that were due to the original loading.

Figure 10.9 shows how two load and stress distributions are superposed to obtain the third. In Fig. 10.9a, a shaft has been loaded to the fully plastic state due to a positive twisting moment ($+M_{fp}$); the associated stress distribution, where the entire section has a shear stress $\tau_{yp}$ acting, is also shown.

In Fig. 10.9b, we have "let go" by adding (superposing) a negative twisting moment ($-M_{fp}$) which just cancels the initial load. Because the fully plastic moment is four-thirds of the yield moment, the maximum shear stress that *would* be associated with $-M_{fp}$ (if the deformation was elastic) is just four-thirds the yield shear stress, as shown. When the two distributions are superposed, we obtain the final *residual* stress distribution, as seen in Fig. 10.9c.

You must realize that the stress never reaches four-thirds of $\tau_{yp}$. That is obviously not possible. At any instant, the actual stress is the superposition of

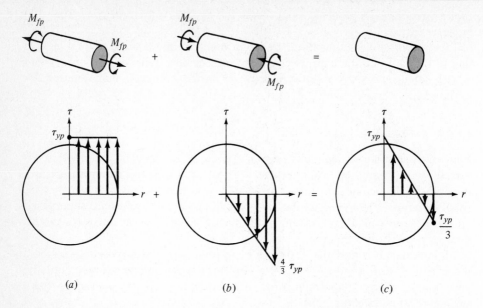

**Figure 10.9** Torsional residual stresses due to *elastic* recovery from the fully plastic condition.

the initial distribution and the recovery stresses. Thus, as we release the load, the stress at the axis does not change; the stress at the outer radius goes from $+\tau_{yp}$ to $-\tau_{yp}/3$; and the stress at three-quarters of the outer radius goes from $\tau_{yp}$ to zero.

Now that we have twisted the shaft to the fully plastic condition and "let go," we have a very interesting piece of metal. It looks the same as when we started—but it is not. If now we apply a positive twisting moment, it will remain elastic and not yield until we have applied a full four-thirds of the initial yield moment; but when it does yield, the entire section will yield at the same time. On the other hand, if we were to apply a negative twisting moment, we would require only two-thirds of the initial yield moment because we already have a stress equal to one-third of the yield stress at the outer radius.

The net result is that we can enhance the elastic behavior of the shaft by properly prestressing it. Clearly if a pretwisted part were twisted in the wrong direction, the result could be most embarrassing!

## 10.6 COMBINED STRESSES

The discussions above relate primarily to the case of pure torsion—i.e., to the case where the only load is a twisting moment. Frequently we also have axial loads, bending loads, internal pressure, or some combination thereof. How can we handle such situations?

So long as the overall deformation of the part is *elastic*, we know that we can superpose the effects of various loads. So we assume that the system is elastic and analyze it accordingly. The process is most straightforward:

1. At any point on the body determine the stresses due to the individual loads relative to an *xyz* coordinate system.
2. Draw an *xyz* stress element and show *all* of the stresses superposed on the single element.
3. Draw the Mohr's stress circle diagram for the element and determine the principal stresses, maximum shear stress, etc.

This gives the total state of stress at that point due to the combined loading or combined stresses. If there is doubt concerning *which* point is the *worst* point on the body, you may have to analyze more than one point. It is worthwhile to give considerable thought to deduce which point is the worst; this can save a lot of time.

**Example 10.2** Consider the propeller shaft on a pleasure boat. The shaft must deliver power to the propeller (torsion), and must transmit the resulting thrust to the boat (axial compression). The shaft is 1.5-in-diameter monel-metal having a tensile yield stress of 75,000 psi. Determine the safety factor relative to yielding for the following conditions:

Maximum power—255 horsepower (hp) at a shaft speed of 2500 rpm
Maximum thrust—4000 lb
Bending—assume a zero bending moment

First we must convert the data to a consistent set of units:

$$255 \text{ hp} = 255(6600) = 1.683 \times 10^6 \text{ in} \cdot \text{lb/s}$$

$$2500 \text{ rpm} = \frac{2500(2\pi)}{60} = 261.8 \text{ rad/s}$$

Power = torque × rotational speed

$$M_t = \frac{1.683 \times 10^6}{261.8} = 6429 \text{ in} \cdot \text{lb}$$

The properties of the area are:

$$r = 0.75 \text{ in} \qquad A = \pi r^2 = 1.77 \text{ in}^2 \qquad I_p = \frac{\pi r^4}{2} = 0.497 \text{ in}^4$$

The stresses are:

$$\sigma_x = \frac{F}{A} = \frac{4000}{1.77} = 2260 \text{ psi} \qquad \text{(compressive)}$$

$$\tau = \frac{M_t r}{I_p} = \frac{6429(0.75)}{0.497} = 9702 \text{ psi}$$

All other stresses are zero.

The *xyz* stress element and the usual Mohr's stress circle are shown in the accompanying drawing. Using the maximum shear-stress yield criterion, we find the maximum applied shear stress to be 9773 psi (clearly, the axial force has little contribution to the maximum shear stress in this instance).

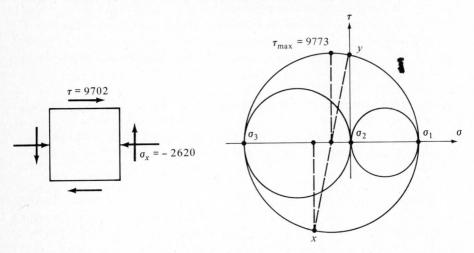

The maximum shear stress at yield ($\tau_{yp}$) is half the tensile yield stress, $75,000/2 = 37,500$. Thus the safety factor is:

$$SF = \frac{37,500}{9773} = 3.84$$

(What would happen if the propeller lost one of its blades?)           ***

## 10.7 TORSIONAL BUCKLING OF HOLLOW SHAFTS

Because the torsional stress distribution varies linearly from the center outward, the core of a solid shaft is not stressed highly; its contribution to the support of the twisting moment is small. Therefore, it is good practice to use hollow shafts whenever we are interested in saving weight or material (very often the case). The design of hollow shafts is perfectly straightforward—we simply use the polar moment of inertia for a hollow shaft instead of a solid one.

If we try to optimize the design of a hollow shaft, we find that we should have a very thin tube of very large diameter. The question arises: how thin and how large? If you make a very thin tube by rolling up a sheet of paper, and then you twist it, you will discover the problem—it buckles!

If we look at the Mohr's stress circle for pure torsion (see Example 10.1), we find that although we have applied only shear stress, we have tensile and compressive stresses at 45° to the axis. The compressive stress is the villain that causes the thin tubes to buckle. (For a thin paper tube can you predict the orientation of the wrinkles that form due to buckling?)

The analysis of buckling (except for the simple columns we considered in Chap. 8) is very complex and well beyond the scope of this text. There is an excellent set of papers which discuss buckling for a variety of shapes and constructions that are of interest to the aircraft industry. They appear in the NACA Technical Notes: G. Gerard and H. Becker, *Handbook of Structural Stability*, NACA Tech. Notes 3781–3786, 1957–58.

In Tech. Note 3783 (August 1957), the authors discuss torsional buckling of thin-walled cylinders and arrive at a "critical shear stress" ($\tau_{cr}$) at which collapse will occur. To avoid incipient buckling, the shear stress should be kept perhaps 20 percent below $\tau_{cr}$.

For very short tubes ($L^2/rt < 12$):

$$\tau_{cr} = \frac{\pi^2 K_t E}{12(1 - \nu^2)} \left(\frac{t}{L}\right)^2 \tag{10.12}$$

where $K_t = 8.95$ for clamped ends (zero axial motion)
$\phantom{where K_t} = 5.35$ for free ends

For intermediate-length tubes, use Eq. (10.12) where:

$$K_t = 0.85\left(\frac{L^2}{rt}\right)^{3/4}(1 - \nu^2)^{3/8}$$

For long tubes [$L/r > 3(r/t)^{1/2}$]:

$$\tau_{cr} = 0.272(1 - \nu^2)^{-3/4}E\left(\frac{t}{r}\right)^{3/2} \tag{10.13}$$

## 10.8 RECTANGULAR CROSS SECTIONS

Analysis of circular sections in torsion was quite simple because of the symmetry: Plane sections remained plane and straight lines in those sections remained straight. For rectangular sections, such is not the case.

If you twist a rubber eraser, or some other compliant rectangular shape, you will see that the sections warp, and that there is no obvious way to describe the deformation. Fortunately, there is a relatively simple solution that we will simply present to you, a solution that assumes that the ends of the rectangular shaft are free to warp. If the ends were restrained against warping (axial motion), the stiffness would be greater and the stresses less.

Figure 10.10 shows a rectangular sectioned element subject to pure torsion. The section is $2a \times 2b$ where $a \geq b$.

The maximum shear stress occurs at the midpoint of the longer side (at a distance $b$ from the axis) and is given by:

$$\tau_{max} = 0.375 \frac{M_t}{ab^2} \left(1 + \frac{0.6b}{a}\right) \tag{10.14}$$

The elastic rotation is most easily written in terms of the rotational spring

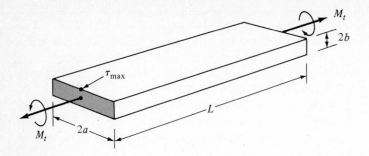

**Figure 10.10** A rectangular cross-sectioned element subject to pure torsion.

constant $K_\phi$:

$$K_\phi = 5.33 \frac{ab^3 G}{L} \left[ 1 - 0.63 \frac{b}{a} \left( 1 - \frac{b^4}{12a^4} \right) \right] \qquad (10.15)$$

where $\qquad K_\phi = \dfrac{M_t}{\phi} \qquad$ in·lb/rad

A bit of thought should convince you that the corners of a rectangular shaft have little effect in twisting. At any corner there can be no shear stress on the outer surfaces. Because $\tau_{xy} = \tau_{yx}$, etc., there can be no shear stress on the cross section at the corner. It is instructive to compare the stresses and stiffnesses for round and square shafts using the equations we have presented:

For a circular solid shaft:

$$\tau_{max} = 0.6366 \frac{M_t}{r^3} \qquad \text{[Eqs. (10.7) and (10.8)]}$$

$$K_\phi = 1.571 \frac{r^4 G}{L} \qquad \text{[Eq. (10.6)]}$$

For a square shaft $(2r \times 2r)$:

$$\tau_{max} = 0.60 \frac{M_t}{r^3} \qquad \text{[Eq. (10.14)]}$$

$$K_\phi = 2.25 \frac{r^4 G}{L} \qquad \text{[Eq. (10.15)]}$$

We see that the stresses are almost the same for the two cases, but that the square shaft is about 40 percent stiffer.

Alternatively, we could compare square and round shafts having equal areas. We would find that the circular shaft was 13 percent stiffer and experienced 26 percent lower stress.

For flat sheets $(a \gg b)$ in torsion, the stresses and stiffness are not greatly affected if the sheets are bent into shapes such as angles, channels, etc. Figure 10.11 shows how a sheet can be formed into various *open* shapes; it also shows

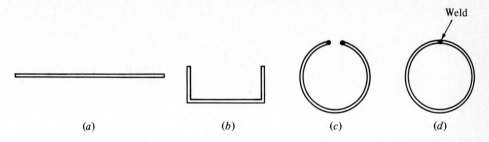

**Figure 10.11** Showing how a flat sheet (*a*) can be formed into various open shapes (*b*) and (*c*) or welded to form a closed shape (*d*).

how it can be welded to form a *closed* shape. Let us compare the properties of the flat sheet (which will be about the same as for the open sections) with those for the closed section. Assume that the sheet is 10 in wide and 0.1 in thick.

For the flat sheet:

$$\tau_{\max} = 30.18 M_t$$

$$K_\phi = 0.0033 \frac{G}{L}$$

For the hollow cylinder made from the same sheet:

$$\tau_{\max} = 0.629 M_t$$

$$K_\phi = 2.53 \frac{G}{L}$$

We see that for this specific case, welding the open form (as in Fig. 10.11*c*) to produce a *closed* form (Fig. 10.11*d*) gives a tremendous improvement in properties:

The stress is reduced by a *factor of* 50.
The stiffness is increased by a *factor of* 700!

The message should be *very* clear: For structures subject to torsion, *always* try to use a closed section.

## 10.9 CLOSED, THIN-WALLED SECTIONS

Because of the excellent torsional properties of closed, tubular sections, we would like to be able to analyze noncircular closed sections as well as circular. There is a relatively simple method that is valid for thin-walled tubes where we can reasonably assume that the shear stresses are approximately constant across the thickness of the tube wall.

Figure 10.12 shows an arbitrary closed, thin-walled tube subject to a

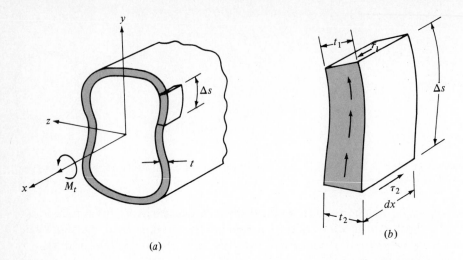

**Figure 10.12** Torsion of a thin-walled closed tube: (*a*) the general geometry; (*b*) an isolated element.

twisting moment $M_t$ along the $x$ axis. The shape of the cross section is arbitrary, and the thickness $t$ can vary with position $s$ around the periphery. Because the wall is thin (compared to the overall section dimensions), we will assume that the shear stress $\tau$ is constant across the thickness of the tube—it can vary with position $s$, but not across the thickness $t$.

Also shown in Fig. 10.12 is a small element of the tube having dimensions $dx$ in the axial direction, and a small, finite length $\Delta s$ along the periphery. At the upper end, the thickness is $t_1$ and the shear stress is $\tau_1$. At the lower end, the thickness can be a different value $t_2$ and the stress $\tau_2$. We assume that the tube is not restrained from warping so that the $x$ faces will not have any tensile or compressive stresses, but will have only shear stresses.

For the element to be in equilibrium, the sum of the forces in the $x$ direction must be zero. The only forces in the $x$ direction are due to the shear stresses $\tau_1$ and $\tau_2$; hence:

For $\Sigma F_x = 0$:
$$t_1\tau_1 \, dx = t_2\tau_2 \, dx$$

$$t_1\tau_1 = t_2\tau_2 = \text{const} = q$$

$$t\tau = q = \text{shear flow} \tag{10.16}$$

Thus, for equilibrium, the product $t\tau$ must be constant at all points on the cross section. The constant $q$ is called "shear flow," emanating from an analogy between torsion of a tubular cross section and fluid flow in a channel of the same shape.

*Note:* Although we will not explore them here, there are various analogies that have been most useful in understanding torsion. In addition to the flow analogy, a "soap film" analogy has been used to determine the torsional properties of various shapes.

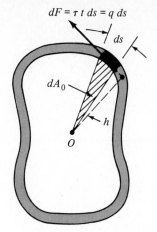

$dF = \tau t\, ds = q\, ds$

$ds$

$dA_0$

$h$

$O$

**Figure 10.13** Showing how the shear force contributes to the twisting moment.

Figure 10.13 shows the shear force $dF$ acting on a small element of sectional area of length $ds$. The moment of that force about any arbitrary point $O$ is simply $dF\, h$, where $h$ is the perpendicular distance from $dF$ to $O$. The sum of all such moments must equal the applied twisting moment $M_t$:

$$M_t = \int_s h\, dF = \int_s h(\tau t\, ds) = \tau t \int_s h\, ds$$

But the shaded area in Fig. 10.13 $(dA_0)$ is just equal to half of the tube periphery $ds$ times the moment arm $h$. Thus $h\, ds = 2\, dA_0$, where $A_0$ is the *total* area enclosed by the tube (or by the centerline of the tube). Do not confuse this with the cross-sectional area $A$ of the tube itself. Thus:

$$M_t = \tau t \int_{A_0} 2\, dA_0 = 2\tau t A_0 = 2qA_0 \qquad (10.17)$$

Solving for the shear stress, we have:

$$\tau = \frac{M_t}{2tA_0} \qquad (10.18)$$

To determine the angle of twist for a closed, thin-walled section requires methods we are not using (energy methods or analogies). We will simply present the result:

$$\phi = \frac{M_t L}{4A_0^2 G} \int_s \frac{ds}{t} \qquad (10.19)$$

where $A_0$, again, is the enclosed area, and the integral is over the entire periphery.

**Example 10.3** A 3-in by 4-in box section is made of aluminum with side walls 0.1 in thick and ends 0.2 in thick (see the accompanying drawing).

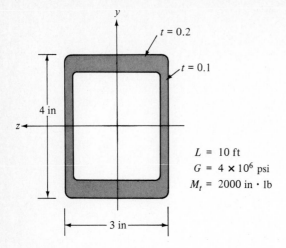

$$L = 10 \text{ ft}$$
$$G = 4 \times 10^6 \text{ psi}$$
$$M_t = 2000 \text{ in} \cdot \text{lb}$$

Estimate the maximum shear stress and the angle of twist when a twisting moment of 2000 in·lb is applied.

To be a bit conservative, use the area inside the tube for $A_0$. This is $A_0 = 2.8(3.6) = 10.08 \text{ in}^2$. The maximum stress will be where $t$ is smallest, i.e., where $t = 0.1$; thus:

$$\tau_{max} = \frac{M_t}{2A_0 t} = \frac{2000}{2(10.08)(0.1)} = 1000 \text{ psi}$$

To compute the angle of twist, we use Eq. (10.19). The integral is easily evaluated by considering 8 in of 0.1-in-thick wall and 6 in of 0.2-in-thick wall ($8/0.1 + 6/0.2 = 80 + 30 = 110$):

$$\phi = \frac{2000(120)(110)}{4(10.08)^2(4 \times 10^6)}$$

$$= 0.0162 \text{ rad} = 0.93°$$

It is instructive to compare this rectangular tube with an "equivalent" circular tube. If we select a tube having the same amount of metal, and the same periphery (circumference), we find:

$$t = 0.143 \text{ in} \qquad \text{(thickness)}$$
$$r = 2.23 \text{ in} \qquad \text{(radius)}$$
$$I_p = 9.9 \text{ in}^4$$
$$\tau_{max} = 450 \text{ psi}$$
$$\phi = 0.006 \text{ rad} = 0.34°$$

Circular tubes are "mighty fine" torsional members!　　　　　　　　***

## 10.10  COMPUTER EXAMPLE

Equations (10.12) and (10.13) for buckling of thin-walled tubes in torsion are readily programmed for computer solution. They are so straightforward that we will simply present an appropriate program (Program Listing 10.1) and an example (Program Output 10.1).

```
10 !"              TORSIONAL BUCKLING"\!
20 !"      Cylindrical Tube, dimensions of lbs and in or N and m"
30 !"              Poisson's rato = .3"\!
40 REM
50 REM  ASSUME POISSON'S RATIO = .3
60 REM
70 INPUT"       Tube length       L = ",L\!
80 INPUT"       Tube radius       R = ",R\!
90 INPUT"       Tube thickness    T = ",T\!
100 INPUT"      Modulus           E = ",E\!
110 REM
120 IF L/R>3*(R/T)^.5 THEN 170
130 K = .85*.97*(L*L/(R*T))^.75
140 IF K<5.35 THEN K=5.35
150 T1=(3.1416)^2*K*E*T*T/(10.92*L*L)
160 GOTO 190
170 T1 = .272*1.07*E*(T/R)^1.5
180 T1=.8*T1
190 I = 6.2832*R*R*R*T
200 M = T1*I/R
210 !"          Tau-crit   (psi, Pa)  ",T1\!
220 !"          Mt-max  (in-lbs, N-m) ",M\!
```

**Program Listing 10.1**

```
          TORSIONAL BUCKLING

Cylindrical Tube, dimensions of lbs and in or N and m
          Poisson's rato = .3

Tube length       L = 100

Tube radius       R = 2

Tube thickness    T = .05

Modulus           E = 30000000

   Tau-crit   (psi, Pa)   27610.477

   Mt-max  (in-lbs, N-m)  34696.43
```

**Program Output 10.1**

## 10.11 PROBLEMS

**10.1** A circular steel shaft 1 in in diameter and 60 in long is subject to a twisting moment $M_t = 5000$ in·lb. Find the maximum shear stress and the angular twist.

**10.2** A 50-mm-diameter steel shaft, made of material having a tensile yield stress $\sigma_{yp} = 400$ MPa, is twisted to the point of yield. Length $= 2.5$ m. How large are the twisting moment and angle of twist?

**10.3** A steel "torsion bar" must support a moment of 10,000 in · lb, and must twist 45° under that load. If the material used has a tensile yield of 90,000 psi, determine the diameter and length of the bar.

**10.4** A hollow steel tube is 50 mm OD, 35 mm ID, and 0.5 m long. The shear modulus is 82 GPa, and the tensile yield stress is 500 MPa. To have a safety factor of 2, how large can the twisting moment be? How large is the angle of twist?

**10.5** A composite circular shaft is made from two cylinders bonded together, as shown in the accompanying illustration. Material 1 is the solid core, while material 2 is the hollow cylinder bonded to the core. A twisting moment $M_t$ twists the assembly. Determine the maximum twisting moment prior to yield for the following conditions:

|                              | Core          | Outer shell   |
| ---------------------------- | ------------- | ------------- |
| Radius, in                   | 1.0           | 1.2           |
| Shear modulus $G$, psi       | $12 \times 10^6$ | $3 \times 10^6$ |
| Tensile yield $\sigma_{yp}$, psi | 60,000    | 30,000        |

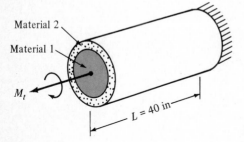

Material 2

Material 1

$M_t$

$L = 40$ in

**Figure P10.5**

**10.6** Gears, being rather complex, must be tested to determine their load-life characteristics. If the life is long (months or years), and the power transmitted high, the process of gear testing can be very difficult and costly. Fortunately, it is possible to build a gear-testing machine that is simple and energy-conserving. In the schematic view shown in the accompanying illustration, four identical gears are mounted on two identical shafts in bearings. A motor at $A$ drives the assembly. In order to "load" the gears, the gear at $A$ is slipped axially out of engagement with $B$ (the gear can slide axially on a "spline" joint); $A$ is then rotated relative to $B$ by some integer number of gear teeth, and they are re-engaged. This process twists the shafts, and the twisting moments must be carried by the gears. Now if the gears are rotated, power will be transmitted around the loop.

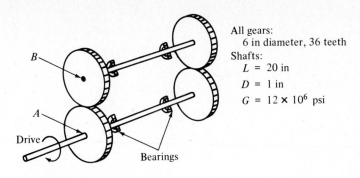

All gears:
  6 in diameter, 36 teeth
Shafts:
  $L = 20$ in
  $D = 1$ in
  $G = 12 \times 10^6$ psi

**Figure P10.6**

In each of the following cases, determine the power (in horsepower and kilowatts) transmitted by the gears:

(*a*) $N = 200$ rpm, $T$ (number of teeth indexed) $= 1$
(*b*) $N = 200$ rpm, $T = 2$
(*c*) $N = 1000$ rpm, $T = 1$
(*d*) $N = 2000$ rpm, $T = 2$

**10.7** Design the shafts for a gear-testing machine like the one described in Prob. 10.6. The machine is to test 3-in-diameter gears having 30 teeth, at a speed of 1800 rpm and power transmission of 0.5 kW. Use 10-in-long steel shafts if possible.

**10.8** A 16-in-long, 1-in-diameter steel rod is bent 90° at $B$, and welded to the wall at $A$, as shown in the accompanying illustration. The rod must support a vertical load of 300 lb at $A$. Where will yielding first occur? How large is the safety factor? What is the deflection of point $C$?

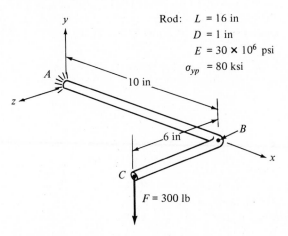

Rod:  $L = 16$ in
      $D = 1$ in
      $E = 30 \times 10^6$ psi
      $\sigma_{yp} = 80$ ksi

10 in

6 in

$B$

$F = 300$ lb

**Figure P10.8**

**10.9** A coil spring is loaded along its axis as shown in the accompanying drawing. Determine the spring constant $k$ in terms of the variables shown. (*Hint*: Consider the contribution to the total motion of $A$ relative to $B$ due to the twist in the small element of length $R\, d\theta$ shown.)

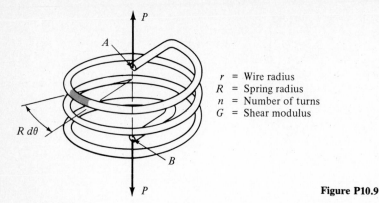

$r$ = Wire radius
$R$ = Spring radius
$n$ = Number of turns
$G$ = Shear modulus

**Figure P10.9**

**10.10–10.12** Piping systems in process plants are usually subject to bending and twisting in addition to pressure loading. Consider pipes made from material with the following properties:

$$\sigma_{yp} = 60,000 \text{ psi} \qquad E = 30 \times 10^6 \text{ psi}$$

$$\sigma_{ult} = 80,000 \text{ psi} \qquad G = 12 \times 10^6 \text{ psi}$$

$$\sigma_{end} = 30,000 \text{ psi} \qquad \alpha = 6 \times 10^{-6} \, ^\circ\text{F}^{-1}$$

**10.10** If the OD = 1.0 in, the ID = 0.9 in, pressure $p$ = 3000 psi, and there is an applied *twisting* moment of 1000 in·lb, determine the safety factor relative to yield.

**10.11** Solve Prob. 10.10 using all the same data except that the applied moment is a *bending* moment of 1000 in·lb.

**10.12** Solve Prob. 10.10 when *both* a 1000-in·lb bending moment and a 1000-in·lb twisting moment are present.

**10.13** We require a very-light-weight drive shaft to transmit 2000 hp at 1200 rpm. We plan to use a hollow titanium tube where $G$ = 6,000,000 psi and $\tau_{yp}$ = 75,000 psi. The minimum safety factor is 2.0, and the shaft must be 80 in long. Determine the tube size based on yield.

**10.14** A steel cantilever beam is built into the wall as shown in the accompanying illustration. When the 400-lb load is applied, how far down does point $A$ move?

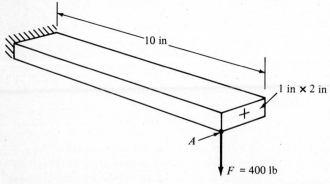

**Figure P10.14**

**10.15–10.17** The crankshaft for a large single-cylinder engine is being studied (see the accompanying drawing). For purposes of stress analysis, it can be modeled as being simply supported in bearings at $A$ and $B$, with the output moment $M_0$ acting at $A$. The three cylindrical portions are solid steel, 1.5 in in diameter; the two radius-arms are sufficiently short and stout to be considered rigid. The driving force $F$, acting at $C$, is at a 30° angle to the vertical. The material properties are:

$$\sigma_{yp} = 75{,}000 \text{ psi} \qquad E = 30 \times 10^6 \text{ psi}$$

$$\sigma_{end} = 35{,}000 \text{ psi} \qquad G = 12 \times 10^6 \text{ psi}$$

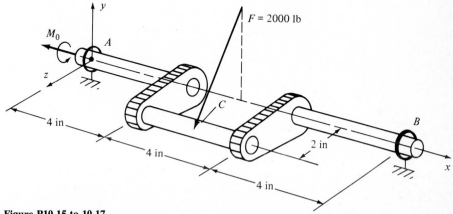

**Figure P10.15 to 10.17**

**10.15** Determine the safety factor relative to yield for points $A$ and $C$.

**10.16** Determine the safety factor relative to fatigue failure at points $A$ and $C$. Assume that $F$ simply oscillates between $+$ and $-2000$ lb, and that $K_f = 2.0$ at both points. For fatigue in torsion (point $A$), consider the maximum shear stress here and in the fatigue test.

**10.17** Estimate the deflection at point $C$.

**10.18** A square, steel, box beam is fabricated by forming a 16-in-wide sheet into the shape shown in the accompanying drawing and then welding the seam. Plates are then welded to the ends for attachment purposes. The steel has a nominal yield stress of 50,000 psi.

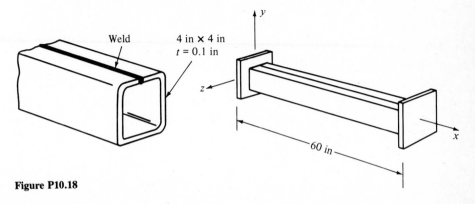

**Figure P10.18**

(*a*) If the weld is only 75 percent as strong as the base metal, what is the largest twisting moment we can apply?

(*b*) Determine the torsional spring constant for the 60-in beam.

**10.19** A thin-walled cylinder of length $L$, radius $r$, and thickness $t$ is made from material having a shear modulus $G$. A twisting moment $M_t$ is applied.

(*a*) Compare the values for shear stress when determined by Eqs. (10.7) and (10.18).

(*b*) Compare the angles of twist when determined by Eqs. (10.6) and (10.19).

# ELEVEN

## MATRIX STRUCTURAL ANALYSIS
## INCLUDING TORSION (THREE-DIMENSIONAL)

## 11.1 OVERVIEW

Now that we can determine the torsional stiffness of simple shapes, we can develop the structural analysis procedure to handle a fairly general, three-dimensional "space frame." This will be a structure, made of uniform elements, subject to axial loads, bending loads, and torsional loads.

In order to limit the complexity of the programs, we again will limit the number of rotational dof at each internal node. We will require that all elements having bending and torsional stiffness rotate together at internal nodes. Thus, for a three-dimensional structure, there will be 3 rotational dof at each node.

At each node, we will now have 6 degrees of freedom, motion on the $x$, $y$, and $z$ directions, and rotation about the three axes. For a complex structure, involving many nodes, the number of active dof can quickly become too large to be comfortably handled by a small microcomputer, and a larger machine may be in order.

Having previously developed the structural analysis procedures in Chaps. 4 and 9, you should have a good insight into what must be done to extend the procedure to this more general case.

At each node we must now input the activity (0, 1) for the 6 dof, and we must input the externally applied forces and moments associated with the 6 dof.
We must input additional element stiffness data. For the programs developed here, we will restrict ourselves to elements having equal bending stiffness in all directions. That is, the Mohr's circle for moment of inertia must be a

point (circular sections, squares, etc.). We will introduce the torsional stiffness in terms of $GI_p$ for a circular section. For other sections, an "equivalent" $GI_p$ can be used (equivalent value for the torsional stiffness per unit length $M_t L/\phi$).

The local stiffness matrices $[k_1]$ will now be $12 \times 12$, to account for the 6 dof at the two nodes. We must develop the terms to fill this rather large matrix.

This chapter will be extremely short because we will address only those aspects that are significantly different from our prior analyses.

## 11.2 DEFINING THE STRUCTURE

Figure 11.1 shows the general three-dimensional element that we will use here. The 6 dof at each node are identified, and are numbered 1 through 12. This is the order they will appear in the local stiffness matrices $[k_1]$. Also shown in the figure are the three direction cosines $C_1$, $C_2$, and $C_3$, which are determined from the $x$, $y$, and $z$ components of the element length $L$. In the programs, of course, we merely input the coordinates of the nodes at each end of an element and the computer will do the necessary calculations for angles, lengths, etc.

As mentioned above, we will limit this analysis to elements that have equal bending stiffness in all directions. We will call these circular sections, but we know that there are many other, equivalent, sections.

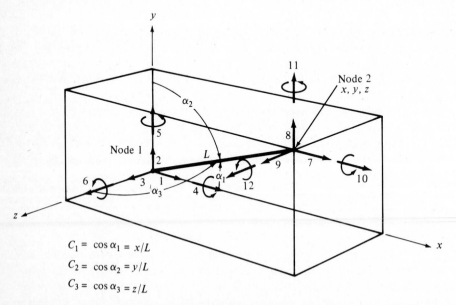

$$C_1 = \cos\alpha_1 = x/L$$
$$C_2 = \cos\alpha_2 = y/L$$
$$C_3 = \cos\alpha_3 = z/L$$

**Figure 11.1** Geometry for calculating the local stiffness matrix $[k_1]$. The six degrees of freedom at each node are numbered 1 through 6 and 1 through 12, as shown.

## 11.3 ESTABLISHING THE STIFFNESS MATRIX

The local stiffness matrix is now $12 \times 12$; the only difficult part of this analysis is establishing the 144 terms from the element stiffness properties. Computing these terms, for the element in Fig. 11.1, requires a great deal of concentrated effort—nothing particularly difficult, but rather taxing on one's patience. Accordingly, we will not subject you to this rather painful process, but will merely present the results.

However, if anyone should *want* to develop any or all of the terms, there are some points to bear in mind:

The meaning of each term is just the same as previously: It is the force or moment, at a node, associated with a given dof, due to a unit displacement or rotation along or about any dof. This is exactly the same as in Chaps. 4 and 9.

When we apply a unit rotation at a node, we must resolve that rotation into bending rotations and twisting rotations relative to the element axis. Although a rotation may *appear* to be a vector quantity (there is an axis of rotation and a magnitude of rotation), it is *not*. Finite rotations cannot be added vectorially (try it if you are skeptical). Fortunately, infinitesimal rotations can be added vectorially; so for small elastic rotations, which are all we ever consider, we can treat the rotations as though they were actually vectors.

The $12 \times 12$ local stiffness matrix can be formed from four different $3 \times 3$ submatrices numbered 1 to 4. Each submatrix, in turn, is formed from six elastic characteristics of the element cross section which we will label $A_1$ to $A_6$ as shown below:

$$A_1 = \frac{EA}{L} \qquad A_4 = \frac{GI_p}{L}$$

$$A_2 = \frac{12EI}{L^3} \qquad A_5 = \frac{4EI}{L}$$

$$A_3 = \frac{6EI}{L^2} \qquad A_6 = \frac{A_5}{2}$$

The four submatrices are:

$$\#1 = \begin{bmatrix} (A_1 - A_2)C_1^2 + A_2 & (A_1 - A_2)C_1C_2 & (A_1 - A_2)C_1C_3 \\ (A_1 - A_2)C_2C_1 & (A_1 - A_2)C_2^2 + A_2 & (A_1 - A_2)C_2C_3 \\ (A_1 - A_2)C_3C_1 & (A_1 - A_2)C_3C_2 & (A_1 - A_2)C_3^2 + A_2 \end{bmatrix}$$

$$\#2 = \begin{bmatrix} 0 & A_3C_3 & -A_3C_2 \\ -A_3C_3 & 0 & A_3C_1 \\ A_3C_2 & -A_3C_1 & 0 \end{bmatrix}$$

$$\#3 = \begin{bmatrix} (A_4 - A_5)C_1^2 + A_5 & (A_4 - A_5)C_1C_2 & (A_4 - A_5)C_1C_3 \\ (A_4 - A_5)C_2C_1 & (A_4 - A_5)C_2^2 + A_5 & (A_4 - A_5)C_2C_3 \\ (A_4 - A_5)C_3C_1 & (A_4 - A_5)C_3C_2 & (A_4 - A_5)C_3^2 + A_5 \end{bmatrix}$$

$$\#4 = \begin{bmatrix} (A_4 - A_6)C_1^2 + A_6 & (A_4 - A_6)C_1C_2 & (A_4 - A_6)C_1C_3 \\ (A_4 - A_6)C_2C_1 & (A_4 - A_6)C_2^2 + A_6 & (A_4 - A_6)C_2C_3 \\ (A_4 - A_6)C_3C_1 & (A_4 - A_6)C_3C_2 & (A_4 - A_6)C_3^2 + A_6 \end{bmatrix}$$

The total local matrix is formed from positive and negative submatrices 1 to 4, as shown below:

$$[k_1] = \begin{bmatrix} +\#1 & +\#2 & -\#1 & +\#2 \\ -\#2 & +\#3 & +\#2 & +\#4 \\ -\#1 & -\#2 & +\#1 & -\#2 \\ -\#2 & +\#4 & +\#2 & +\#3 \end{bmatrix}$$

If you ever feel an urgent need for some good intellectual exercise, derive all of the above!

## 11.4 SOLVING THE EQUATIONS

Our previous method of solution, using the gaussian elimination technique, is perfectly appropriate here. We need change nothing. As mentioned earlier, the number of active dof can be quite large, and the time for solution may become significant.

**Example 11.1** "Steps" are made of 0.5-in-diameter steel and welded to the side of a ship as shown. Determine the forces and deflections when the single 500-lb force is applied at node 3 as shown. Second, determine the forces and deflections when 500-lb loads are applied at both nodes 2 and 3.

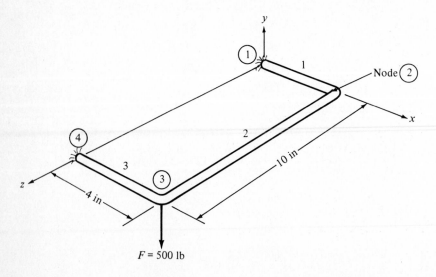

Before we can run a program used to solve this problem, we must compute the elastic properties of the cross section. For steel, let:

$$E = 30 \times 10^6 \text{ psi}$$
$$G = 12 \times 10^6 \text{ psi}$$

For $r = 0.25$ in:

$$EA = 0.597 \times 10^{-6} \text{ lb}$$
$$EI = 0.092 \times 10^{-6} \text{ lb} \cdot \text{in}^2$$
$$GI_p = 0.074 \times 10^{-6} \text{ lb} \cdot \text{in}^2$$

The numbering (arbitrary) for the elements and nodes is shown in the sketch. The input to and output from Program Listing 11.1 are reproduced in Program Output 11.1 for the two sets of loadings. (The computation requires about three-quarters of a minute.) ***

```
Number of ELEMENTS = 3
Number of NODES   = 4
```

Input data for each NODE:
    For degrees of freedom, 'active'=1; 'nonactive'=0
    '?' at end of line means hit CR if OK, N if not

| Deg of freedom | | | | | | | Applied forces (lb) and moments (in-lb) | | | | | | Coordinates (in) | | |
|---|---|---|---|---|---|---|---|---|---|---|---|---|---|---|---|
| # x | y | z | Ax | Ay | Az | Fx | Fy | Fz | Mx | My | Mz | x | y | z |
| 1 0 | 0 | 0 | 0 | 0 | 0 | 0 | 0 | 0 | 0 | 0 | 0 | 0 | 0 | 0? |
| 2 1 | 1 | 1 | 1 | 1 | 1 | 0 | 0 | 0 | 0 | 0 | 0 | 4 | 0 | 0? |
| 3 1 | 1 | 1 | 1 | 1 | 1 | 0 | -500 | 0 | 0 | 0 | 0 | 4 | 0 | 10? |
| 4 0 | 0 | 0 | 0 | 0 | 0 | 0 | 0 | 0 | 0 | 0 | 0 | 0 | 0 | 10? |

Input data to define ELEMENT location, stiffness, and initial conditions
    Note multipliers for E$A, E$I, and Temp-Strain

| Element No. | Located Between Nodes | E$A $(10^{-6})$ (lbs.) | E$I $(10^{-6})$ (lb-in^2) | G$Ip $(10^{-6})$ (lb-in^2) | Delta-Zero (in.) | Temp. Strain $(10^6)$ |
|---|---|---|---|---|---|---|
| 1 | 1 2 | .597 | .092 | .074 | 0 | 0 ? |
| 2 | 2 3 | .597 | .092 | .074 | 0 | 0 ? |
| 3 | 3 4 | .597 | .092 | .074 | 0 | 0 ? |

**Program Output 11.1**

There are 12 degrees of freedom
Wait for computation

| Node | Nodal displacements (in) | | | Nodal rotations (rad) | | |
|------|--------|--------|--------|--------|--------|--------|
| | x | y | z | Ax | Ay | Az |
| 1 | .00000 | .00000 | .00000 | .00000 | .00000 | .00000 |
| 2 | .00000 | -.02043 | .00000 | .00562 | .00000 | -.00961 |
| 3 | .00000 | -.09551 | .00000 | .00562 | .00000 | -.03387 |
| 4 | .00000 | .00000 | .00000 | .00000 | .00000 | .00000 |

| El. | Node | Nodal forces (lbs.) | | | Nodal moments (in-lbs) | | |
|-----|------|------|--------|------|--------|------|--------|
| | | Fx | Fy | Fz | Mx | My | Mz |
| 1 | 1 | .0 | 20.8 | .0 | -104.0 | .0 | 262.7 |
| 1 | 2 | .0 | -20.8 | .0 | 104.0 | .0 | -179.5 |
| 2 | 2 | .0 | 20.8 | .0 | -104.0 | .0 | 179.5 |
| 2 | 3 | .0 | -20.8 | .0 | -104.0 | .0 | -179.5 |
| 3 | 3 | .0 | -479.2 | .0 | 104.0 | .0 | 179.5 |
| 3 | 4 | .0 | 479.2 | .0 | -104.0 | .0 | 1737.3 |

Run other loadings for this structure? (CR/N)

| Node # | Applied forces (lb) and moments (in-lb) | | | | | | |
|--------|------|------|----|----|----|----|---|
| | Fx | Fy | Fz | Mx | My | Mz | |
| 2 | 0 | -500 | 0 | 0 | 0 | 0 | ? |
| 3 | 0 | -500 | 0 | 0 | 0 | 0 | ? |

| Element | Delta-Zero | T-Strain #(10^6) | |
|---------|-----------|-----------|---|
| 1 | 0 | 0 | ? |
| 2 | 0 | 0 | ? |
| 3 | 0 | 0 | ? |

Wait for computation

**Program Output 11.1** (continued)

|      | Nodal displacements (in) |         |         | Nodal rotations (rad) |         |         |
|------|--------------------------|---------|---------|-----------------------|---------|---------|
| Node | x                        | y       | z       | Ax                    | Ay      | Az      |
| 1    | .00000                   | .00000  | .00000  | .00000                | .00000  | .00000  |
| 2    | .00000                   | -.11594 | .00000  | .00000                | .00000  | -.04348 |
| 3    | .00000                   | -.11594 | .00000  | .00000                | .00000  | -.04348 |
| 4    | .00000                   | .00000  | .00000  | .00000                | .00000  | .00000  |

|     |      | Nodal forces (lbs.) |        |     | Nodal moments (in-lbs) |     |        |
|-----|------|---------------------|--------|-----|------------------------|-----|--------|
| El. | Node | Fx                  | Fy     | Fz  | Mx                     | My  | Mz     |
| 1   | 1    | .0                  | 500.0  | .0  | -.0                    | .0  | 2000.0 |
| 1   | 2    | .0                  | -500.0 | .0  | .0                     | .0  | .0     |
| 2   | 2    | .0                  | .0     | .0  | -.0                    | .0  | .0     |
| 2   | 3    | .0                  | -.0    | .0  | -.0                    | .0  | -.0    |
| 3   | 3    | .0                  | -500.0 | .0  | .0                     | .0  | -.0    |
| 3   | 4    | .0                  | 500.0  | .0  | -.0                    | .0  | 2000.0 |

**Program Output 11.1** (continued)

## 11.5 STRUCTURAL ANALYSIS PROGRAM LISTING

```
20  !"                    STRUCTURAL ANALYSIS"
30  !"                 (finite element-method)" \!
60  !"              3-DIMENSIONAL STRUCTURES"
70  !"                 Axial loading    (E*A)"
80  !"                 Bending          (E*I)"
90  !"                 Twisting         (G*Ip)" \ !
110 REM
120 !" ***********************************************************************"\!\!
130 REM
140 REM
150 REM   1.  DEFINE THE STRUCTURE
160 REM          a. Number of elements and nodes
170 REM          b. Input element data
180 REM          c. Input node data
190 REM          d. Input element data
200 REM          e. Determine and number the active degrees of freedom (1,2,--M)
210 REM          f. Dimension arrays
220 REM          g. Establish the applied force (moment) vector (f)
```

**Program Listing 11.1**

```
230                 GOSUB 800
240 REM
250 REM
260 REM    2.  ESTABLISH THE GLOBAL STIFFNESS MATRIX  (k)
270 REM        For each element,
280 REM        a. Determine element constants and store in array
290 REM        b. Establish local stiffness matrix (k1)
300 REM        c. Put appropriate terms into global matrix (k)
310                 GOSUB 1710
320 REM
330 REM
340 REM    3.  ESTABLISH THE INITIAL FORCE VECTOR  {f0}
350 REM        For each element,
360 REM        a. Calculate initial forces due to temperature or misfit
370 REM           and put into vector {f0} and final element force
380 REM           array E1
390 REM        b. Subtract initial forces from the applied forces
400 REM             {f}={f}-{f0}
410                 GOSUB 2090
420 REM
430 REM
440 REM    4.  SOLVE FOR DISPLACEMENTS {u} BY GAUSSIAN ELIMINATION
450 REM             (k)*{u}={f}
460 REM        a. Triangularize stiffness matrix {k}
470 REM        b. Calculate displacements {u}
480 REM        c. Put displacements into nodal array N1
490                 GOSUB 2360
500 REM
510 REM
520 REM    5.  DETERMINE FORCES ON EACH ELEMENT  (at both nodes)
530 REM        For each element\
540 REM        a. Determine nodes N1 AND N2
550 REM        b. Establish local displacement vector {u1}
560 REM        c. Establish local stiffness matrix (k1)
570 REM        d. Calculate forces due to displacements {f1}
580 REM             {f1}=(k1)*{u1}
590 REM        e. Add to initial forces in array E1
600                 GOSUB 2740
610 REM
620 REM
630 REM    6.  PRINT OUT RESULTS
640 REM        a. Nodal motion
650 REM        b. Forces on each element at both ends
```

**Program Listing 11.1** (continued)

```
660                     GOSUB 3080
670 REM
680 REM
690 REM    7.  FURTHER CALCULATIONS
700 REM         a. Other loadings for this structure?
710                     GOTO 3650
720 REM
730 REM
740 REM  ******************************************************************************
750 REM
760 REM    1.  DEFINE THE STRUCTURE
770 REM
780 REM         **** a. Number of elements and nodes ****
790 REM
800 !"             Input data to define the problem"\!\!
810 INPUT"            Number of ELEMENTS = ",I9      \REM I=ELEMENTS
820 INPUT"            Number of NODES    = ",J9      \REM J=NODES
830 !\!
840 REM         **** b. Dimension arrays ****
850 REM
860 DIM N(J9,15)    \REM NODAL INPUT DATA
870 DIM N1(J9,6)    \REM NODAL DISPLACEMENTS FOR OUTPUT
880 DIM E(I9,7)     \REM ELEMENT INPUT DATA
890 DIM E1(I9,14)   \REM ELEMENT FORCES FOR OUTPUT
900 DIM K0(I9,9)    \REM ELEMENT CONSTANTS
910 DIM M(12)       \REM GLOBAL DOF NUMBERS FOR LOCAL ELEMENT DOF
920 DIM K1(12,12)   \REM LOCAL ELEMENT STIFFNESS MATRIX
930 DIM C(3,3)      \REM USED TO CALCULATE K1(12,12)
940 DIM C1(3)       \REM LOCAL ELEMENT DIRECTION COSINES
950 DIM F1(12)      \REM LOCAL FORCE VECTOR
960 DIM U1(12)      \REM LOCAL DISPLACEMENT VECTOR
970 REM
980 REM         **** c. Input nodal data ****
990 REM
1000 !"Input data for each NODE:"
1010 !"     For degrees of freedom, 'active'=1; 'nonactive'=0"
1020 !"     '?' at end of line means hit CR if OK, N if not" \!
1030 !"  Deg of freedom    Applied forces (lb) and moments (in-lb)   Coordinates (in)" \ !
1040 !
1050 !" # x y z Ax Ay Az    Fx     Fy     Fz    Mx    My    Mz     x     y     z"\!
1060 FOR J=1 TO J9
1070 !J,\ !TAB(3), \ INPUT1"",N(J,1) \ INPUT1" ",N(J,2) \ INPUT1" ",N(J,3)
1080 INPUT1" ",N(J,4) \ INPUT1" ",N(J,5) \ INPUT1" ",N(J,6)
```

**Program Listing 11.1** (continued)

```
1090 !TAB(21), \ INPUT1" ",N(J,7) \ !TAB(28), \ INPUT1" ",N(J,8)
1100 !TAB(35), \ INPUT1" ",N(J,9) \ !TAB(42), \ INPUT1" ",N(J,10)
1110 !TAB(49), \ INPUT1" ",N(J,11)\ !TAB(56), \ INPUT1" ",N(J,12)
1120 !TAB(63), \ INPUT1" ",N(J,13) \ !TAB(69), \ INPUT1" ",N(J,14)
1130 !TAB(75), \ INPUT1" ",N(J,15)
1140 INPUT Z$
1150 IF Z$="N" THEN 1070
1160 NEXT \ !\!
1170 REM
1180 REM          **** d. Input element data
1190 REM
1200 !"Input data to define ELEMENT location, stiffness, and initial conditions"
1210 !"       Note multipliers for E*A, E*I, and Temp-Strain"
1220 !\!
1230 !"Element  Located     E*A      E*I      G*Ip     Delta-    Temp."
1240 !"  No.    Between   *(10^-6)  *(10^-6) *(10^-6) Zero     Strain"
1250 !"         Nodes     (lbs.)   (lb-in^2) (lb-in^2) (in.)    *(10^6)"
1260 FOR I=1 TO I9
1270 !"   ",I,TAB(11),\INPUT1"",E(I,1)\!TAB(14),\INPUT1"",E(I,2)
1280 !TAB(20), \ INPUT1" ",E(I,3)
1290 !TAB(30), \ INPUT1" ",E(I,4)
1300 !TAB(40), \ INPUT1" ",E(I,5)
1310 !TAB(50), \ INPUT1" ",E(I,6)
1320 !TAB(60), \ INPUT1" ",E(I,7)
1330 E(I,3)=E(I,3)*10^6 \ E(I,7)=E(I,7)*10^-6 \ E(I,4)=E(I,4)*10^6 \ E(I,5)=E(I,5)*10^6
1340 !" ", \ INPUT Z$
1350 IF Z$="N" THEN 1270
1360 NEXT \!\!\!
1370 REM
1380 REM          **** e. Determine and number active degrees of freedom ****
1390 REM
1400 FOR J=1 TO J9
1410 FOR K=1 TO 6
1420 IF N(J,K)=0 THEN 1440
1430 M=M+1 \ N(J,K)=M
1440 NEXT
1450 NEXT
1460 !"         There are ",M," degrees of freedom"
1470 !"             Wait for computation"\!\!
1480 REM
1490 REM          **** f. Dimension arrays ****
1500 REM
1510 DIM K(M,M)      \REM GLOBAL STIFFNESS MATRIX
```

**Program Listing 11.1** (continued)

```
1520 DIM F(M)          \REM APPLIED FORCE VECTOR
1530 DIM F0(M)         \REM INITIAL FORCE VECTOR
1540 DIM U(M)          \REM DISPLACEMENT VECTOR
1550 REM
1560 REM        **** g. Establish the applied force vector {f} ****
1570 REM
1580 FOR J=1 TO J9 \ FOR K=1 TO 6
1590 IF N(J,K)=0 THEN 1620
1600 M=N(J,K)
1610 F(M)=N(J,K+6)
1620 NEXT
1630 NEXT
1640 RETURN
1650 REM
1660 REM **********************************************************************
1670 REM
1680 REM    2.  ESTABLIST THE GLOBAL STIFFNESS MATRIX (k)
1690 REM
1700 REM        **** a. Determine and store element constants ****
1710 REM
1720 FOR U=1 TO M \ FOR V=1 TO M \ K(U,V)=0 \ NEXT \ NEXT
1730 REM
1740 FOR I=1 TO I9
1750 N1=E(I,1) \ N2=E(I,2)                \REM N1 IS NODE AT LOCAL ORIGIN
1760 REM
1770 FOR K=1 TO 6 \ M(K)=N(N1,K) \ M(K+6)=N(N2,K) \ NEXT
1780 X=N(N2,13)-N(N1,13)
1790 Y=N(N2,14)-N(N1,14)
1800 Z=N(N2,15)-N(N1,15)
1810 L=SQRT(X*X+Y*Y+Z*Z)
1820 C1=X/L
1830 C2=Y/L
1840 C3=Z/L
1850 K0(I,1)=E(I,3)/L
1860 K0(I,2)=E(I,4)*12/(L*L*L)
1870 K0(I,3)=K0(I,2)*L/2
1880 K0(I,4)=E(I,5)/L
1890 K0(I,5)=E(I,4)*4/L
1900 K0(I,6)=C1 \ K0(I,7)=C2 \ K0(I,8)=C3 \ K0(I,9)=L
1910 REM
1920 REM        **** b. Establish local stiffness matrix (k1) ****
1930 REM
```

**Program Listing 11.1** (continued)

```
1940 GOSUB 3260           \REM LOCAL STIFFNESS MATRIX SUBROUTINE
1950 REM
1960 REM        **** c. Put appropriate terms into global matrix (k) ****
1970 REM
1980 FOR U=1 TO 12\ FOR V=1 TO 12
1990 M1=M(U) \ M2=M(V)
2000 IF M1=0 THEN 2020 \ IF M2=0 THEN 2020
2010 K(M1,M2)=K(M1,M2)+K1(U,V)
2020 NEXT \ NEXT
2030 NEXT
2040 RETURN
2050 REM
2060 REM *************************************************************************
2070 REM
2080 REM    3.  ESTABLIST THE INITIAL FORCE VECTOR {f0}
2090 REM
2100 REM        **** a. Calculate initial forces, put in {f0} and E1 ****
2110 REM
2120 FOR I=1 TO I9
2130 N1=E(I,1) \ N2=E(I,2)                \REM N1 IS NODE AT LOCAL ORIGIN
2140 FOR K=1 TO 6 \ M(K)=N(N1,K) \ M(K+6)=N(N2,K) \ NEXT
2150 F=-K0(I,1)*(E(I,6)+E(I,7)*K0(I,9))
2160 IF F=0 THEN 2230
2170 FOR U=1 TO 3
2180 F0(M(U))=F0(M(U))-F*K0(I,U+5) \ E1(I,U+2)=-F*K0(I,U+5)
2190 NEXT
2200 FOR U=7 TO 9
2210 F0(M(U))=F0(M(U))+F*K0(I,U-1) \ E1(I,U+2)= F*K0(I,U-1)
2220 NEXT
2230 NEXT
2240 REM
2250 REM        **** b. Subtract initial forces {f}={f}-{f0} ****
2260 REM
2270 FOR K=1 TO M \ F(K)=F(K)-F0(K) \ NEXT
2280 RETURN
2290 REM
2300 REM *************************************************************************
2310 REM
2320 REM    4.  SOLVE FOR DISPLACEMENTS (u) BY GAUSSIAN ELIMINATION
2330 REM
2340 REM        **** a. Triangularize global matrix (k) ****
2350 REM
2360 FOR I=1 TO M
```

**Program Listing 11.1** (continued)

```
2370  FOR J=I+1 TO M
2380    Z=-K(J,I)/K(I,I)
2390     FOR K=I TO M
2400      K(J,K)=K(J,K)+Z*K(I,K)
2410     NEXT
2420     F(J)=F(J)+Z*F(I)
2430    NEXT
2440  NEXT
2450 REM
2460 REM         **** b. Calculate displacement vector {u} ****
2470 REM
2480 FOR I=M TO 1 STEP-1
2490 Z=0
2500 FOR J=I+1 TO M
2510   Z=Z+U(J)*K(I,J)
2520 NEXT
2530 U(I)=(F(I)-Z)/K(I,I)
2540 NEXT
2550 REM
2560 REM          c. Put displacements in nodal array N1 ****
2570 REM
2580 M1=1
2590 FOR J=1 TO J9 \ FOR K=1 TO 6
2600 IF N(J,K)<>M1 THEN 2630
2610 N1(J,K)=U(M1)
2620 M1=M1+1
2630 NEXT \ NEXT
2640 RETURN
2650 REM
2660 REM ***********************************************************************
2670 REM
2680 REM
2690 REM    5.  DETERMINE FORCES ON EACH ELEMENT
2700 REM
2710 REM
2720 REM         **** a. Determine nodes N1 and N2 ****
2730 REM
2740 FOR I=1 TO I9
2750 N1=E(I,1) \ N2=E(I,2)
2760 E1(I,1)=N1 \ E1(I,2)=N2
2770 REM
2780 REM         **** b. Establish local displacement vector {u1} ****
2790 REM
```

**Program Listing 11.1** (continued)

```
2800 FOR K=1 TO 6 \ M1=N(N1,K) \ U1(K)=U(M1) \ NEXT
2810 FOR K=1 TO 6 \ M1=N(N2,K) \ U1(K+6)=U(M1) \ NEXT
2820 REM
2830 REM          $$$$ c. Establish local stiffness matrix (k1) $$$$
2840 REM
2850 GOSUB 3260          \REM LOCAL STIFFNESS MATRIX SUBROUTINE
2860 REM
2870 REM          $$$$ d. Calculate local forces {f1} due to {u1} $$$$
2880 REM          $$$$ e. Add {f1} to initial forces in E1 $$$$
2890 REM
2900 FOR U=1 TO 12
2910 F1(U)=0
2920 FOR V=1 TO 12
2930 F1(U)=F1(U)+K1(U,V)$U1(V)
2940 NEXT
2950 E1(I,U+2)=E1(I,U+2)+F1(U)
2960 NEXT
2970 NEXT
2980 RETURN
2990 REM
3000 REM
3010 REM $$$$$$$$$$$$$$$$$$$$$$$$$$$$$$$$$$$$$$$$$$$$$$$$$$$$$$$$$$$$$$$$$$$$$$$$$$$$$$$$$$$$$$$$
3020 REM
3030 REM    6. PRINT OUT THE RESULTS
3040 REM
3050 REM          $$$$ a. Nodal displacements $$$$
3060 REM          $$$$ b. Forces at both ends of each element $$$$
3070 REM
3080 !"       Nodal displacements (in)        Nodal rotations  (rad)"
3090 !"Node     x         y         z         Ax        Ay        Az"\!
3100 FOR J=1 TO J9
3110 !J,    "    ",%10F5, N1(J,1),N1(J,2),N1(J,3),N1(J,4),N1(J,5),N1(J,6)
3120 NEXT
3130 !\!\!
3140 Z3=0
3150 !"               Nodal forces    (lbs.)        Nodal moments (in-lbs)"
3160 !"E1.   Node       Fx        Fy        Fz        Mx        My        Mz"
3170 !
3180 FOR I=1 TO I9
3190 !%2I,I,%10I, E1(I,1), %10F1, E1(I,3),E1(I,4),E1(I,5),E1(I,6),E1(I,7),E1(I,8)
3200 !%2I,I,%10I,  E1(I,2), %10F1,E1(I,9),E1(I,10),E1(I,11),E1(I,12),E1(I,13),E1(I,14)
3210 NEXT
3220 !\!\!INPUT "OK? (CR/N) ",Z$
```

**Program Listing 11.1** (continued)

```
3230 IF Z$="N" THEN 3080
3240 !\!\!\RETURN
3250 REM
3260 REM  ‡‡‡‡‡‡‡‡‡‡‡‡‡‡‡‡‡‡‡‡‡‡‡‡‡‡‡‡‡‡‡‡‡‡‡‡‡‡‡‡‡‡‡‡‡‡‡‡‡‡‡‡‡‡‡‡‡‡‡‡‡‡‡‡‡‡‡‡‡‡‡‡
3270 REM
3280 REM                    LOCAL STIFFNESS MATRIX ROUTINE
3290 REM
3300 FOR U=1 TO 12\FOR V=1 TO 12\ K1(U,V)=0 \ NEXT\NEXT
3310 FOR K=1 TO 3 \ C1(K)=K0(I,5+K) \ NEXT
3320 U9=0 \ V9=0 \ A7=K0(I,1)-K0(I,2) \ A8=K0(I,2) \ GOSUB 3530
3330 U9=6 \ V9=6 \ GOSUB 3530
3340 U9=0 \ A7=-A7 \ A8=-A8 \ GOSUB 3530
3350 U9=6 \ V9=0 \ GOSUB 3530
3360 U9=3 \ V9=3 \ A7=K0(I,4)-K0(I,5) \ A8=K0(I,5) \ GOSUB 3530
3370 U9=9 \ V9=9 \ GOSUB 3530
3380 U9=3 \ A7=-K0(I,4)-K0(I,5)/2 \ A8=K0(I,5)/2 \ GOSUB 3530
3390 U9=9 \ V9=3 \ GOSUB 3530
3400 REM ‡‡‡‡‡‡‡‡‡‡‡‡‡‡‡‡‡‡‡‡‡‡‡‡‡
3410 C(1,2)=K0(I,8) \ C(1,3)=-K0(I,7) \ C(2,1)=-C(1,2) \ C(2,3)=K0(I,6)
3420 C(3,1)=-C(1,3) \ C(3,2)=-C(2,3) \ A3=K0(I,3)
3430 U9=0 \ V9=3 \ GOSUB 3580
3440 V9=9 \ GOSUB 3580
3450 U9=3 \ V9=6 \ GOSUB 3580
3460 U9=9 \ V9=6 \ GOSUB 3580
3470 U9=3 \ V9=0 \ A3=-A3 \ GOSUB 3580
3480 U9=6 \ V9=9 \ GOSUB 3580
3490 U9=9 \ V9=0 \ GOSUB 3580
3500 U9=6 \ V9=3 \ GOSUB 3580
3510 RETURN
3520 REM ‡‡‡‡‡‡‡‡‡‡‡‡‡‡‡‡‡‡‡‡‡‡‡‡‡
3530 FOR U=1 TO 3 \ FOR V=1 TO 3
3540 A9=A7‡C1(U)‡C1(V)
3550 IF U=V THEN A9=A9+A8
3560 K1(U+U9,V+V9)=A9
3570 NEXT \ NEXT \ RETURN
3580 FOR U=1 TO 3 \ FOR V=1 TO 3 \K1(U+U9,V+V9)=A3‡C(U,V) \ NEXT \ NEXT
3590 RETURN
3600 REM
3610 REM ‡‡‡‡‡‡‡‡‡‡‡‡‡‡‡‡‡‡‡‡‡‡‡‡‡‡‡‡‡‡‡‡‡‡‡‡‡‡‡‡‡‡‡‡‡‡‡‡‡‡‡‡‡‡‡‡‡‡‡‡‡‡‡‡‡‡‡‡‡‡‡‡‡
3620 REM
3630 REM    7.  FURTHER CALCULATIONS
3640 REM
3650 REM          ‡‡‡‡ a. Other loadings for this structure? ‡‡‡‡
```

**Program Listing 11.1** (continued)

```
3660 REM
3670 !\ INPUT "Run other loadings for this structure? (CR/N) ",Z$
3680 IF Z$="N" THEN CHAIN "2.01$"
3690 FOR I=1 TO I9 \ FOR K=3 TO14 \ E1(I,K)=0 \ NEXT \ NEXT
3700 FOR U=1 TO M \ FO(U)=0 \ NEXT
3710 !\!
3720!"                    Applied forces (lb) and moments (in-lb)"
3730!"Node #              Fx    Fy    Fz    Mx    My    Mz"
3740 FOR J=1 TO J9
3750 FOR K=1 TO 6 \ IF N(J,K)<>0 THEN EXIT 3770 \ NEXT
3760 GOTO 3860
3770 !"    ",J,
3780!TAB(21), \ INPUT1" ",N(J,7) \ !TAB(28), \ INPUT1" ",N(J,8)
3790!TAB(35), \ INPUT1" ",N(J,9) \ !TAB(42), \ INPUT1" ",N(J,10)
3800!TAB(49), \ INPUT1" ",N(J,11)\ !TAB(56), \ INPUT1" ",N(J,12)
3810 FOR K=1 TO 6 \ M1=N(J,K)
3820 F(M1)=N(J,K+6)
3830 NEXT
3840 !"    ", \ INPUT Z$
3850 IF Z$="N" THEN 3770
3860 NEXT
3870 !\!
3880 !"Element   Delta-Zero   T-Strain $(10^6)"\!
3890 FOR I=1 TO I9
3900 !%7I,I,\ INPUT1"          ",E(I,6)
3910 !TAB(27),
3920 INPUT1"",E(I,7) \ !"   ", \ INPUT Z$
3930 E(I,7)=E(I,7)$10^-6
3940 IF Z$="N" THEN 3900
3950 NEXT
3960 !\!"             Wait for computation"\!\!
3970 GOTO 310
```

**Program Listing 11.1** (continued)

## 11.6 PROBLEMS

**11.1–11.3** Consider the general three-dimensional element shown in Fig. 11.1. The local stiffness matrix $[k_1]$ is a $12 \times 12$ matrix. In terms of the angles and the element stiffness parameters:

**11.1** Determine the (1, 1) term.

**11.2** Determine the (4, 4) term.

**11.3** Determine the (4, 5) term.

**11.4** A sign post, made from 2-in steel pipe is loaded at $A$ as shown in the accompanying drawing.

(a) Using the analytic techniques developed in Chaps. 8 and 10 calculate the deflection at A "by hand."

(b) Using the appropriate FEM program, determine the deflection at A and compare it with the answer to part (a).

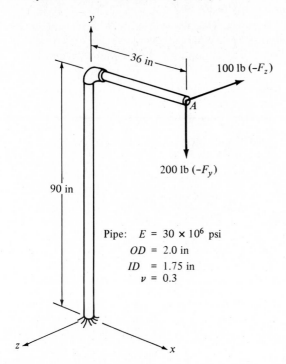

Pipe:  $E = 30 \times 10^6$ psi
$OD = 2.0$ in
$ID = 1.75$ in
$\nu = 0.3$

**Figure P11.4**

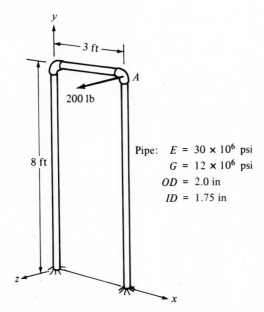

Pipe:  $E = 30 \times 10^6$ psi
$G = 12 \times 10^6$ psi
$OD = 2.0$ in
$ID = 1.75$ in

**Figure P11.5**

**11.5** A simple pipe frame, embedded in a concrete base, is loaded at $A$ in the $z$ direction, as shown in the accompanying illustration. Determine:
   (a) The deflection at $A$
   (b) The maximum *bending* stress
   (c) The yield stress required for a safety factor of 2.0

**11.6** The space frame shown in the accompanying drawing is made from aluminum tubing having a tensile yield stress of 40,000 psi. What is the largest force $F$ that can be allowed for a safety factor of 2? When $F_{max}$ is applied, what is the deflection of $A$ relative to $B$?

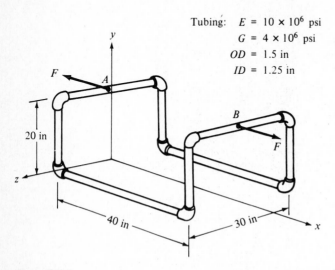

Tubing:
$E = 10 \times 10^6$ psi
$G = 4 \times 10^6$ psi
$OD = 1.5$ in
$ID = 1.25$ in

**Figure P11.6**

**11.7–11.9** To solve Probs. 11.7 to 11.9 (which will be very time-consuming), start with the bicycle frame shown in the accompanying drawing, which is made for aluminum tubing welded together. For modeling purposes assume the following:

Distance between nodes 5 and 6 is fixed (5 in)
Rear wheel axis is fixed in space
Axis of the front wheel cannot move up or down

The element section sizes are given in the accompanying table.

| Element(s) | OD, in | ID, in |
|---|---|---|
| 1 | 1.25 | 1.00 |
| 2, 3, 4 | 1.0 | 0.80 |
| 5, 6 | 0.5 | 0.40 |
| 7, 8 | 0.75 | 0.60 |

For this aluminum, $\sigma_{yp} = 70,000$ psi; $E = 10^7$ psi, and $G = 3.8 \times 10^6$ psi.

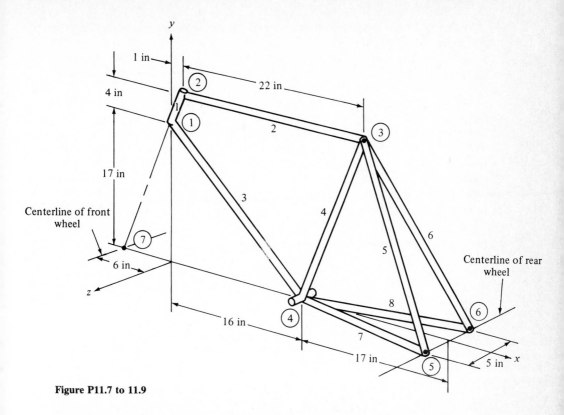

**Figure P11.7 to 11.9**

**11.7** Using the same format as in Example 11.1, code the bicycle structure. Note that the front-wheel support is at the wheel axis, node 7, which is not on the frame itself. Determine the weight of the frame.

**11.8** When a bicycle rider is "standing up and pedaling hard," the forces and moments applied by the handlebars and the pedal shaft could be as follows:

At node 2:

$$F_y = +200 \text{ lb}$$

$$M_x = -2400 \text{ in} \cdot \text{lb}$$

At node 4:

$$F_x = +800 \text{ lb}$$

$$F_y = -400 \text{ lb}$$

$$M_x = +2400 \text{ in} \cdot \text{lb}$$

$$M_y = +1600 \text{ in} \cdot \text{lb}$$

Convince yourself that these are not unreasonable "worst" conditions. Determine, using the appropriate FEM program:

(*a*) The point having the highest stress relative to yield. How large is the safety factor?

(*b*) The relative rotation, about the *x* axis, of nodes 2 and 4. Too much "springiness" can be annoying to the rider.

**11.9** Having worked Prob. 11.8, redesign all elements except No. 1 to have a common safety factor of 3.0. To do this, assume (although it is not so) that any diameter tubing is available, and that the ratio of inner to outer diameter is always 0.8.

Using the FEM program, rework the problem to determine whether, with the new stiffnesses, the safety factors are actually close to 3.0.

Determine the weight of this newly designed frame.

**11.10–11.11** The FEM programs provide output in terms of the force and moment components on each element at both nodes. There are other outputs that can be desirable, and which can be output automatically. Develop the logic and write an algorithm.

**11.10** Output the axial force, and the maximum bending and twisting moments, for each element.

**11.11** Output the maximum shear stress for each element.

# TWELVE

## BENDING OF BEAMS II

## 12.1 OVERVIEW

In Chap. 8 we studied the bending of uniform beams that were loaded only at their ends. We learned how to characterize the cross section of a beam in terms of the moments of inertia and the product of inertia (of the area), how to compute the stresses due to the bending moments, and how to determine the beam deflection. These developments were sufficient for the solution of many practical problems, and were incorporated into the structural analysis programs.

Here we will extend the developments of Chap. 8 into a more comprehensive treatment of beams in bending. First, we will remove the restriction that the beam be loaded only at its ends; we will consider point loads along the beam as well as loads that are distributed along the beam (such as a weight load).

Whenever we apply point loads or point moments along the length of a beam, we are faced with the problem of integrating discontinuous functions. A set of "singularity functions" will be developed which will permit us to treat some discontinuous functions as though they were continuous.

We noted in Chap. 8 that some thin-walled beams, if loaded through the centroid, will twist. Here we will determine the point (shear center) where the beam can be loaded without twisting.

The postyield behavior of beams will be introduced, and we will consider the limiting load characteristics of structures in terms of "limit analysis."

As we start this chapter, we will assume a good understanding of the material in Chap. 8.

## 12.2 GENERALIZED BEAM LOADING

Figure 12.1 shows a beam subject to a completely generalized loading in the $xy$ plane. The loading is *distributed* along the beam, so the units for the loading will be force/length (lb/in or N/m). We will use the symbol $q(x)$ to represent the value of the loading at any position $x$, just as was done in Sec. 2.6. In accordance with our usual sign conventions, *external* loadings (force or moment) in the positive $x$, $y$, or $z$ directions will be considered positive. Thus a downward loading, due to the weight of a beam, will be a negative $q(x)$.

In much of our work we have been concerned with the effects of "point loads" and "point moments." In terms of a distributed loading $q(x)$, a point load must have an *infinite* value; that is, a finite load, divided by zero beam length, gives a $q(x)$ of infinity, as shown in Fig. 12.2 (a "singularity" at the point of load application).

The value of $q(x)$ goes to infinity as $\Delta x$ goes to zero. On the other hand, the resultant of the $q(x)$ associated with a point load must be equal to the point

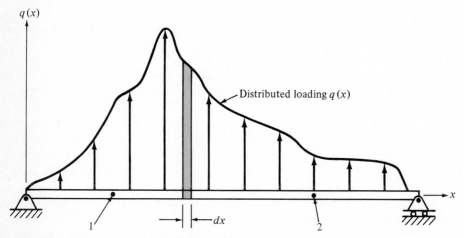

**Figure 12.1** Beam subject to a generalized loading.

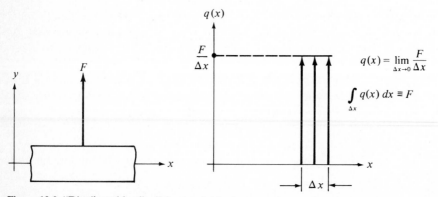

**Figure 12.2** "Distributed loading" for a point force.

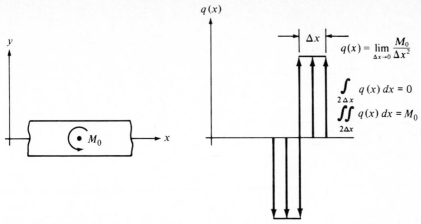

**Figure 12.3** "Distributed loading" for a point moment.

load itself. Therefore, the integral of $q(x)\,dx$ across a point load simply equals the load.

Figure 12.3 shows the equivalent distributed loading for a point moment $M_0$. Here we have the distributed loading associated with two point forces a distance $\Delta x$ apart which produce the "couple" $M_0$. As $\Delta x$ goes to zero, the point forces go to infinity, and they become zero distance apart. The first integral of $q(x)$ is the net force, which is zero. The second integral, however, just equals the applied moment $M_0$.

## 12.3 EQUILIBRIUM RELATIONS (LOADING, SHEAR, AND BENDING)

The manner in which the loading is distributed, and the manner in which the beam is supported, determine the shear force $V$ and the bending moment $M_b$ at all points along the beam. Because the loading *varies* with $x$, we will use our usual approach and isolate a differential beam element of length $dx$, as shown in Figs. 12.1 and 12.4. By writing the equilibrium equations for the beam element, we can establish very useful relations between $q(x)$, $V$, and $M_b$.

For $\Sigma F_y = 0$: $\qquad q(x)\,dx + (V + dV - V) = 0$

$$\frac{dV}{dx} = -q(x) \tag{12.1}$$

For $\Sigma M_z = 0$: $\quad (M_b + dM_b) - M_b + \dfrac{(V + dV)\,dx}{2} + \dfrac{V\,dx}{2} = 0$

$$\frac{dM_b}{dx} = -V \tag{12.2}$$

Equations (12.1) and (12.2) are the *equilibrium relations* for a beam, and they provide a great deal of information.

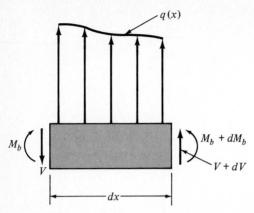

**Figure 12.4** Isolation of a beam element for consideration of equilibrium.

Recall that in Chap. 8 we constructed plots of both $V$ and $M_b$ as functions of position $x$ along the beam; these were the "shear force and bending moment diagrams." The equilibrium relations tell us how those diagrams relate to the loading $q(x)$ and to each other:

The *slope* of the shear-force diagram must equal (minus) the magnitude of the loading $q(x)$.

The *slope* of the bending moment diagram must equal (minus) the magnitude of the shear force $V$.

We can write the equilibrium relations in the integral form as well as the derivative form. We merely integrate the two equations between any two points along the beam (points 1 and 2 in Fig. 12.1). The result is:

$$V_2 - V_1 = -\int_1^2 q(x)\, dx = \text{area under } q(x) \text{ between 1 and 2} \qquad (12.3)$$

$$M_{b_2} - M_{b_1} = -\int_1^2 V(x)\, dx = \text{area under } V(x) \text{ between 1 and 2} \qquad (12.4)$$

These equations show that:

The *difference* in shear force between points 1 and 2 must equal (minus) the area under the loading diagram between the same two points.

When there is a point load, there is a singularity in the loading $q(x)$ which produces a discontinuity in the shear exactly equal to the load. If the load is positive, the discontinuity is negative.

The *difference* in bending moment between points 1 and 2 must equal (minus) the area under the shear diagram between the same two points.

Just as a point external load produces a discontinuity in the shear diagram, a point externally applied moment will give a discontinuity in the bending moment diagram. If the external moment is positive, the discontinuity will be negative.

To summarize: Moving from left to right, at a point force $+F$, the shear

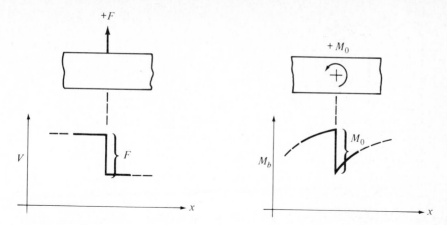

**Figure 12.5** Showing discontinuities in the shear force and bending moment diagrams due to point loads and point moments.

force *decreases* by $F$; at a point moment $+M_0$, the bending moment *decreases* by $M_0$ as shown in Fig. 12.5.

**Example 12.1** Consider the simply supported beam shown in the accompanying drawing, which carries a uniformly distributed weight $W$. In part (b) of the drawing the beam is isolated, showing the two support forces $R_A$ and $R_B$ and the distributed loading $q(x) = -W/L = -w_0$. So long as the beam is determinate, we can solve for the support forces. In this case they both equal $W/2$ or $w_0 L/2$.

To draw the shear-force diagram shown in part (c), we must determine the shear force $V$ at $x = 0$ and then use Eqs. (12.1) and (12.3) to guide us as we sketch in the total curve. The shear force at $x = 0$ is *always* equal to the magnitude of the support point force at $x = 0$ ($R_A$). If the support force is upwards, as in this problem, the shear force is negative.

Having determined $V$ at $x = 0$, Eq. (12.1) gives us the *slope* of $V(x)$ at all points along the beam. Because $q(x)$ is constant, negative, the slope is constant, positive, as shown. Using Eq. (12.3) we can compute the change in $V$ between any two points in terms of the area under the loading diagram. The total area $A_1$ is just equal to the change in $V$ from $x = 0$ to $x = L$. Clearly the shear force is zero in the center of the beam.

To establish the bending moment diagram, in similar fashion, we determine $M_b$ at $x = 0$ and then use Eqs. (12.2) and (12.4) to guide us along the curve. $M_b$ at $x = 0$ is *always* equal to any externally provided point moment at $x = 0$. In this case, with a pin-joint support, there can be no moment.

Having determined $M_b$ at $x = 0$, Eq. (12.2) gives the *slope* of the $M_b$ curve. As $x$ grows from zero to $L$, the shear force goes from a large negative value, through zero, to a large positive value. Accordingly, the moment diagram starts with a large positive slope, reaches zero slope at the center of the beam, and ends with a large negative slope.

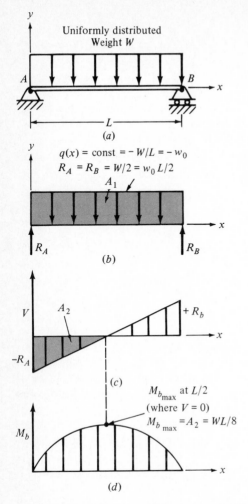

Because we are usually concerned with maxima of the bending moment diagrams, we are particularly interested in points where $V = 0$. In this case, we can readily calculate the maximum moment using Eq. (12.4) where point 1 is at the origin, and point 2 is at $x = L/2$. The maximum moment is simply equal to the area $A_2$.   ***

In Example 12.1, we saw that the bending moment went through a maximum where the shear force went through zero. This is to be expected because the shear force is just the derivative of the moment. Frequently, a beam will have several maxima and minima; if the beam cross section is symmetrical and uniform along the beam, the point having the largest-magnitude ($+$ or $-$) moment will be the weakest point. If the beam is not uniform or symmetrical, all of the maxima (minima) will have to be considered. Maxima and minima will *always* occur where the shear force is zero *or* where a point external moment is applied.

## 12.4 SINGULARITY FUNCTIONS

In Chap. 8, we determined the slope and deflection of a beam through two integrations of the bending moment equation [Eqs. (8.13)]. We have just seen that the bending moment equation can be obtained through two integrations of the loading relation $q(x)$.

Thus, if we can write the loading equation $q(x)$, and integrate it four times, we can directly determine the beam deflection equation. Of course, we must introduce the beam stiffness $EI$, and determine four constants of integration. When the stiffness is constant along the beam, it can be taken outside of the integrals. The resulting set of equations is shown as Eqs. (12.5):

$$q(x) = \text{function of } x$$

$$-V(x) = \int q(x)\, dx + C_1$$

$$M_b(x) = \int -V(x)\, dx + C_2 \qquad (12.5)$$

$$EI\phi(x) = \int M_b(x)\, dx + C_3$$

$$EIv(x) = \int EI\phi(x)\, dx + C_4$$

As long as the loading equation $q(x)$ is "continuous," we can use Eqs. (12.5) directly. However, we saw that point loads and point moments give rise to singularities in $q(x)$ and discontinuities in the shear force and bending moment relations.

One way to handle beams with singularities in the loading equation is to divide the beam into segments such that $q(x)$ is continuous over each segment. Doing so gives four constants of integration to be determined for each segment—a very considerable chore.

A rather elegant way to handle problems involving singularities is through a set of "singularity functions": functions that permit us to write and to integrate discontinuous functions as easily as continuous functions.

The family of singularity functions is:

$$f_n(x) = \langle x - a \rangle^n \qquad (12.6)$$

where $a$ is a constant and $n = -2, -1, 0, 1, 2, \ldots$. We define the meaning of the pointed brackets $\langle\ \rangle$ as follows:

When the argument $x - a$ is zero or negative, the value of the function $f(x)$ is zero and $\langle x - a \rangle$ equals zero. When the argument is positive, then the value of $\langle x - a \rangle$ just equals $x - a$:

If $x - a \leqq 0$ then $\langle x - a \rangle = 0$ and $f(x) = 0$

If $x - a > 0$ then $\langle x - a \rangle = x - a$ \qquad (12.7)

Figure 12.6 shows this relationship.

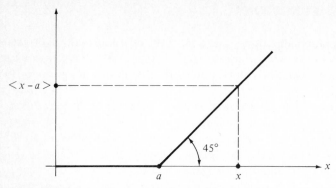

**Figure 12.6** Showing the meaning of pointed brackets ⟨ ⟩.

*Note*: Singularity functions are used in various disciplines with various names and notations (i.e., Dirac delta function, unit impulse, unit doublet, etc.). The notation in Eq. (12.6) is due to Macauley, who can be credited with introducing the method for beam analysis.

Let us study singularity functions by first considering the function for $n = 0$, then the derivatives of that function, and finally its integrals.

Figure 12.7a shows the function for $n = 0$. For $x < a$ the total function is zero, by definition [Eqs. (12.7)]. For $x > a$, the quantity $x - a$ is positive, and any positive number to the zero power equals 1. This function is often called a "unit step function," for obvious reasons. The function steps from a value of 0 for $x < a$ to a value of 1 for $x > a$.

Now consider the derivative of the step function as shown in Fig. 12.7b. The derivative of a curve equals the slope of the curve. The slope of the step function is zero everywhere except at $x = a$, where it is infinite as shown. If we compare Figs. 12.7b and 12.2, it is clear that the derivative of the step function can properly represent the distributed loading $q(x)$ for a unit point force in the positive $y$ direction.

Note that in Fig. 12.7b the functional notation for the derivative of the step function is $\langle x - a \rangle_{-1}$ rather than the usual convention $\langle x - a \rangle^{-1}$. A *subscripted* exponent (as opposed to the usual superscript) is a *flag* to indicate that this function is *different* in that it has a nonconventional integration law.

Normally, the integration of $\langle x - a \rangle^{-1} \, dx$ would yield a natural logarithm form; here, as shown in Fig. 12.7, it must integrate into the unit step function. That is:

$$\int \langle x - a \rangle_{-1} \, dx = \langle x - a \rangle^0 \tag{12.8}$$

In similar fashion, Fig. 12.7c shows the second derivative of the step function. When compared with Fig. 12.3, it is clear that this function can properly represent an externally applied point moment of *negative* sign. Again, because the integration law is peculiar, we use the subscripted exponent. The unit moment (or unit doublet) must integrate to the unit force (or unit impulse):

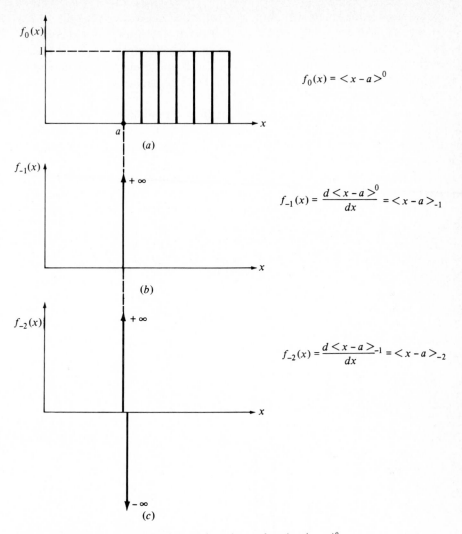

$$f_0(x) = <x - a>^0$$

$$f_{-1}(x) = \frac{d<x-a>^0}{dx} = <x-a>_{-1}$$

$$f_{-2}(x) = \frac{d<x-a>_{-1}}{dx} = <x-a>_{-2}$$

**Figure 12.7** Showing the derivatives of the unit step function $\langle x - a \rangle^0$.

$$\int \langle x - a \rangle_{-2} \, dx = \langle x - a \rangle_{-1} \tag{12.9}$$

We see then, that the three functions shown in Fig. 12.7 can represent three types of loading on a beam:

1. A uniformly distributed load of magnitude $+1$ that "starts" at $x = a$ and exists for all $x > a$
2. A point load of magnitude $+1$ acting at $x = a$
3. A point moment of magnitude $-1$ acting at $x = a$

The integral of the loading for the point moment is that for a point load; the integral of the point load is the step function.

Now consider the singularity functions for $n > 0$ (successive integrals of the step function). These are all well behaved in that they integrate according to the normal rules. Accordingly, they will all have superscript exponents. Figure 12.8$a$ shows the unit step function, and Fig. 12.8$b$ and $c$ the first two integrals. The step becomes a "ramp," the ramp becomes a second-order curve, etc. As shown, the integrals are perfectly normal.

To use any of the singularity functions in a beam-loading equation, they must, of course, be multiplied by the magnitude of the actual load. Also, you must be very careful regarding signs; an error here will propagate throughout the entire analysis.

The step function and its integrals provide functions that "start" at $x = a$, and continue for all $x > a$. Frequently we have need to "stop" a function, as in a uniformly distributed loading which acts only in the central portion of a beam.

We have no means to "stop" a function once it "starts" at $x = a$; however, we can effectively "stop" it by "starting" a new function of proper sign and magnitude to just cancel the original function.

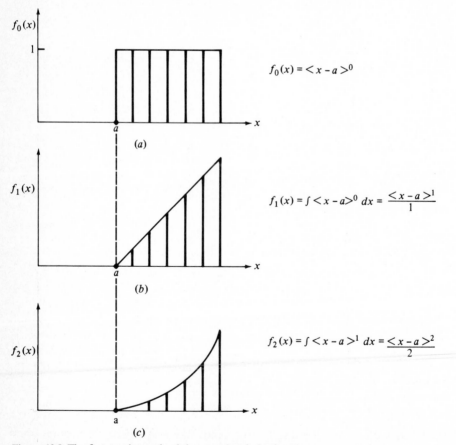

$$f_0(x) = <x - a>^0$$

$$f_1(x) = \int <x - a>^0 \, dx = \frac{<x - a>^1}{1}$$

$$f_2(x) = \int <x - a>^1 \, dx = \frac{<x - a>^2}{2}$$

(a)

(b)

(c)

**Figure 12.8** The first two integrals of the step singularity function.

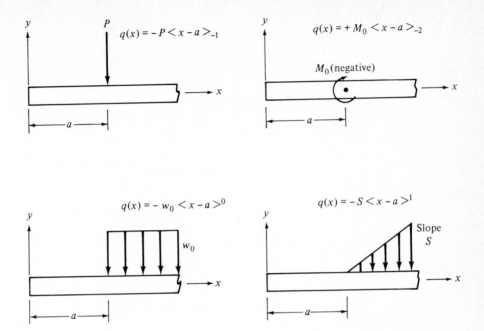

**Figure 12.9** Showing the proper use of singularity functions to represent typical beam loads.

Figure 12.9 shows the loading equations $q(x)$ that would be used to represent a variety of loads.

To solve any beam problem using the singularity functions, we must write the full loading equation, integrate four times (if the deflection is desired), and satisfy the boundary conditions (through the constants of integration).

At this point we have a choice of two procedures when we write the basic loading equation:

1. We can incorporate *all* loading terms, including the forces and moments acting at each end of the beam.
2. We can include *only* loads that act *between* the ends of the beam (at $0 < x < L$).

For most purposes, it is clear that the second procedure is better to use: The equations are shorter, there is less chance for error, and we gain greater insight. Accordingly, we will *not* include any end loads (typically support loads) in the loading equations; the four constants of integration take care of this deficiency. The examples to follow will demonstrate just how this works.

**Example 12.2** A uniform beam of length $L$ and stiffness $EI$ is simply supported at the ends, as shown. A point load $F$ (downward) acts at the center of the beam. Find the deflection under the load.

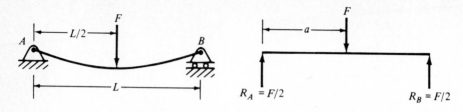

This is a very simple problem, but it will serve to demonstrate the power of the singularity function method. Because this problem is statically determinate, we can solve for the support forces and moments at $A$ and $B$. Clearly:

$$R_A = \frac{F}{2} \qquad M_A = 0$$

$$R_B = \frac{F}{2} \qquad M_B = 0$$

We now write the loading equation $q(x)$, considering *only* loads that act *between* the ends of the beam, of which there is only one, the load $F$ at $x = a = L/2$. After writing $q(x)$, we carry out the four integrations as follows:

$$q(x) = -F\langle x - a\rangle_{-1} \tag{1}$$

$$-V(x) = -F\langle x - a\rangle^0 + C_1 \tag{2}$$

$$M_b(x) = -F\langle x - a\rangle^1 + C_1 x + C_2 \tag{3}$$

$$EI\phi(x) = \frac{-F\langle x - a\rangle^2}{2} + \frac{C_1 x^2}{2} + C_2 x + C_3 \tag{4}$$

$$EIv(x) = \frac{-F\langle x - a\rangle^3}{6} + \frac{C_1 x^3}{6} + \frac{C_2 x^2}{2} + C_3 x + C_4 \tag{5}$$

The first integration above followed the special singularity function integration rule, while the others are normal. Now, all we need do is to determine the four constants of integration. Because this is really the *only* effort in using this method, we will consider the steps in some detail:

In determining the constants, we will note some simplifying relations that will add considerable insight to the problems. These relations are *almost* always true: they are true *if* the loading function and its integrals are zero when $x = 0$. The exceptions are loads that vary with $\sin x$, $\cos x$, $e^x$, etc. These exceptions are not often seen, but should be borne in mind. The *almost* true relations will be identified as "*almost*" always true.

From Eq. (2), $C_1$ is *almost* always equal to $-V$ at $x = 0$; and $-V$ at $x = 0$ equals $R_A$. Therefore, $C_1$ *almost* always equals the support force at $x = 0$. If the support force is upward (positive), then $C_1$ is also positive. For this case, $C_1 = +F/2$.

From Eq. (3), $C_2$ is *almost* always equal to $M_b$ at $x = 0$; but $M_b$ at $x = 0$ equals $-M_0$ where $M_0$ is the externally applied moment at $x = 0$. If the support moment $M_0$ is positive, then $C_2$ is negative. For this case, as with *almost* any pin-jointed end support, $C_2 = 0$.

From Eq. (4), $C_3$ is *almost* always equal to the slope of the beam at $x = 0$. If the beam is built in at $x = 0$, then $C_3 = 0$; otherwise it is generally not known and must be determined. For this case, we do not know the slope at $x = 0$ from the displacement constraints.

From Eq. (5), $C_4$ is *almost* always equal to the displacement of the beam at $x = 0$. If the beam is supported at $x = 0$, then $C_4 = 0$. For this case, $C_4 = 0$.

For this problem we have solved for three of the constants, $C_1$, $C_2$, and $C_4$. In order to solve for the remaining constant(s), there will always be a sufficient number of boundary constraints that have not yet been used. (If the problem has symmetry, as does this one, there will be excess, redundant constraints.) Here, the slope is zero at $x = L/2$ and the displacement is zero at $x = L$. From Eq. (4) or (5) we find $C_3 = -FL^2/16$. Summarizing:

$$C_1 = \frac{F}{2}$$

$$C_2 = 0$$

$$C_3 = -\frac{FL^2}{16}$$

$$C_4 = 0$$

Introducing the constants into Eq. (5) gives the deflection at $x = L/2$:

$$\delta_{L/2} = -\frac{FL^3}{48EI}$$

Now, as part of this same example, consider the same beam, with the same load, $F$, but supported in different ways. Shown in the illustration on p. 386 are some of the many ways it could be supported at its ends.

The loading *between* the ends has not changed; therefore Eqs. (1) to (5) above are still appropriate. The only differences are in the displacement constraints at the ends. Those affect only the constants of integration. In each case illustrated here, we have indicated the information that would be used to determine the constants and therefore the deflection.

If you wish to explore any of the loadings shown in the illustration, you can check your results from the following:

For (*a*): $$\delta_{L/2} = -\frac{FL^3}{24EI}$$

For (*b*): $$\delta_{L/2} = -\frac{7}{768}\frac{FL^3}{EI} \qquad C_1 = \frac{11F}{16} \qquad C_2 = -\frac{3FL}{16}$$

For (*c*): $$\delta_{L/2} = -\frac{1}{192}\frac{FL^3}{EI} \qquad C_2 = -\frac{FL}{8}$$

For (*d*): $$\delta_{L/2} = -\frac{7}{768}\frac{FL^3}{EI} - \frac{5\delta_L}{16} \qquad C_1 = \frac{11F}{16} + \frac{3EI\delta_L}{L^3} \qquad C_2 = -\frac{3FL}{16} - \frac{3EI\delta_L}{L^2}$$

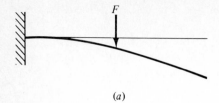

$$C_1 = F \qquad C_2 = -FL/2$$
$$C_3 = 0 \qquad C_4 = 0$$

(a)

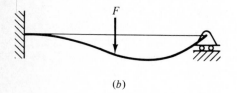

$$C_1 = ? \qquad C_2 = ?$$
$$C_3 = 0 \qquad C_4 = 0$$
$$\text{at } x = L \qquad M_b = 0, v = 0$$

(b)

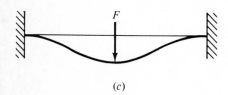

$$C_1 = F/2 \qquad C_2 = ?$$
$$C_3 = 0 \qquad C_4 = 0$$
$$\text{at } x = L/2 \qquad \phi = 0$$
$$\text{at } x = L \qquad \phi = 0, v = 0$$

(c)

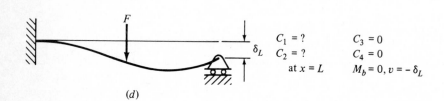

$$C_1 = ? \qquad C_3 = 0$$
$$C_2 = ? \qquad C_4 = 0$$
$$\text{at } x = L \qquad M_b = 0, v = -\delta_L$$

(d)

**Example 12.3** A beam of length $3a$ and stiffness $EI$ is built in at the left-hand end, as shown. It is loaded by a point force $W$ at the end and another load $W$ distributed over the central portion. Determine the deflection at the free end.

From equilibrium of the entire beam, the support force $R_A = 2W$ and the support moment $M_0 = 9Wa/2$.

The loading equation, for the loading *between* the ends of the beam, involves only two terms, one to "start" the distributed load at $x = a$, and one to "stop" it at $x = 2a$:

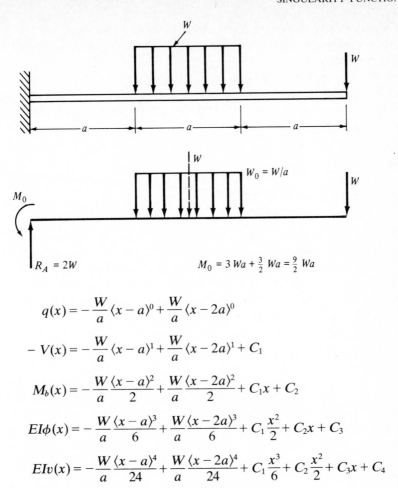

$$q(x) = -\frac{W}{a}\langle x - a\rangle^0 + \frac{W}{a}\langle x - 2a\rangle^0$$

$$-V(x) = -\frac{W}{a}\langle x - a\rangle^1 + \frac{W}{a}\langle x - 2a\rangle^1 + C_1$$

$$M_b(x) = -\frac{W}{a}\frac{\langle x - a\rangle^2}{2} + \frac{W}{a}\frac{\langle x - 2a\rangle^2}{2} + C_1 x + C_2$$

$$EI\phi(x) = -\frac{W}{a}\frac{\langle x - a\rangle^3}{6} + \frac{W}{a}\frac{\langle x - 2a\rangle^3}{6} + C_1\frac{x^2}{2} + C_2 x + C_3$$

$$EIv(x) = -\frac{W}{a}\frac{\langle x - a\rangle^4}{24} + \frac{W}{a}\frac{\langle x - 2a\rangle^4}{24} + C_1\frac{x^3}{6} + C_2\frac{x^2}{2} + C_3 x + C_4$$

Here we can solve for the constants by inspection:

$$C_1 = R_A = 2W \qquad\qquad C_3 = 0$$

$$C_2 = -M_0 = -\frac{9Wa}{2} \qquad C_4 = 0$$

And the deflection at the free end is:

$$-\frac{285}{24}\frac{Wa^3}{EI} \qquad\qquad\qquad ***$$

From Examples 12.2 and 12.3, we see how the use of singularity functions facilitates the solution of beam problems. In both examples we were asked to determine the deflection at a specific point on the beam. Frequently, we must determine the *maximum* deflection, or the *maximum* bending moment, when the location of the maximum is not known. From our past work, we know that to locate a maximum, we set the derivative of the function equal to zero and solve for the location $x$.

There can be confusion as to the handling of the $\langle x - a \rangle$ terms when solving for the location of a maximum; if $x$ turns out to be less than $a$, we should not have included that term when we solved for $x$. The procedure is as follows:

1. Assume that the maximum is located within some specific portion of the beam, perhaps $a > x > b$.
2. Solve for $x$ ignoring any terms involving $\langle x - b \rangle$ which must be negative.
3. If the value of $x$ is in the zone you had assumed, all well and good; if not, try a new zone.

One word of caution: When determining the maximum bending moment using singularity functions, always sketch the shear-force and bending moment diagrams. Otherwise you may miss the maximal maximum.

Similarly, it is always useful to sketch your best estimate of the beam shape *before* you start to work on the problem. We all seem to have quite good intuition regarding the general shape of a loaded beam.

## 12.5 SHEAR CENTER

In our consideration of shear effects in beams (Chap. 8), we restricted our analyses to beams where the plane of loading was also a plane of beam symmetry. This restriction ensured that the beam bent downward (in the plane of loading) and did not twist. We now want to consider the effects of relaxing the symmetry requirement.

Figure 12.10 shows a thin-walled channel beam laid on its side. The

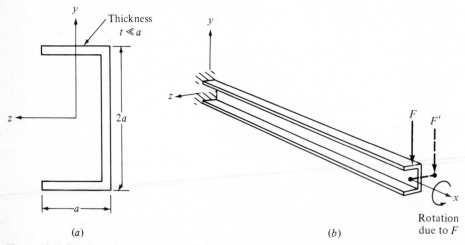

*(a)*        *(b)*

**Figure 12.10** Bending of a nonsymmetrical thin-walled channel section.

dimensions are $a \times 2a$, and the thickness is $t$. If we load the beam as a cantilever, where the load passes through the centroid, we will find that it will not only bend, but it will also twist about the $x$ axis (the rotation will be a positive rotation about $x$).

In general we do not want beams to twist, and the situation shown is not satisfactory. Clearly, if we also applied a twisting moment, of proper sign and magnitude, we could remove the twist due to bending.

This is precisely what we shall do; and we shall apply the twisting moment simply by moving the force $F$ sidewise to $F'$, as shown in Fig. 12.10. The point where $F'$ is applied is called the "shear center"; if we load the beam through this point, it will bend without twisting, as desired.

To analyze this problem, we will:

1. Force the beam to bend *without* twisting
2. Determine the shear stresses and the resultant shear forces
3. Find how to load the beam such that the above forces are produced by the loading.

If we "force" a beam to bend straight downward without twisting, then the geometry of deformation will be the same as for a symmetrical beam, and the stresses can be calculated by the standard equations. Thus the bending stresses will be determined by $\sigma_x = My/I$, and the shear stresses by $\tau = VQ/Ib$.

First, determine the moment of inertia about the $z$ axis in the usual way, making approximations that are appropriate for a thin-walled beam (for the horizontal portions, ignore the value of $I_{yy}$ about its own centroidal axis when compared to the parallel-axis term):

$$I = 2(at)a^2 + \frac{t(2a)^3}{12} = \frac{8}{3}a^3t$$

Using the geometry illustrated in Fig. 12.11, we will determine the shear-

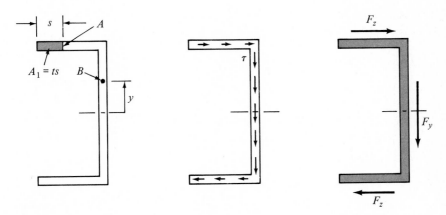

**Figure 12.11** Calculation of shear stress and resultant forces for the channel beam.

stress distribution in the upper (and lower) flange and the resultant horizontal shear force $F_z$. In a parallel way we will obtain the vertical shear force $F_y$.

In the upper flange, at a distance $s$ from the left-hand edge (point $A$):

$$Q = \bar{y}_1 A = a(ts) \qquad \tau = \frac{VQ}{It} = \frac{3}{8} \frac{Vs}{a^2 t}$$

The resultant force $F_z$, obtained by integrating the $\tau\, dA$ over the flange area, is:

$$F_z = \int_A \tau\, dA = \tfrac{3}{16} V$$

The shear force on the lower flange is the same magnitude, in the opposite direction.

In the vertical web, at point $B$,

$$Q = \sum_i \bar{y}_i A_i = a(at) + 0.5(a+y)[t(a-y)]$$

The web shear stress, calculated using $Q$, can be integrated over the web area to get $F_y$. It is not surprising that:

$$F_y = V$$

Looking at Fig. 12.12, we see that in order to force the beam to bend downward without twisting, we must apply not only a vertical force $V$ in the web, but also equal and opposite shear forces $F_z$ in the flanges. Externally, the latter two forces form a couple of magnitude $2aF_z$.

The applied force $F$ will produce both the shear force $V$ and the couple if it is offset a distance $d$ as shown. For this particular geometry:

$$d = \frac{2aF_z}{F}$$

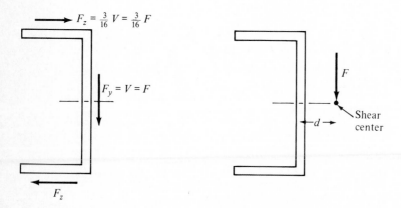

**Figure 12.12** Showing the determination of the shear center.

But
$$F_z = \frac{3V}{16} = \frac{3F}{16}$$

Thus
$$d = \tfrac{3}{8}a$$

We see, then, that if we had applied $F$ through the centroid (which might seem to be correct), the beam would twist, but if we apply $F$ through the shear center, the resulting twisting moment keeps the beam from twisting (or if you prefer, straightens it again).

**Example 12.4** An angle beam has thickness $t$ and two sides of length $a$, as shown. We want the beam to carry a vertical force $F$ and move straight downward without twisting. How must we apply the force $F$?

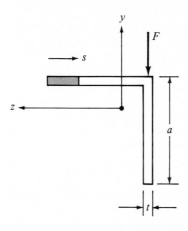

The procedure will be the following: ($a$) locate the centroid, ($b$) calculate the moment of intertia $I$ ($I_{yy}$), ($c$) find the shear flow and the resulting shear force in the horizontal leg, ($d$) find the shear flow and the resulting shear force in the vertical leg, and ($e$) determine how to apply the load so that the shear force requirements are satisfied.

Using the standard method for step ($a$), we find that the centroid is down and in a distance $a/4$ from the juncture of the two legs or flanges.

For step ($b$), we use the methods described in Chap. 8 to compute $I_{yy}$ about the centroidal axis; we find $I_{yy} = \frac{5}{24}a^3t$.

For step ($c$), determine $Q$ at a distance $s$ from the left-hand edge:

$$Q = \frac{ats}{4} \qquad q = \frac{VQ}{I}$$

Therefore
$$q = \frac{6Vs}{5a^2} \qquad F_z = \int_0^a q\,ds = \tfrac{3}{5}V$$

Thus the resultant horizontal shear force in the upper leg equals three-fifths of the shear force $V$. For smooth shear flow, because the vertical shear force is downward, the horizontal force $F_z$ must be to the right as shown in the illustration below.

For step $(d)$, now determine $Q$ for the area above a height $y$ on the vertical leg:

$$Q = \frac{t}{2}\left(\frac{9}{16}a^2 - y^2\right) \qquad q = \frac{12V}{5a^3}\left(\frac{9}{16}a^2 - y^2\right)$$

$$F_y = \int_{-3a/4}^{a/4} q\,ds = V \qquad \text{(not surprising)}$$

The resultant shear force in the vertical leg just equals the shear force $V$, as it must.

For step $(e)$, as shown in the illustration below, the applied loading must produce a vertical force equal to $V$ in the vertical leg, and must *also* produce a horizontal force of $0.6V$ in the horizontal leg. Clearly, in order that the beam move straight downward, we must apply both the vertical and the horizontal force components at the intersection of the two legs.

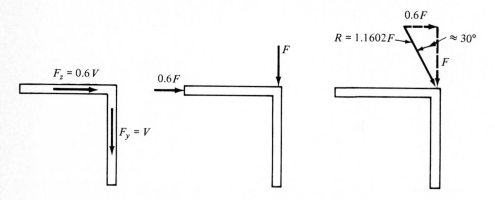

The center of shear is at the intersection; if we apply any force there (as contrasted to the centroid), the beam will not twist. However, in general, the beam will not move in the direction of the force.

We can generalize the above result by noting that for *any* 90° angle section, the shear center must be at the intersection of the two flanges.

For this example, are there directions such that the force and deflection are collinear?

*\*\*\**

**Example 12.5** Although we are calling this a separate example, it is really a continuation of Example 12.4. It is meant to demonstrate , for any who may be confused or concerned, that the various approaches we have seen relative to asymmetrical beam sections are, in fact, consistent.

We will examine the same "angle iron" discussed above (relative to shear center), from two different analytical approaches:

1. Use the Mohr's circle for moments of inertia to determine the principal axes of inertia (magnitude and direction). Then load the beam along the principal axes with the equivalent of $F$ and $0.6F$ as determined above, and find the resultant deflection (magnitude and direction).

2. Given the principal moments of inertia and their directions, calculate the principal compliances. Draw the Mohr's circle for compliance, and find the total deflection when we apply the *resultant* of $F$ and $0.6F$ to the beam.

   In regard to approach 1, the cross-sectional area and the Mohr's circle diagram are shown in the illustration below. We already know (Example 12.4) that the centroid is down $a/4$, and that the moment of inertia $I_{yy} = 5a^3t/24$. Because of symmetry, $I_{zz} = I_{yy}$ (in fact, because of symmetry we *know* that the principal axes must be at 45° as shown). To plot the circle, we need only find $I_{yz}$, the product of inertia.

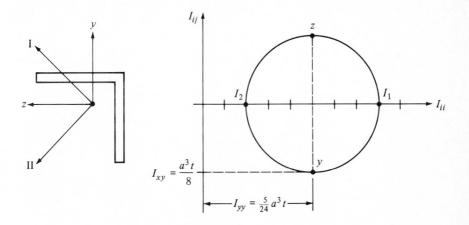

   If we consider the two legs separately, we find that the centroid of one is at $y = z = +a/4$, and the other area is at $y = z = -a/4$. The product of inertia then is:

$$I_{yz} = \int_A yz \, dA = \sum_i \bar{y}_i \bar{z}_i A_i = \frac{a}{4}\frac{a}{4} \, at + \left(-\frac{a}{4}\right)\left(-\frac{a}{4}\right) at$$

$$= +\frac{a^3t}{8}$$

   To use the Mohr's circle sign convention, we note that we have a $yz$ set of axes instead of $xy$. Thus the rule that would normally be applied to the $x$ axis is now applied the equivalent axis, the $y$ axis. Because $I_{yz}$ is positive, we plot $y$ downward and $z$ upward, as shown in the illustration. The principal axes are at 45°, and the principal moments of inertia are:

$$I_1 = \frac{5}{24} a^3 t + \frac{1}{8} a^3 t = \frac{a^3 t}{3}$$

$$I_2 = \frac{5}{24} a^3 t - \frac{1}{8} a^3 t = \frac{a^3 t}{12}$$

$$\frac{I_1}{I_2} = 4.0$$

We know that we can calculate the total beam deflection $\delta$ by using our usual beam equations to calculate the components $\delta_1$ and $\delta_2$ along the two principal directions. First we determine the components of load along the 1 and 2 directions, and then compute the corresponding deflections.

Shown below are a force diagram and the corresponding deflection diagram. The applied resultant load is the force $R$ composed of the initial load $F$ and the additional horizontal force of $0.6F$ (Example 12.4 showed that we must apply this extra load if we want the beam to bend straight downward). The components of $R$ along the principal directions are shown in the illustration.

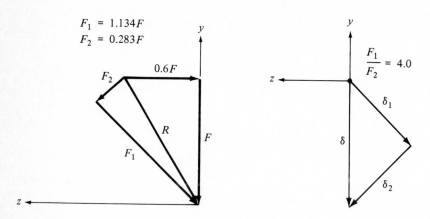

$$F_1 = 1.134F$$
$$F_2 = 0.283F$$
$$\frac{F_1}{F_2} = 4.0$$

Thus we find that the ratio of the forces in the principal directions equals the ratio of the principal moments of inertia. Clearly the principal deflections are equal and the beam does, in fact, move straight downward.

The magnitudes of the deflections are:

$$\delta_1 = \frac{FL^3}{3EI} = \frac{1.134FL^3}{3Ea^3 t/3} = \frac{1.134FL^3}{Ea^3 t}$$

$$\delta_2 = \frac{FL^3}{3EI} = \frac{0.283FL^3}{3Ea^3 t/12} = \frac{1.134FL^3}{Ea^3 t}$$

$$\delta = \sqrt{\delta_1^2 + \delta_2^2} = 1.6 \frac{FL^3}{Ea^3 t} \qquad \text{(straight down)}$$

In regard to approach 2, if we wish to use the Mohr's circle for compliance, we can calculate the principal compliances ($C_1$ and $C_2$), draw the circle, and determine the compliances in the direction of the resultant applied force $R$. The principal compliances are simply the inverse of the stiffnesses in the principal directions. Again, using the standard equations for a cantilever beam loaded in a principal direction:

$$C_1 = \frac{1}{k_1} = \frac{L^3}{3EI_1} = \frac{L^3}{Ea^3t}$$

$$C_2 = \frac{1}{k_2} = \frac{L^3}{3EI_2} = 4\frac{L^3}{Ea^3t}$$

The Mohr's circle for compliance is shown in the drawing below, together with a diagram for the force and deflection geometry. The axes associated with the resultant load $R$ have been labeled $x'$ and $y'$. The $x'$ axis is 14.04° clockwise from the 1 axis in real space. Therefore, on the Mohr's circle, it is twice as far (28.08°) in the clockwise direction. Note that the 1 axis, which has the greatest stiffness, has the *least* compliance.

Using the usual sign convention for Mohr's circle, because $x'$ is plotted *up*, $C_{x'y'}$ must be negative. The compliances associated with a force in the $x'$ direction are:

$$C_{x'x'} = 1.177\frac{L^3}{Ea^3t}$$

$$C_{x'y'} = -0.706\frac{L^3}{Ea^3t} = C_{yx}$$

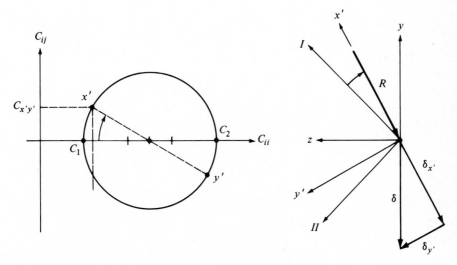

The applied force $R$ (equaling $1.1662F$) is applied in the negative $x'$ direction. If we calculate the deflections, we find:

$$\delta_{x'} = F_{x'}C_{x'x'} = -1.372 \frac{FL^3}{Ea^3t}$$

$$\delta_{y'} = F_{x'}C_{y'x'} = +0.823 \frac{FL^3}{Ea^3t}$$

$$\delta = \sqrt{\delta_{x'}^2 + \delta_{y'}^2} = 1.6 \frac{FL^3}{Ea^3t} \qquad \text{(straight down)} \qquad\qquad *** $$

## 12.6 YIELD AND POSTYIELD BEHAVIOR

When we studied torsion of a circular member, where the shear strain varied linearly from the axis outward, we found that yield occurred first at the outer layers. If we continued to twist the cylinder beyond the initial yield condition, the twisting moment increased as more and more of the cross section yielded. We finally reached the "fully plastic" condition where the entire section was at the yield stress.

In bending, where the normal strain varies linearly from the neutral axis outward, yield occurs first in the material farthest from the neutral axis. If we continue to bend the beam beyond the initial yield condition, the bending moment will increase as more and more of the cross section reaches the yield stress. Eventually, the entire section becomes plastic.

Qualitatively, then, torsion and bending are quite similar in their yield and postyield behavior. We will observe the same effects here except that the stresses and strains are "normal" rather than shear due to a twisting moment.

If we consider (as we did for torsion) an elastic, perfectly plastic material (where the stress-strain curve is horizontal following yield), the stress distribution on a section will vary as shown in Fig. 12.13.

In Fig. 12.13a, the outermost "fibers" have just reached the yield stress in

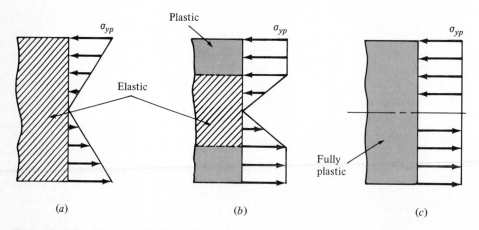

**Figure 12.13** Bending stress distributions: (a) at yield; (b) following yield; (c) fully plastic.

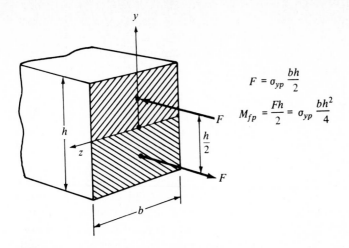

$$F = \sigma_{yp} \frac{bh}{2}$$

$$M_{fp} = \frac{Fh}{2} = \sigma_{yp} \frac{bh^2}{4}$$

**Figure 12.14** Determination of the fully plastic bending moment ($M_{fp}$).

tension ($\sigma_{yp}$) and the bending moment is the yield moment ($M_{yp}$). In Fig. 12.13$b$, the outer layers are plastic with an elastic core. Finally, in Fig. 12.13$c$, the entire section is in the plastic condition, and the bending moment has reached the "fully plastic bending moment" ($M_{fp}$) or "limit moment" ($M_L$).

The analysis of the fully plastic condition is very easy because the stress is constant over the section. Consider the rectangular section beam shown in Fig. 12.14: Above the neutral axis, the stress is a constant compressive stress of $-\sigma_{yp}$; below the NA the stress is constant at $+\sigma_{yp}$. The two resultant forces $F$ act at the centroids of the upper and lower halves of the section, and the bending moment ($M_{fp}$) equals the couple formed by these forces.

We can find the ratio of the fully plastic moment ($M_{fp}$) to the yield moment ($M_{yp}$) just as we did for torsion. From the elastic relations:

$$M_{yp} = \frac{\sigma_{yp}I}{y} = \sigma_{yp} \frac{bh^3/12}{b/2} = \sigma_{yp} \frac{bh^2}{6}$$

Thus
$$\frac{M_{fp}}{M_{yp}} = 1.5$$

If we were to consider an I beam, we would find that the fully plastic moment was only slightly larger than the yield moment.

If the cross section is not symmetrical about the $z$ axis, we must determine the location of the neutral axis. You may recall that the neutral axis was determined by the condition of zero force in the $x$ direction. In the fully plastic condition, where stresses are either $+\sigma_{yp}$ or $-\sigma_{yp}$, there must be equal *area* above and below the NA.

For an ideally plastic material, once the fully plastic condition is reached, the beam can continue to bend with no increase in bending moment. Large angles of rotation can be reached before fracture in many ductile materials. This situation is aptly called a "plastic hinge" because it is equivalent to a hinge that rotates with constant friction torque.

### 12.6.1 Residual Stresses

If we bend a beam into the plastic regime and then "let go," there will be some degree of elastic "springback," and there will be residual stresses. We can determine the residual stress distribution, as we did for torsion, by superposing the *elastic* stress distribution associated with the springback on the *plastic* stress distribution that existed prior to "letting go" of the beam.

Figure 12.15 shows the superposition required to determine the residual stresses in a rectangular beam. We start by applying a positive $M_{fp}$ as shown in Fig. 12.15a. To "let go" or release the beam, we add an oppositely directed moment of magnitude $M_{fp}$ as shown in Fig. 12.15b. The stress distributions shown are the fully plastic distribution and the *elastic* distribution due to a moment equal to $M_{fp}$.

Because we saw that the limit moment $M_{fp}$ was just three-halves of the yield moment $M_{yp}$ (for a rectangular beam), the maximum elastic stress in Fig. 12.15b is just three-halves of the yield stress $\sigma_{yp}$. Now when we superpose the two distributions, we find the distribution shown in Fig. 12.15c; the residual stresses in the outer layers equals half the yield stress—a large residual stress.

Having established the residual stress pattern shown, the beam will respond quite differently to bending moments of different sign. If we apply a

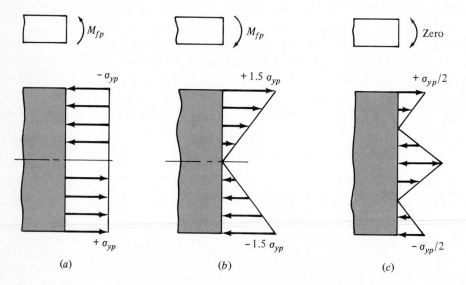

**Figure 12.15** Showing the development of residual stresses in a bent rectangular beam.

positive moment, there will be no yielding until the moment equals 1.5 times the original yield moment ($M_b = 1.5M_{yp}$). If we apply a negative moment, however, the residual stresses will add to the bending stresses and yielding will occur at 0.5 times the yield moment ($M_b = 0.5M_{yp}$). This represents a $3:1$ ratio of yield moments!

## 12.7 LIMIT ANALYSIS

We saw in Sec. 12.6 that a property of every ductile beam is its fully plastic bending moment ($M_{fp}$). This is also called the "limit moment" ($M_L$). This is the moment that cannot be exceeded. When the limit moment is reached, unrestrained bending can occur (plastic hinge).

In any structure made from beams, if the loading is increased sufficiently, the structure will collapse (undergo very large, catastrophic deformations). The "collapse mechanism" is intimately associated with the occurrence of plastic hinges in the beams of the structure. The "collapse load" is the set of loads that will produce complete collapse. The process of determining the collapse loads, or limit loads, is called "limit analysis."

For simple structures it is quite easy to determine the collapse mechanism and the collapse load. Consider the cantilever beam shown in Fig. 12.16a. The load $P$ produces a maximum bending moment at the wall. In Fig. 12.16b, the load has become large enough to produce a plastic hinge at the wall, and collapse can occur.

The limit load $P_L$ can be determined by isolating the beam at the instant of collapse; the bending moment at the left-hand end is *known* to be the limit moment $M_L$. Clearly,

$$P_L = \frac{M_L}{L}$$

For the cantilever beam, which was statically determinate, a single plastic hinge was required for collapse. If, however, we add redundant supports, we will require more plastic hinges before collapse can occur. Figure 12.17 shows a centrally loaded cantilever beam with a (redundant) support at the end. The

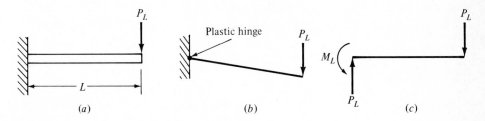

**Figure 12.16** Showing collapse of a cantilever beam.

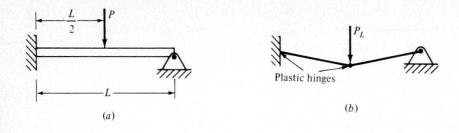

(a)

(b)

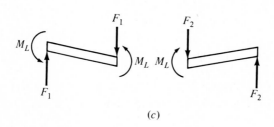

(c)

**Figure 12.17** Collapse of a statically indeterminate beam.

beam cannot collapse until two hinges form. Now we can isolate the two halves of the beam as shown. From moment equilibrium of the two parts we have:

Left half: $\qquad\qquad 2M_L = \dfrac{F_1 L}{2} \qquad F_1 = \dfrac{4M_L}{L}$

Right half: $\qquad\qquad M_L = \dfrac{F_2 L}{2} \qquad F_2 = \dfrac{2M_L}{L}$

thus: $\qquad\qquad P_L = F_1 + F_2 = \dfrac{6M_L}{L}$

As structures become more complex, there are generally more than one collapse mechanism, that is, there are more than one set of plastic hinges that could cause collapse. The true collapse mechanism is that one which requires the smallest limit load. The approach should be clear: Analyze all possibilities to determine the minimum limit load.

## 12.8 PROBLEMS

**12.1–12.10** Isolate each beam and show all loads acting on it. Then draw the shear-force and bending moment diagrams directly under the loading diagram. Label all important points. Be careful with signs.

**12.1**

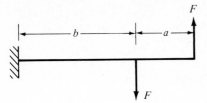

**Figure P12.1**

**12.2**

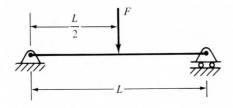

**Figure P12.2**

**12.3**

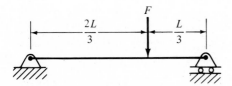

**Figure P12.3**

**12.4**

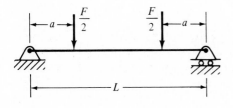

**Figure P12.4**

**12.5**

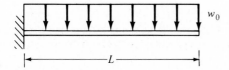

**Figure P12.5**

**12.6**

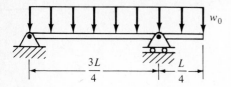

**Figure P12.6**

**12.7**

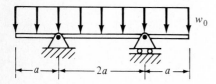

**Figure P12.7**

**12.8**

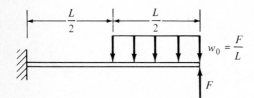

**Figure P12.8**

**12.9**

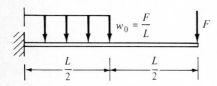

**Figure P12.9**

**12.10**

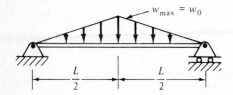

**Figure P12.10**

**12.11** Determine the axial force, shear force, and bending moment as functions of $\theta$.

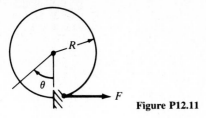

**Figure P12.11**

**12.12** For the angle bracket, determine the bending and twisting moments for points $A$, $B$, and $C$. Note that the bending moment at $C$ is identical to the twisting moment at $B$.

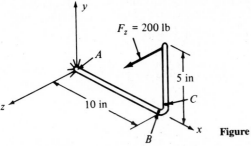

**Figure P12.12**

**12.13** Show the bending and twisting moments at point $A$.

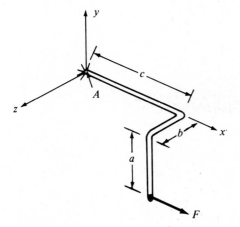

**Figure P12.13**

**12.14** Determine bending and twisting moments as functions of $\theta$.

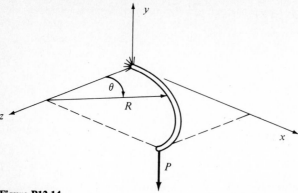

**Figure P12.14**

**12.15–12.20** In many structures in bending it is possible to design the system to minimize the effects of bending; the stresses can be minimized or the deflections can be minimized. For the beams illustrated in Probs. 12.15 to 12.20, determine the location $x$ of the supports such that the maximum bending moment is as small as possible.

**12.15**

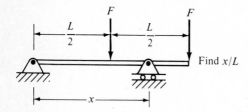

Find $x/L$

**Figure P12.15**

**12.16**

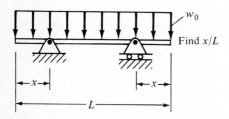

Find $x/L$

**Figure P12.16**

**12.17**

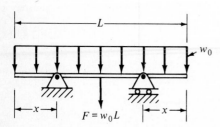

**Figure P12.17**

**12.18**

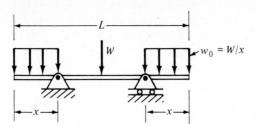

**Figure P12.18**

**12.19**

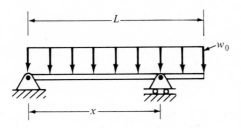

**Figure P12.19**

**12.20**

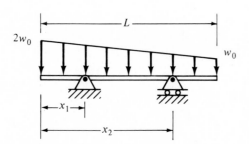

**Figure P12.20**

**12.21** A uniform vertical beam falls over under the action of gravity, rotating about point $O$, as shown. Locate the position $x$ of the maximum bending moment.

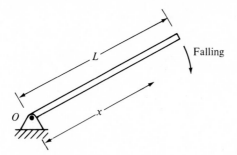

**Figure P12.21**

**12.22–12.32** Problems 12.22 to 12.32, involving uniform beams $(EI)$ of length $L$ under various loads, are to be solved using *singularity functions.* In each case, write the loading equation $q(x)$, integrate it four times to obtain $-V$, $M_b$, $EI\phi$, and $EIv$. Use the boundary conditions to evaluate the four constants of integration. Determine the magnitude and location of both the maximum bending moment and the maximum deflection.

**12.22**

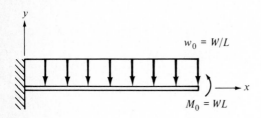

$w_0 = W/L$

$M_0 = WL$   **Figure P12.22**

**12.23**

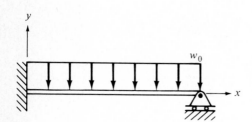

$w_0$

**Figure P12.23**

**12.24**

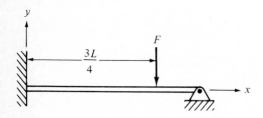

$F$

$\dfrac{3L}{4}$

**Figure P12.24**

**12.25**

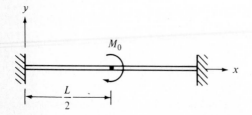

$M_0$

$\dfrac{L}{2}$

**Figure P12.25**

**12.26**

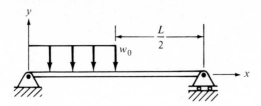

**Figure P12.26**

**12.27**

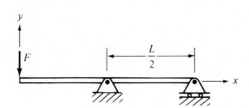

**Figure P12.27**

**12.28**

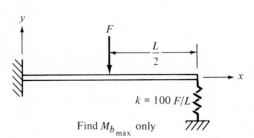

$k = 100\,F/L$

Find $M_{b_{\max}}$ only

**Figure P12.28**

**12.29**

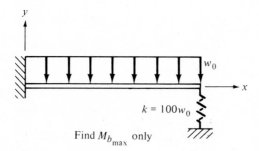

$k = 100w_0$

Find $M_{b_{\max}}$ only

**Figure P12.29**

**12.30**

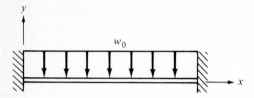

Figure P12.30

**12.31**

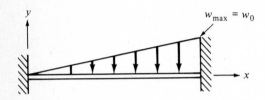

Figure P12.31

**12.32**

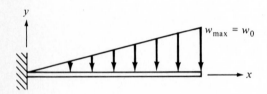

Figure P12.32

**12.33** For the thin-walled channel section shown, determine the location of the shear center ($s$) such that the channel will bend without twisting.

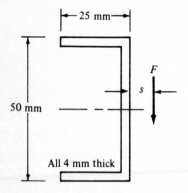

Figure P12.33

**12.34–12.38** Each of the beams illustrated in Probs. 12.34 to 12.38 is 10 in long, 1 in high, and loaded, as a cantilever, by a 50-lb force $P$, as shown. In each case, determine the maximum bending stress and the deflection of the free end.

**12.34**

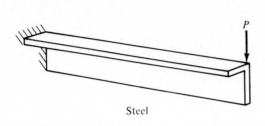

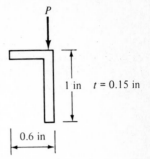

1 in     $t = 0.15$ in

0.6 in

**Figure P12.34**

**12.35**

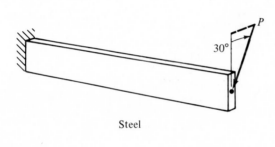

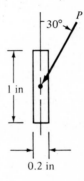

30°

1 in

0.2 in

**Figure P12.35**

**12.36**

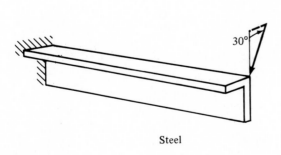

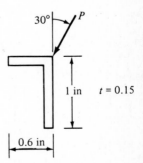

30°

1 in     $t = 0.15$

0.6 in

**Figure P12.36**

**12.37**

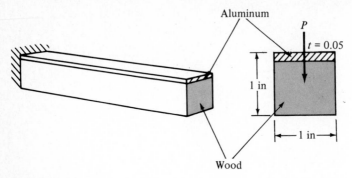

**Figure P12.37**

**12.38**

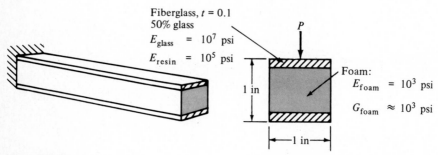

**Figure P12.38**

**12.39** An aluminum flagpole is made by rolling 0.25-in-thick sheet stock into a tapered cylinder and welding the seam. The pole is 40 ft high and tapers linearly from a 5-in

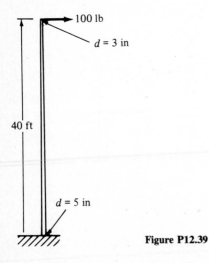

**Figure P12.39**

diameter at the base to a 3-in diameter at the top, as shown in the accompanying drawing. How far will the top of the pole deflect when a 100-lb horizontal force acts at the top?

**12.40–12.42** Determine the fully plastic bending moment (or limit moment $M_L$) for the cross section. Assume that the material is an ideal plastic with a tensile yield stress $\sigma_{yp}$.

**12.40**

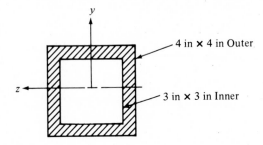

4 in × 4 in Outer

3 in × 3 in Inner

**Figure P12.40**

**12.41**

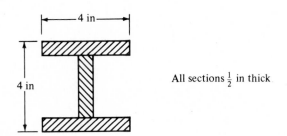

4 in

4 in

All sections $\frac{1}{2}$ in thick

**Figure P12.41**

**12.42**

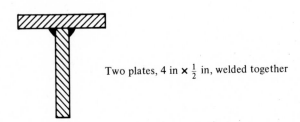

Two plates, 4 in × $\frac{1}{2}$ in, welded together

**Figure P12.42**

**12.43–12.45** The beam of length $L$ has a known limit bending moment $M_L$. Determine the limit load $P_L$ in terms of $M_L$ and $L$.

**12.43**

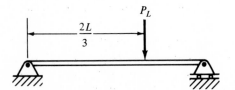

$\dfrac{2L}{3}$

$P_L$

**Figure P12.43**

**12.44**

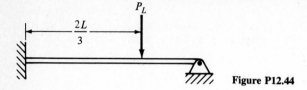

**Figure P12.44**

**12.45**

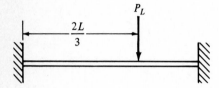

**Figure P12.45**

# THIRTEEN
## COMPUTER APPLICATION EXAMPLES

### 13.1 OVERVIEW

In the earlier chapters, except for the introduction of computerized structural analysis, and the computer examples at the end of most chapters, the emphasis has been on the *understanding* of relatively simple concepts and simple problems. The methods were simple, the equations few, and the computation minimal. Pencil, paper, and calculator (or sliderule) were sufficient.

There are many real situations, however, where the computational effort would be so great that the "old ways" would not suffice. In those cases, either we avoid proper analysis (frequently by oversimplification of the problem), or we must resort to computer methods.

For problems of considerable complexity, there is a growing set of computer analyses that are commercially available. However, in order to use sophisticated software, a considerable learning effort is usually required before useful results can be obtained; and it is often difficult to know whether the results are valid.

For problems of more moderate complexity, a perfectly viable approach is to develop computer analyses that are specifically devoted to a given problem. We will see that the analyses developed in the earlier chapters of this book, coupled with a modest capability in computer programming, will provide you with a considerable competence for solving a wide variety of complex problems.

In this chapter, we will present a set of examples on the use of computers for problems relevant to this text. We will present each problem, develop the necessary analytic background to understand it, and give a program for solving it.

The various analyses and programs will illustrate *how* we can approach rather complex problems, and will provide useful programs that can be altered to suit other problems. Here we will give more detail on logic and programming than we have in the programs given in earlier chapters.

## 13.2 TAPERED BEAM

Whenever the cross section of a beam varies along its length, the moment of inertia ($I_{yy}$) varies and must be included within the integrals. In this example, we will consider a uniformly tapered beam (a shape that can be handled by standard integration for simple loadings). However, the method we will present is sufficiently robust to handle *any* cross section with *any* loading.

Basically, the method involves dividing the beam into a number of segments ($N$), writing the pertinent equations for each segment, and carrying out the integrations numerically. The method could be called "numerical analysis," "finite-element analysis," or "finite-difference analysis." The name is not important, the method is. As we proceed, it should become clear that the method is applicable to a wide variety of problems.

By treating a problem having an exact solution, we can explore the errors introduced via the finite-segment approach.

Consider the solid, rectangular, tapered beam shown in the first illustration below. It is cantilevered from the left-hand end at $x = 0$, where the dimensions are $B \times H$. At $x = L$, the beam has tapered down to a height of $KH$ and a width of $K_1 B$, that is, there are two "taper ratios" $K$ and $K_1$.

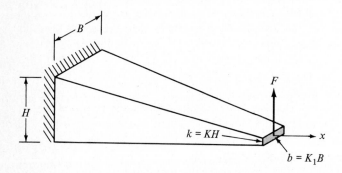

For this type of analysis, we divide the length $L$ into $N$ uniform segments of length $L/N$. It is assumed that for any segment, the bending moment $M_b$ is constant and the section stiffness $EI$ is also constant. The geometry for the beam, and for the $i$th element are shown in the second accompanying illustration.

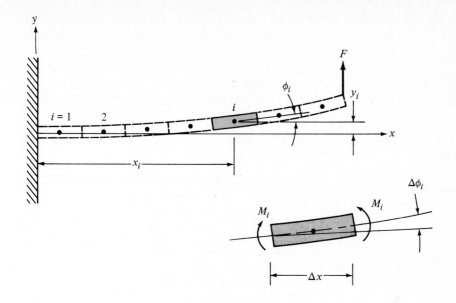

We will write the pertinent equations in terms of the problem variables $x$, $M_b$, $I_{yy}$, $E$, $F$, $\phi$, and $y$. However, we frequently will prefer to obtain output in "dimensionless" form—numbers that can later be applied to the same problem with different numerical input. Dimensionless output can be obtained simply by setting the input parameters equal to 1. In this program, the user is given the option of dimensional or dimensionless output.

Referring to Program Listing 13.1 for "TAPERED BEAM," we see that most of it is devoted to input and output. The actual body of the program is contained in lines 280 to 370 and is really very simple. Consider those lines:

280　X is the position of the center of the $i$th element.

290　M is the bending moment at $x = X$

300　I0 is the moment of inertia ($I_0$) at the wall ($x = 0$).

310　I1 is the moment of inertia at $x$, adjusted for the taper in the vertical direction.

320　I1 is now adjusted for any horizontal taper, and now represents $I_{yy}$ at any position $x$.

330　P1 is the change in slope across the element $i$. This is the familiar "$\phi = ML/EI$" for any simple beam element. This change in slope is the basis for the remaining deformation calculations in this and subsequent beam examples.

340　P is the slope at $i$. We simply sum the slope increments P1.

350　Y is the deflection at $i$. The increment in Y is just the slope at $i$ times the length of the element ($L/N$).

360　S (Sigma), the maximum normal bending stress ($My/I$).

370     T ($\tau$), the shear stress at the neutral axis. For a rectangular section, $\tau$ equals three-halves the mean shear stress (use different equations for other sections).

The above relations are evaluated for the $N$ elements and the results are printed out. Very simple, but very powerful.

```
10 !"        TAPERED BEAM (tapered height and width)" \ !
40 REM
50 REM
60 !"Solution for beam cantilevered at left end (x=0)"
70 !"Vertical load F at right end (x=L)."\!
80 !"Height tapers from H at x=0, to K*H at x=L."
90 !"Width tapers from B at x=0, to K1*B at x=L."\!
100 !"For zero taper, set K or K1 equal to '1'"\!\!
110 INPUT"     Number of elements N= ",N \ ! \ !
120 INPUT"     Height taper rato K = ",K \!\!
130 INPUT"     Width  taper rato K1= ",K1 \!
140 INPUT"     Dimensionless ratios?  (CR/N) ",Z$ \ !\!
150 IF Z$<>"N" THEN 230
160 INPUT"        Length L  (in) = ",L
170 INPUT"        Width  B  (in) = ",B
180 INPUT"        Height H  (in) = ",H
190 INPUT"        Modulus E (psi)= ",E
200 INPUT"        Force  F  (lbs)= ",F \ !\!
210 GOTO 250
220 REM
230 L=1 \ B=12 \ H=1 \ E=1 \ F=1
240 REM
250 !"     X     Moment    Slope     Defl.    Sigma      Tau" \ !
260 REM
270 FOR I=1 TO N
280 X=(I-.5)*L/N              \REM Position x
290 M=F*(L-X)                 \REM Bending moment Mb
300 I0=B*H^3/12               \REM Moment of inertia at x=0
310 I1=I0*(1-(1-K)*X/L)^3     \REM Adjust Io for height taper
320 I1=I1*(1-(1-K1)*X/L)      \REM Adjust Iyy for width taper
330 P1=M*L/(N*I1*E)           \REM Delta-phi (change of slope)
340 P=P+P1                    \REM Slope at i (radians)
350 Y=Y+P*L/N                 \REM Deflection in y-direction
360 S=M*(1-(1-K)*X/L)*H/(2*I1)\REM Max. normal stress  'Sigma'
370 T=3*F/(2*B*(1-(1-K)*X/L)) \REM Max. shear stress    'Tau'
380 IF Z$<>"N" THEN 420
```

**Program Listing 13.1**

```
390 !%10F2,X,M, %10F4,P ,Y , %10F0,S,T
400 NEXT
410 ! \ ! \ ! \ GOTO 450
420  !% 10F3,X,M,P,Y,S,T
430 NEXT
440 ! \ ! \ ! \ END
```

**Program Listing 13.1** (continued)

Consider first, the case of a straight beam $(K = K_1 = 1)$. We know the solution, so we can compare the computer results for various numbers $N$ of elements with the exact solution. In Program Output 13.1*a*, only the results for $x = L$ are shown.

*Note*: In Program Listing 13.1, in order to obtain dimensionless parameters, the width $B$ is set at 12 rather than 1. This is to make the output dimensionless relative to $I_0$ $(BH^3/12)$ rather than to $BH^3$.

```
      TAPERED BEAM (tapered height and width)

Solution for beam cantilevered at left end (x=0)
Vertical load F at right end (x=L).

Height tapers from H at x=0, to K*H at x=L.
Width tapers from B at x=0, to K1*B at x=L.

For zero taper, set K or K1 equal to '1'

      Height taper rato K = 1

      Width  taper rato K1= 1

      Dimensionless ratios?  (CR/N)
```

| N | Slope | Defl. | Error |
|-----|-------|-------|-------|
| 10 | .500 | .358 | .073 |
| 20 | .500 | .346 | .037 |
| 30 | .500 | .342 | .025 |
| 40 | .500 | .340 | .019 |
| 50 | .500 | .338 | .015 |
| 60 | .500 | .337 | .012 |
| 70 | .500 | .337 | .011 |
| 80 | .500 | .336 | .009 |
| 90 | .500 | .336 | .008 |
| 100 | .500 | .336 | .007 |

**Program Output 13.1*a***

The dimensionless numbers for slope and deflection should be $\frac{1}{2}$ and $\frac{1}{3}$ ($\phi = FL^2/2EI$) and ($\delta = FL^3/3EI$). We see that the slope is correct (to three decimal places), and that the deflection is in error by a factor less than $1/N$. The error of about $1/N$ is typical for such analyses, so it is hardly worthwhile using $N$ greater than 20 to 50.

The program has the option of using real dimensions as well as dimensionless quantities. Typical output is shown in Program Output 13.1*b*.

```
        TAPERED BEAM (tapered height and width)

Solution for beam cantilevered at left end (x=0)
Vertical load F at right end (x=L).

Height tapers from H at x=0, to K#H at x=L.
Width tapers from B at x=0, to K1#B at x=L.

For zero taper, set K or K1 equal to '1'

     Number of elements N= 10

     Height taper rato K = .3

     Width  taper rato K1= .3

     Dimensionless ratios?  (CR/N) N

       Length L  (in) = 10
       Width  B  (in) = 1
       Height H  (in) = 1
       Modulus E (psi)= 10000000
       Force  F  (lbs)= 100
```

| X | Moment | Slope | Defl. | Sigma | Tau |
|------|--------|-------|-------|-------|------|
| .50 | 950.00 | .0013 | .0013 | 6343. | 155. |
| 1.50 | 850.00 | .0029 | .0042 | 7114. | 168. |
| 2.50 | 750.00 | .0048 | .0091 | 8014. | 182. |
| 3.50 | 650.00 | .0072 | .0163 | 9062. | 199. |
| 4.50 | 550.00 | .0102 | .0266 | 10267. | 219. |
| 5.50 | 450.00 | .0140 | .0406 | 11607. | 244. |
| 6.50 | 350.00 | .0188 | .0594 | 12973. | 275. |
| 7.50 | 250.00 | .0247 | .0840 | 13996. | 316. |
| 8.50 | 150.00 | .0314 | .1154 | 13548. | 370. |
| 9.50 | 50.00 | .0361 | .1515 | 7980. | 448. |

**Program Output 13.1*b***

We see then, that a very simple program, using numerical integration (summation), can solve rather difficult problems:

An arbitrary bending moment function can be introduced in a DATA statement, and the proper value selected at each $i$ in line 290. For a point moment $M_0$ equal to $FL$ applied at $x = L$, simply set $M = FL$ in line 290.

An arbitrary moment of inertia can also be input as data and recalled in lines 310 to 320.

If you change the section, re-evaluate the stresses.

## 13.3 RING ANALYSIS

The tapered beam example above was particularly simple because of two characteristics: (*a*) The beam was straight, and (*b*) the problem was statically determinate. The transition from straight to curved beams just requires us to take account of both horizontal and vertical components of the deflection increments (as long as the beam height $h$ is small relative to the initial radius of beam curvature).

For this example, we will demonstrate a different method for computing the deflection increments, a method that appears particularly well suited to curved beams.

The transition to an indeterminate beam is more substantial, because we now cannot *a priori* state the value of the bending moment at each location along the beam. As we develop this example, we will see that, just as in the analytic treatment of beams, we must utilize the support displacement constraints in order to obtain a solution.

As in the prior example, we will demonstrate the methods with a problem that can be solved exactly: a ring of mean radius $R$, thickness $T$, width $B$, and loaded, along a diameter by a force $F$, as shown in the first illustration below.

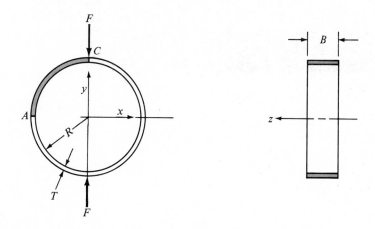

We wish to determine the ring compression in the vertical direction, and the bending moment at all points around the ring (for $T \ll R$).

We first note that an isolation of the entire ring provides no information. Being statically indeterminate, we must "cut" the ring and isolate an appropriate portion or portions. Because of symmetry about the $x$ and $y$ axes, we can isolate the 90° segment $AC$ shown shaded. There can be zero change in slope at both $A$ and $C$, so we can consider $AC$ to be built in at $A$ as shown in the next illustration below. As in the previous example, we divide the segment into $i = 1, 2, \ldots, N$ elements.

At $C$, because we are considering just part of the ring, the applied force will be $F/2$, and the deflection $\delta/2$. In order to maintain the slope equal to zero at $C$, we must apply an *unknown* moment $M_0$. We will determine $M_0$ such that the slope does remain zero.

The second free body below is a portion of the 90° segment determined by the angle $\theta$. At $B$, we see a support force $R_B$ and the bending moment $M_b$. We want to determine $M_b$ so that we can determine the change of slope at $B$ and hence the total deformation of the ring.

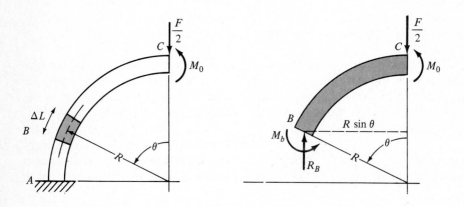

Summing the moments about $B$, we have:

$$M_b = \frac{F}{2} R \sin \theta - M_0$$

The change in slope, across element $i$ at $B$, is:

$$\Delta\phi = \frac{M_b \, \Delta L}{EI} = \frac{\Delta L}{EI} \left( \frac{F}{2} R \sin \theta - M_0 \right)$$

The total change in slope, from $A$ to $C$, is just the summation of the $N$ incremental changes:

$$\phi = \sum_i \Delta\phi_i = \frac{\Delta L}{E} \sum_i \left( \frac{FR}{2I_i} \sin \theta_i - \frac{M_0}{I_i} \right)$$

But the total change of slope must be zero, so that the summation above must also be zero:

$$M_0 \sum_i \frac{1}{I_i} = \frac{FR}{2} \sum_i \frac{\sin \theta_i}{I_i}$$

We should note that for this particular problem, where $I_{yy}$ is constant, it can be taken outside of the summations. $M_0$ could be determined analytically to be $FR/\pi$. However, we are setting up a procedure that will accommodate variable $I_{yy}$, and will set up the program accordingly, as shown in Program Listing 13.2.

```
10 !"              RING ANALYSIS"\!
50 REM
60 !" Cylindrical 'ring' with vertical load."\!
70 !" Vertical load F acts along a diameter"
80 !" Average ring radius = R"
90 !" Ring width = B"
100 !" Ring thickness = T; ( T can vary with angular position if desired)"\!\!
110 INPUT"        Number of Elements N = ",N\!
120 INPUT"        Dimensionless ratios?  (CR/N) ",Z$ \!\!
130 IF Z$<>"N" THEN 220
140 INPUT"        Ring radius       R = ",R \!
150 INPUT"        Ring width        B = ",B \!
160 INPUT"        Ring thickness    T = ",T \!
170 INPUT"        Modulus           E = ",E \!
180 INPUT"        Force             F = ",F \!\!
190 REM
200 GOTO 240
210 REM
220 R=1 \ B=1 \ T=1 \ E=1 \ F=1
230 REM
240 !"    Theta    Moment"\!
250 A1 = 1.5708/N                  \REM  Delta theta (A = angle)
260 I0 = B*T*T*T/12                \REM  Iyy Could vary with theta)
270 L1 = R*A1                      \REM  Element length
280 REM
290 FOR I=1 TO N
300 A  = (I-.5)*A1                 \REM  Theta, radians
310 S1 = S1+1/I0
320 S2 = S2+SIN(A)/I0
330 NEXT
340 REM
350 M0 = F*R*S2/(2*S1)             \REM  Moment at theta = 0
```

**Program Listing 13.2**

```
360 REM
370 FOR I=1 TO N
380   A = (I-.5)*A1
390   M = F*R*SIN(A)/2-M0          \REM  Bending moment
400   P1= M*L1/(E*I0)              \REM  Delta phi
410   Y = Y + 2*P1*R*SIN(A/2)*COS(A/2)  \REM  Deflection
420   !%10F2, 57.296*A, M
430 J=J+1
440 IF J<>20 THEN 470
450 J=0
460 INPUT" CR to continue ",Z1$
470 NEXT
480 !\!
490 IF Z$<>"N" THEN 540
500 !"  Deflection  delta = ",%10F5,2*Y \!
510 !"  Spring constant k = ",%10F1,F/(2*Y) \!
520 !"  End moment     Mo = ",%10F3,M0 \!\!\GOTO 570
530 REM
540 !"  Deflection  delta = ",%8F5,2*Y,"        Dimensionless ratios" \!
550 !"  Spring constant k = ",%8F5,.5*F/Y \!
560 !"  End moment     Mo = ",%8F5,M0 \!\!\END
```

**Program Listing 13.2** (continued)

Again, most of the program involves input, output, and comments. The calculations are contained in lines 250 to 410, as follows:

250    A1 is the increment in angle $\theta$ ($\Delta\theta$).
260    I0 is the moment of inertia $I_{yy}$ for element $i$ (constant for the example).
270    L1 is the increment of beam length $R\,\Delta\theta$.
300    A is the angle $\theta$ for element $i$ (rad).
310    S1 is the summation of $1/I_{yy}$.
320    S2 is the summation of $\sin\theta/I_{yy}$.
350    M0 is the end moment $M_0$ required for zero slope.
390    M is the bending moment ($M_b$) acting on element $i$.
400    P1 is the increment of slope at $i$ ($ML/EI$).
410    Y is the vertical component of the deflection at point $C$. The associated analysis is described below.

For this problem, because the beam is curved, we use a different method for computing the deflection from the incremental changes in slope.

In the sketch on page 423 we show the deflection of point $C$ due to the rotation of a single element at the angle $\theta$. When *only* that element rotates, the rest of the ring remains *rigid*. The portion between $B$ and $C$ simply rotates, as a rigid body, about $B$. Thus point $C$ moves in an arc about $B$.

Because the deformations are small, the motion of $C$ to $C'$ can be approximated by a straight line perpendicular to $BC$. We can then calculate the vertical and horizontal components of the total motion at $C$. This vertical component, due to the rotation of a single element, is the quantity summed in line 410.

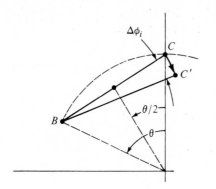 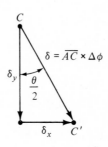

An example of dimensionless output is shown in Program Output 13.2a. With only 10 elements, the deflection error is less than 2 percent. We could note that the bending moment goes through zero at an angle near 40°. The true angle for zero moment is 39.6°. So again, we obtain good precision with little effort.

Also shown, in Program Output 13.2b, is the output for an example involving real dimensions.

```
        RING ANALYSIS

Cylindrical 'ring' with vertical load.

Vertical load F acts along a diameter
Average ring radius = R
Ring width = B
Ring thickness = T; ( T can vary with angular position if desired)


    Number of Elements N = 10

    Dimensionless ratios?  (CR/N) Y
```

**Program Output 13.2a**

```
Theta    Moment

 4.50    -.28
13.50    -.20
22.50    -.13
31.50    -.06
40.50     .01
49.50     .06
58.50     .11
67.50     .14
76.50     .17
85.50     .18
```

Deflection   delta =   1.76962        Dimensionless ratios

Spring constant k =    .56509

End moment      Mo =   .31864

**Program Output 13.2a** (continued)

### RING ANALYSIS

Cylindrical 'ring' with vertical load.

Vertical load F acts along a diameter
Average ring radius = R
Ring width = B
Ring thickness = T; ( T can vary with angular position if desired)

Number of Elements N = 20

Dimensionless ratios?   (CR/N) N

Ring radius       R = .5

Ring width        B = .5

Ring thickness    T = .1

**Program Output 13.2b**

```
    Modulus          E = 30000000
    Force            F = 100

  Theta   Moment
   2.25   -14.94
   6.75   -12.98
  11.25   -11.04
  15.75    -9.13
  20.25    -7.27
  24.75    -5.45
  29.25    -3.70
  33.75    -2.03
  38.25     -.44
  42.75     1.05
  47.25     2.44
  51.75     3.71
  56.25     4.87
  60.75     5.89
  65.25     6.78
  69.75     7.54
  74.25     8.14
  78.75     8.60
  83.25     8.91
  87.75     9.06
CR to continue

  Deflection  delta =     .00148

  Spring constant k =    67362.0

  End moment     Mo =     15.920
```

**Program Output 13.2b** (continued)


## 13.4 EQUIVALENT BEAM

There are occasions when we would like to use a *nonuniform* beam with one of the structural analysis programs. But those programs were written for uniform beams only. If we can replace the nonuniform beam with an *equivalent uniform beam*, then we can use the programs. In this example, we will limit ourselves to making the beams "equivalent" in bending only.

In this and foregoing chapters we have seen how to obtain stiffness parameters for a wide variety of beams; how can we use these quantities to define an equivalent beam?

For use in the FEM programs, the equivalent beam must undergo the equivalent deflection and rotation at an end when a point force and a point moment are applied at the end. That is, four parameters must be identical for the real and the equivalent beam:

$$K_1 = \frac{\phi}{M} \qquad K_2 = \frac{\delta}{M} \qquad K_{2'} = \frac{\phi}{F} \qquad K_3 = \frac{\delta}{F}$$

The second and third parameters are identical, so that only three conditions must be met. If you think about it, you will realize that we cannot design a single equivalent beam to satisfy three requirements. By varying $L$ and $I_{yy}$, we can only satisfy two conditions. However, if we place two uniform beams end-to-end as shown in the accompanying illustration, we can satisfy *four* conditions.

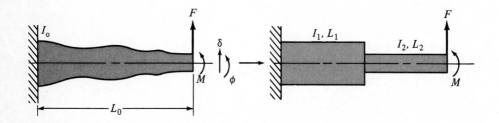

We can write the three stiffness parameters $K$ for the equivalent beam using superposition and the equations from Chap. 8 as follows:

$$K_1 = \frac{L_1}{EI_1} + \frac{L_2}{EI_2}$$

$$K_2 = \frac{L_1^2}{2EI_1} + \frac{L_1 L_2}{EI_1} + \frac{L_2^2}{2EI_2}$$

$$K_3 = \frac{L_1^3}{3EI_1} + \frac{L_1^2 L_2}{EI_1} + \frac{L_1 L_2^2}{EI_1} + \frac{L_2^3}{3EI_2}$$

The three $K$'s can be replaced with their dimensionless equivalents:

$$Z_1 = K_1 \frac{EI_0}{L_0}$$

$$Z_2 = K_2 \frac{EI_0}{L_0^2}$$

$$Z_3 = K_3 \frac{EI_0}{L_0^3}$$

Because we can set four parameters, but need only three, we can be arbitrary relative to one. For this analysis, the moment of inertia of beam element 1 ($I_1$) is set equal to that for the original beam at $x = 0$ ($I_0$); that is, $I_1 = I_0$.

We can now determine three quantities, which are chosen to be dimensionless ratios:

$$R_1 = \frac{L_1}{L_0} \qquad R_2 = \frac{L_2}{L_0} \qquad R_3 = \frac{I_0}{I_2}$$

A bit of manipulation will show that:

$$Z_1 = R_1 + R_2 R_3$$

$$Z_2 = \frac{R_1^2}{2} + R_1 R_2 + \frac{R_2^2 R_3}{2}$$

$$Z_3 = \frac{R_1^3}{3} + R_1^2 R_2 + R_1 R_2^2 + \frac{R_2^3 R_3}{3}$$

We now simply solve for the three $R$'s knowing the three $Z$'s. This sounds simple, but is not (at least for the author).

The short program below solves these equations by "brute-force" methods:

Assume a value for $R_1$.
Use the first and second equations to solve for $R_2$ and $R_3$ in terms of $R_1$.
See if the third equation is satisfied; if not, change $R_1$.

The program very quickly gives the required values.

Program Listing 13.3 and Program Output 13.3 involve a specific set of $Z$ values.

```
10 !"               EQUIVALENT BEAM"\!
20 !"   Two cantilever beams in series, length and Iyy adjusted to"
30 !"   match generalized cantilever stiffness parameters. "
40 !"   The input quantities are dimensionless stiffnesses." \!
50 REM
60 INPUT" Phi divided by MLo/EIo   = Z1 = ",Z1\!
70 INPUT" Defl    ''       MLo^2/EIo = Z2 = ",Z2\!
80 INPUT" Defl    ''       FLo^3/EIo = Z3 = ",Z3 \!
90 REM
100 R1=0
110 R2=(2#Z2-R1#R1)/(R1+Z1)
120 R3 = (Z1-R1)/R2
130 D = R1^3/3 + R1#R1#R2 + R1#R2#R2 + R2^3#R3/3 - Z3
140 IF D<0 THEN 160
150 GOTO 190
160 R1=R1+.01
170 GOTO 110
180 REM
190 !%# 10F5
200 !"    R1 = ",R1
210 !"    R2 = ",R2
220 !"    R3 = ",R3
```

**Program Listing 13.3**

EQUIVALENT BEAM

Two cantilever beams in series, length and Iyy adjusted to
match generalized cantilever stiffness parameters.
The input quantities are dimensionless stiffnesses.

Phi divided by MLo/EIo   = Z1 = 2.375

Defl   ''      MLo^2/EIo = Z2 = .887

Defl   ''      FLo^3/EIo = Z3 = .510


      R1 =     .60000
      R2 =     .47529
      R3 =    3.73453

**Program Output 13.3**


## 13.5 SHAFT WHIP

All rotating devices (indeed, all mechanical devices) are subject to the phenomenon of *resonance* at certain *critical speeds* (natural frequencies). Most people are aware of the severe vibration in a washing machine when it goes into its spin cycle. As the machine speeds up, it vibrates more and more, and then, beyond a critical speed, the vibration dies away, and the machine runs smoothly. If the machine were run at or near the critical speed for any length of time, it might well "self-destruct"—the amplitude of vibration would grow until something (or someone) was damaged.

Rotating shafts, no matter how they are supported, will exhibit a series of critical speeds. The resonance characteristics of uniform shafts are well understood, and equations for critical speeds can be found in most books on vibration.

However, if the shafts are *not* uniform, or if the supports are not simple supports, solutions may not be available.

Here we will demonstrate how the methods we have been developing can be applied to an otherwise very difficult problem. We will consider the simple case of a uniform shaft, cantilevered from the left-hand end in bearings. As usual, we will divide the shaft into $N$ elements.

We will permit various initial eccentricities at any $i$, so we can investigate crooked shafts. We will also allow for a spring support at any position $i$. A spring support provides a radial restoring force proportional to the beam runout at that position.

The overall geometry is shown in the two diagrams below. At each element, the actual radius of rotation $r_i$ is the sum of the eccentricity $e_{0i}$ and the shaft elastic deflection $u_i$.

Acting on each element, in *addition* to the forces due to the rest of the beam (not shown in the diagrams), is a spring force $F = k_i r_i$.

Because the center of the mass element is rotating in a circle of radius $r_i$, it is accelerated inward radially with the centripital acceleration $a = r\omega^2$ ($\omega$ is the rotational speed in radians/second). Thus, the d'Alembert's force, which equals $ma$ in magnitude, acts radially outward ($F = mr\omega^2$).

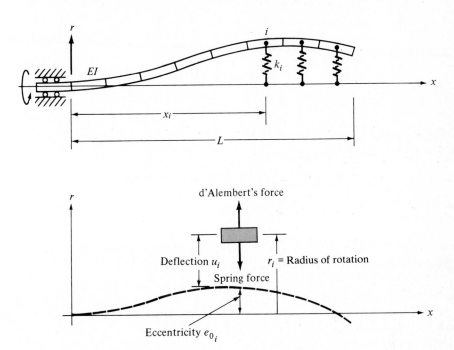

Thus we have a rather complex situation where the external forces acting on the shaft (beam) are a function of the beam deflection, somewhat akin to the buckling problem. However, let us use a different approach.

In this case, we have forces applied to every element, which will affect the deflection of every element. Let us define a set of "influence coefficients" ($C_{ij}$) which give the deflection at $i$ ($u_i$), due to a force at $j$ ($F_j$), as shown in the illustration on page 430.

At any element $i$, the total deflection is the sum of the contributions from the $j = 1$ to $N$ forces:

$$u_i = \sum_i C_{ij} F_j$$

Now, if we determine the forces $F_j$ (due to the springs and the d'Alembert's forces) we should be able to compute the deflections.

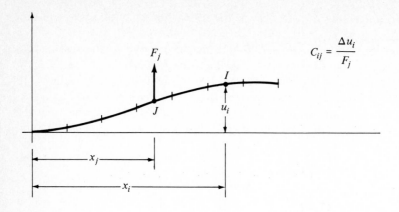

The force is:

$$F_j = m_j\omega^2 r_j - k_j r_j$$

But

$$r_j = e_{0j} + u_j$$

And so

$$u_i = \sum_j C_{ij}(m_j\omega^2 - k_j)(e_{0j} + u_j)$$

The summation can be split into two parts, one of which can be carried out first because it does not depend on the $u_j$:

$$e_i = \sum_j C_{ij}k_{1j}e_{0j}$$

where $k_{1j} = m_j\omega^2 - k_j$.

Then we can write:

$$u_i = e_i + \sum_j C_{ij}k_{1j}u_j$$

We can solve for $u_i$ in terms of the following matrix equation:

$$[k]\{u\} = \{e\}$$

where $k_{ii} = 1 - C_{ii}k_{1i}$

$k_{ij} = -C_{ij}k_{1j}$

At this point, if we knew the $C_{ij}$, we could solve the above matrix equation using gaussian elimination or other methods. The $C_{ij}$ can be obtained through superposition and the simple cantilever beam equations given in Chap. 8. We find that for $i > j$,

$$C_{ij} = \frac{\Delta L^3}{EI}\left[\frac{(j - 0.5)^3}{3} + \frac{(j - 0.5)^2(i - j)}{2}\right]$$

These equations are implemented in Program Listing 13.4. The gaussian elimination section is taken directly from the FEM programs.

This program, which is a bit more ambitious than the previous ones, takes some time to run. However, good accuracy is obtained with a few elements—five seem ample.

```
10 !"                 SHAFT WHIP"\!
20 !"              Filed under WHIP",\!
30 REM
40 !"      Cantilevered shaft rotating at n rpm"\!
50 !"      Can vary initial eccentricity (E0(I))"
60 !"      Can attach 'spring' K2(I) ιo ground at any I"
70 !"      Modulus and density are for steel"\!
80 INPUT"      Number of elements    N = ",N\!
90 INPUT"      Length of shaft (in)  L = ",L\!
100 INPUT"      Radius of shaft (in)  R = ",R\!
110 INPUT"      Initial RPM           N1 = ",N1\!
120 INPUT"      Speed increment (rpm)  = ",N2\!\!
130 !"      RPM        Radius of rotation at"
140 !"            I = 1      2      3      4      5      6"\!
150 REM
160 DIM C(N,N), K(N,N), F(N),U(N), K2(N), K1(N), R(N), E(N), E0(N), F1(N)
170 REM
180 M1 = 3.1416*R*R*L*.284/(386*N)        \REM  Mass per element
190 E1 = 30*10^6*3.1416*R^4/4             \REM  E*I
200 REM  K2(N) is the spring constant at any I.  Two examples below
210 K2(N)=100000000                       \REM  'Fixed' support at x=L
220 K2(N)=0                               \REM  No support at x=L
230 X2=(L/N)^3/(E1)
240 REM
250 FOR I=1 TO N
260  E0(I)=.01                            \REM  Initial eccentricity at i"
270 NEXT
280 REM
290  FOR I=1 TO N
300   FOR J=1 TO I
310    C(I,J) = X2*( (J-.5)^3/3 + (I-J)*(J-.5)^2/2 )
320    C(J,I)=C(I,J)                      \REM  Influence coef.
330   NEXT
340  NEXT
350 REM
360 !%10I,N1,
370 W=N1*6.2832/60                        \REM  Radians/second
```

**Program Listing 13.4**

```
380 K0=M1*W*W
390 FOR J=1 TO N
400  K1(J)=K0-K2(J)
410  E(J)=0
420 NEXT
430 FOR I=1 TO N
440  FOR J=1 TO N
450   E(I) = E(I)+C(I,J)*K1(J)*E0(J)  \REM  Vector {e}
460  NEXT
470 NEXT
480 REM
490  FOR I=1 TO N
500   FOR J=1 TO N
510    K(I,J)=-C(I,J)*K1(J)
520    IF I=J THEN K(I,J)=K(I,J)+1   \REM  Matrix (k)
530   NEXT
540  NEXT
550 REM
560 REM
570 REM
580 FOR I=1 TO N
590 U(I)=0
600 NEXT
610 REM ************************************************************************
620 REM
630 REM       SOLVE FOR DISPLACEMENTS (u) BY GAUSSIAN ELIMINATION
640 REM
650 REM                   (k)*{u} = {e}
660 REM
670 REM       **** a. Triangularize matrix (k) ****
680 REM
690 FOR I=1 TO N
700  FOR J=I+1 TO N
710   Z=-K(J,I)/K(I,I)
720    FOR K=I TO N
730    K(J,K)=K(J,K)+Z*K(I,K)
740    NEXT
750    E(J) = E(J) + Z*E(I)
760   NEXT
770 NEXT
780 REM
790 REM       **** b. Calculate displacement vector {u} ****
800 REM
```

**Program Listing 13.4** (continued)

```
810 FOR I=N TO 1 STEP-1
820 Z=0
830 FOR J=I+1 TO N
840   Z=Z+U(J)*K(I,J)
850 NEXT
860 U(I) = (E(I)-Z)/K(I,I)          \REM  Deflection at I
870 NEXT
880 REM
890 FOR I=1 TO N
900   R(I)=U(I)+E0(I)                \REM  Radius of rotation
910   !%10F5,R(I),
920 NEXT
930 !
940 N1=N1+N2
950 GOTO 360
```

**Program Listing 13.4** (continued)

To run the program, we must input a speed (rpm) which we know is below the critical point. We next input the speed increment by which the speed will be increased stepwise. For each speed, the program will calculate the shape of the shaft. A resonance is identified by large vibration amplitude followed by a change in sign of the amplitude. This corresponds to a phase change of 180°.

Two runs are shown below. In Program Output 13.4a, the shaft is simply cantilevered. All spring constants are zero. The resonance is between 650 and

```
     SHAFT WHIP

     Filed under WHIP
Cantilevered shaft rotating at n rpm

Can vary initial eccentricity (E0(I))
Can attach 'spring' K2(I) to ground at any I
Modulus and density are for steel

Number of elements    N = 5

Length of shaft (in)  L = 72

Radius of shaft (in)  R = 1

Initial RPM          N1 = 500

Speed increment (rpm)   = 25
```

**Program Output 13.4a**

| RPM | Radius of rotation at | | | | | |
|-----|-------|-------|-------|-------|-------|---|
|     | I = 1 | 2 | 3 | 4 | 5 | 6 |
| 500 | .01038 | .01300 | .01736 | .02267 | .02836 | |
| 525 | .01048 | .01383 | .01940 | .02620 | .03349 | |
| 550 | .01063 | .01503 | .02238 | .03137 | .04101 | |
| 575 | .01086 | .01694 | .02713 | .03961 | .05300 | |
| 600 | .01130 | .02046 | .03585 | .05475 | .07506 | |
| 625 | .01235 | .02901 | .05707 | .09163 | .12878 | |
| 650 | .01868 | .08044 | .18488 | .31373 | .45244 | |
| 675 | .00396 | -.03918 | -.11239 | -.20294 | -.30054 | |
| 700 | .00763 | -.00941 | -.03841 | -.07438 | -.11320 | |
| 725 | | | | | | |

**Program Output 13.4a** (continued)

SHAFT WHIP

Filed under WHIP

Cantilevered shaft rotating at n rpm

Can vary initial eccentricity (E0(I))
Can attach 'spring' K2(I) to ground at any I
Modulus and density are for steel

Number of elements    N = 5

Length of shaft (in)  L = 72

Radius of shaft (in)  R = 1

Initial RPM           N1 = 3000

Speed increment (rpm)  = 100

| RPM | Radius of rotation at | | | | | |
|-----|-------|-------|-------|-------|-------|---|
|     | I = 1 | 2 | 3 | 4 | 5 | 6 |
| 3000 | .01209 | .02158 | .02594 | .01764 | .00000 | |
| 3100 | .01274 | .02550 | .03188 | .02220 | .00000 | |
| 3200 | .01374 | .03149 | .04100 | .02912 | .00000 | |
| 3300 | .01553 | .04222 | .05742 | .04126 | .00000 | |
| 3400 | .01990 | .06849 | .09789 | .07129 | .00000 | |
| 3500 | .04728 | .23327 | .35139 | .26111 | .00002 | |
| 3600 | -.01478 | -.14044 | -.22384 | -.16887 | -.00001 | |
| 3700 | .00037 | -.04928 | -.08362 | -.06426 | -.00000 | |
| 3800 | .00378 | -.02876 | -.05205 | -.04058 | -.00000 | |
| 3900 | .00538 | -.01918 | -.03727 | -.02973 | -.00000 | |
| 4000 | .00658 | -.01209 | -.02675 | -.02257 | -.00000 | |
| 4100 | | | | | | |

**Program Output 13.4b**

675 rpm. An exact solution gives 655 rpm. If a run is made with 10 elements, and a speed increment of 1 rpm, we find the resonance between 655 and 656 rpm.

For Program Output 13.4b, the spring constant at $I = N$ is $10^8$ lb/in. This is equivalent to a fixed support. We find the resonance to be between 3500 and 3600 rpm. An exact solution gives 2892 rpm. For this case, the beam shape is more complex than in the previous example, so we really should use more elements to obtain better precision. If we make the same run with 10 elements, the resonance is located between 3100 and 3200 rpm; a considerable improvement. Given patience, good accuracy can be achieved.

Certainly these methods would not be used when exact solutions exist; however, the above should demonstrate how rather complex problems can be handled.

## 13.6 RECIPROCATING DRIVE COMPONENTS

In the world of mechanical engineers and chemical engineers there are a great many pumps: those for pumping gases, fluids, and slurries; small pumps, large pumps; high-pressure and low-pressure pumps; high-volume and low-volume pumps; reciprocating and rotary pumps.

High-pressure pumps generally involve reciprocating pistons, in cylinders with valves, much like the piston-cylinder-valve mechanisms in an internal combustion engine. In fact, if the maximum pressures are of the order of 1000 psi or less, the mechanism might well look much like an engine.

For some chemical processes, however, gases and liquids may require pressurization to 30,000 psi or more. Here the fluid pressure begins to approach the yield strength of metals, and the entire structural design must be very carefully carried out. Any device of a reciprocating nature, and operating at high stress levels, is a natural candidate for fatigue failure.

Consider a portion of a hypothetical, high-pressure machine, as shown schematically on page 436. A crankshaft *OA* is driven at 300 rpm by a 3500-hp electric motor. The "throw" for the crank *OA* is 5 in, giving a "stroke" of 10 in. The mechanism drives two pistons at *D* and *E*. In order to carry the driving force around the cylinder at *D*, a framework of crossheads and struts is used as shown. During each cycle, the total alternating force applied to the frame is 200,000 lb, each strut carrying ±100,000 lb. For this example, we will consider the structural integrity of this framework.

The crossheads are very stiff cast-steel beams; however, because the loads are very high, they will bend as shown in the drawing on page 436. This beam bending will cause bending in the struts in addition to the axial forces in the struts. Of course, the bending of the beams is somewhat limited by the bending stiffness of the struts.

The weak point in any mechanism like this is apt to be at the threads in the bolted connection between the struts and the crossheads. Generally, there are

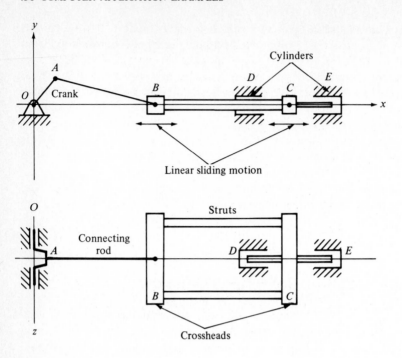

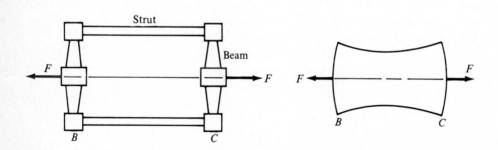

large average (mean) loads, large alternating loads, and a fatigue strength reduction factor $K_f$ of about 4.

In order to reduce the *alternating* load applied to the threads, a bolt-sleeve combination is often used. The bolt is prestressed in tension, putting the sleeve in compression. Then, when an alternating load is applied, the sleeve carries a portion of it.

The details of construction are shown in the illustration on page 437. The beam consists of relatively rigid center and end sections, with a square, hollow, tapered section in between. The beam wall thickness $t$ tapers, as well as the outer dimensions.

The strut rod and sleeve are cylindrical as shown. The threads are very carefully "rolled" 3″-12 UN (unified standard screw thread, nominally 3 in in diameter, having 12 threads per inch). The diameter at the "root" of the thread is 2.898 in. Typically, the "worst" point is at the root of the first thread in engagement (point $F$ in the drawing). Convince yourself that the worst point is at $F$ and not at $G$.

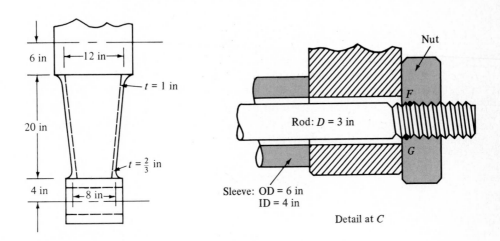

The bolt material is AISI 4340 with the following properties:

$$\sigma_{yp} = 100,000 \text{ psi}$$

$$\sigma_{end} = 50,000 \text{ psi}$$

$$\sigma_{ult} = 130,000 \text{ psi}$$

$$E = 30 \times 10^6 \text{ psi}$$

$$K_f = 4.0$$

We will ask several questions concerning this structure, each of which constitutes an "example" problem:

1. In order to use the structural analysis programs (FEM), find an appropriate equivalent beam to replace the tapered beam actually used.
2. Model the structure and determine the input for a FEM program.
3. Assume that a prestress of 15,000 psi is desired in the bolt. Find the required initial bolt stretch ($\delta_0$). (We cannot "pretension" this bolt by tightening with a wrench. It must be stretched by a hydraulic jack arrangement before tightening the nut.)
4. With the 15,000-psi prestress, apply ±200,000 lb to the structure. Find the

nominal mean and alternating stresses at $F$. Assuming that the prestress is *actually* 15,000 psi, determine the safety factor (relative to fatigue). Consider ways to improve the safety factor.

Almost everything connected with answering these problems has already been explained earlier in the text. Accordingly, the discussion here will be brief.

Because there are numerous pages of computer output for the answers, they are assembled together at the end of the example.

### 13.6.1 Problem 1—Finding the Equivalent Beam

We will first use Program Listing 13.1 ("TAPERED BEAM") (Sec. 13.2) to determine the pertinent dimensionless stiffness parameters for the actual beam. Then we will use Program Listing 13.3 ("EQUIVALENT BEAM") (Sec. 13.4) to find the equivalent pair of uniform beams.

We will consider the ends of the beam to be relatively stiff, so that all compliance occurs in the central, tapered section.

The moment of inertia at the large end of the beam is

$$I_0 = \frac{12^4 - 10^4}{12} = 895 \text{ in}^4$$

Because the cavity inside the beam has the same "taper ratio" as the outside, we can use the program TAPER with $K = \frac{8}{12} = 0.667$, $I_0 = 895$, and $L_0 = 20$. Running the program as written, we obtain stiffness quantities for the beam loaded with a *force* at its end. The results are (see Program Output 13.5 at the end of this section):

Slope: $\qquad$ $\phi$ divided by $\dfrac{FL^2}{EI_0} = 0.875$

Deflection: $\qquad$ $\delta$ divided by $\dfrac{FL^3}{EI_0} = 0.522$

To obtain equivalent quantities for a *moment* at the end, we simply change a program statement to make the bending moment constant along the beam, of magnitude $FL = 1$. This can be done by writing at line 290:

$$M = F*L$$

The results, shown in Program Output 13.6 are:

Slope: $\qquad$ $\phi$ divided by $\dfrac{ML}{EI_0} = 2.374$

Deflection: $\qquad$ $\delta$ divided by $\dfrac{ML^2}{EI_0} = 0.935$

Theoretically, the 0.875 and the 0.935 should be equal. We will replace them with the mean value, 0.905.

We can now load the "EQUIVALENT BEAM" program, input the dimensionless stiffness parameters $(Z_1, Z_2, Z_3)$, and output the dimensionless beam ratios $(R_1, R_2, R_3)$ (see Program Output 13.7 at the end of this section):

$$Z_1 = 2.374 \qquad R_1 = 0.570 \qquad L_1 = 20(0.570) = 11.4$$

$$Z_2 = 0.905 \qquad R_2 = 0.505 \qquad L_2 = 20(0.505) = 10.1$$

$$Z_3 = 0.522 \qquad R_3 = 3.576 \qquad I_2 = \frac{895}{3.576} = 250$$

Thus (for bending purposes) we can replace the tapered beam with two uniform beams end to end. The large end of the tapered beam is replaced with a beam 11.4 in long with $I_{yy} = 895$. This is joined to a 10.1-in beam with $I_{yy} = 250$.

## 13.6.2 Problem 2—Modeling the Structure

The structure has a high degree of symmetry, so we can properly model only a portion of it. The slope of the strut must be zero at its midpoint, and the slope of the beam must be zero where the external force is applied. Accordingly, we model a quarter of the structure, as shown in the illustration below.

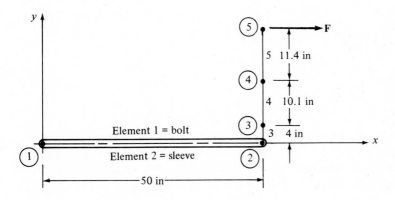

Node 1 represents the center of the strut. In the model, the rod (1), and the sleeve (2) are joined here. Only motion in the $y$ direction is permitted. Node 5 represents the end of the tapered beam; we permit motion only in the $x$ direction.

The pertinent numerical values are given in the following table.

Element 3 is relatively stiff, so we will simply introduce large values for the area and $I_{yy}$. Since there is no axial force on beams 1 and 2, we will give them arbitrary, large areas in order that the programs can function.

| Element | Area, in$^2$ | $I_{yy}$, in$^4$ |
|---|---|---|
| 1 (bolt) | 7.07 | 3.98 |
| 2 (sleeve) | 15.71 | 51.05 |
| 3 (spacer) | ?? | ?? |
| 4 (beam 2) | ?? | 250 |
| 5 (beam 1) | ?? | 895 |

The input data are shown in the first run of the structural analysis program, given in Program Output 13.8a, at the end of this section.

### 13.6.3 Problem 3—Prestressing

To prestress the bolt in tension, we must make it shorter than it would normally be. In the FEM programs, this is input as a negative $\delta_0$ ("delta-zero") for the element that represents the bolt.

Although we do not know how large $\delta_0$ should be, it is very easy to find out. Because all effects in the FEM programs are elastic effects, the pretension will be proportional to $\delta_0$. So we can choose any $\delta_0$ find the pretension, and adjust $\delta_0$ accordingly.

In the first run, shown in Program Output 13.8b, $\delta_0 = -0.025$ in, the pretension is 73,098 lb. We want $15,000(7.07) = 106,000$ lb. Therefore, set $\delta_0 = -0.025(106,000)/73,098 = -0.036$ in.

The second FEM run, shown in Program Output 13.8c, is for $\delta_0 = -0.036$. The resulting pretension is 105,261 lb—close enough.

### 13.6.4 Problem 4—Fatigue Analysis

The next two runs, shown in Program Outputs 13.8d and 13.8e, are for the prestressed system, with an applied load of $\pm100,000$ lb. These represent the design load conditions. We see that the strut (element 1) undergoes variations in both axial loads (136,300- to 74,221-lb tension) and bending moments ($-4869$ to $+4869$ in·lb).

We are interested in the "worst" conditions, which will occur at the root of the thread, at a diameter of 2.898 in. The area $A$ and $I_{yy}$ there are:

$$A = 6.60 \text{ in}^2 \qquad I_{yy} = 3.46 \text{ in}^4$$

The stresses are $\sigma_x = F/A + (-My/I)$:

At $+100,000$ lb:          $\sigma_x = 20,652 + 2093 = 22,691$ psi

At $-100,000$ lb:          $\sigma_x = 11,246 - 2903 = 9153$ psi

The mean stress $\sigma_m = 15,992$ psi and the alternating stress $\sigma_a = 6769$ psi.

Now to see if these represent a safe set of stresses: The maximum total stress is 22,691 psi. Multiplying this by the value of $K_f$ gives 90,764. This is less than the yield stress; therefore, we must apply $K_f$ to both $\sigma_m$ and $\sigma_a$.

The Goodman diagram for this material, corrected on both axes for $K_f = 4.0$, is shown in the illustration below. At a mean stress of 15,922 psi, the diagram suggests a maximum allowable alternating stress of 6376 psi. This is *less* than the applied alternating stress: hence:

$$\text{Safety factor} = \frac{6376}{6769} = 0.94 \qquad \text{(for a known mean stress)}$$

We would like to see a safety factor of at least 1.5, and probably 2 to 2.5. This design does not appear to be satisfactory!

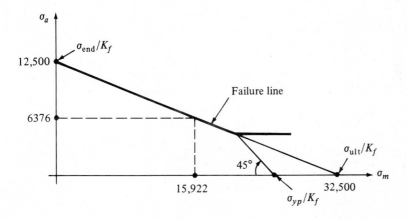

Consider how we might improve the design: A full 30 percent of the alternating stresses are due to bending. If we could make the beam infinitely stiff, the safety factor would increase, but only to 1.3.

We could reduce the alternating tensile load by (*a*) reducing the diameter of the bolt, and/or (*b*) increasing the area of the sleeve. Try the following:

1. Double the wall thickness of the box beam (by reducing the cavity size, not increasing the external dimensions).
2. Reduce the bolt diameter to 2.5 in, but maintain the 3-in-diameter threaded portion (relieve the central portion of the bolt). Assume that we still want to *prestress* the bolt to 15,000 psi.
3. Increase the sleeve outer diameter to 7 in.

The aggregate effect of these changes is to raise the safety factor from 0.94 to 2.7. This looks a bit better!

TAPERED BEAM (tapered height and width)

Filed under TAPER

Solution for beam cantilevered at left end (x=0)
Vertical load F at right end (x=L).

Height tapers from H at x=0, to K\$H at x=L.
Width tapers from B at x=0, to K1\$B at x=L.

For zero taper, set K or K1 equal to '1'

Number of elements N= 20

Height taper rato K = .6667

Width  taper rato K1= .6667

Dimensionless ratios?  (CR/N)

Find: $\dfrac{\phi}{FL^2/EI_0}$  $\dfrac{\delta}{FL^3/EI_0}$

| X | Moment | Slope | Defl. | Sigma | Tau |
|------|--------|-------|-------|-------|------|
| .025 | .975 | .050 | .003 | .500 | .126 |
| .075 | .925 | .102 | .008 | .499 | .128 |
| .125 | .875 | .153 | .015 | .497 | .130 |
| .175 | .825 | .206 | .026 | .494 | .133 |
| .225 | .775 | .259 | .039 | .490 | .135 |
| .275 | .725 | .312 | .054 | .484 | .138 |
| .325 | .675 | .365 | .072 | .476 | .140 |
| .375 | .625 | .419 | .093 | .466 | .143 |
| .425 | .575 | .472 | .117 | .455 | .146 |
| .475 | .525 | .524 | .143 | .440 | .149 |
| .525 | .475 | .575 | .172 | .423 | .152 |
| .575 | .425 | .625 | .203 | .402 | .155 |
| .625 | .375 | .673 | .237 | .378 | .158 |
| .675 | .325 | .718 | .273 | .349 | .161 |
| .725 | .275 | .759 | .311 | .315 | .165 |
| .775 | .225 | .797 | .350 | .276 | .169 |
| .825 | .175 | .828 | .392 | .230 | .172 |
| .875 | .125 | .853 | .435 | .176 | .176 |
| .925 | .075 | .869 | .478 | .113 | .181 |
| .975 | .025 | .875 | .522 | .041 | .185 |

**Program Output 13.5**

Other Tapered Beam Problems? (Y/CR)

TAPERED BEAM (tapered height and width)

Filed under TAPER

Solution for beam cantilevered at left end (x=0)
Vertical load F at right end (x=L).

Height tapers from H at x=0, to K#H at x=L.
Width tapers from B at x=0, to K1#B at x=L.

For zero taper, set K or K1 equal to '1'


Number of elements N= 20


Height taper rato K = .6667


Width  taper rato K1= .6667

Dimensionless ratios? (CR/N)

[line 290, $M = F*L$]


| X | Moment | Slope | Defl. | Sigma | Tau |
|------|--------|-------|-------|-------|------|
| .025 | 1.000 | .052 | .003 | .513 | .126 |
| .075 | 1.000 | .107 | .008 | .539 | .128 |
| .125 | 1.000 | .166 | .016 | .568 | .130 |
| .175 | 1.000 | .230 | .028 | .599 | .133 |
| .225 | 1.000 | .298 | .043 | .632 | .135 |
| .275 | 1.000 | .372 | .061 | .667 | .138 |
| .325 | 1.000 | .451 | .084 | .705 | .140 |
| .375 | 1.000 | .536 | .111 | .746 | .143 |
| .425 | 1.000 | .628 | .142 | .791 | .146 |
| .475 | 1.000 | .728 | .178 | .839 | .149 |
| .525 | 1.000 | .836 | .220 | .890 | .152 |
| .575 | 1.000 | .953 | .268 | .947 | .155 |
| .625 | 1.000 | 1.080 | .322 | 1.008 | .158 |

**Program Output 13.6** (continued on next page)

| | | | | | |
|---|---|---|---|---|---|
| .675 | 1.000 | 1.219 | .383 | 1.074 | .161 |
| .725 | 1.000 | 1.370 | .451 | 1.146 | .165 |
| .775 | 1.000 | 1.535 | .528 | 1.225 | .169 |
| .825 | 1.000 | 1.716 | .614 | 1.312 | .172 |
| .875 | 1.000 | 1.915 | .710 | 1.407 | .176 |
| .925 | 1.000 | 2.133 | .816 | 1.511 | .181 |
| .975 | 1.000 | 2.374 | .935 | 1.626 | .185 |

**Program Output 13.6** (continued)

```
                    EQUIVALENT BEAM

   Two cantilever beams in series, length and Iyy adjusted to
   match generalized cantilever stiffness parameters.
   The input quantities are dimensionless stiffnesses.

 Phi divided by MLo/EIo   = Z1 = 2.374

 Defl  ''       MLo^2/EIo = Z2 = .905

 Defl  ''       FLo^3/EIo = Z3 = .522


      R1 =    .57000 ⎫
      R2 =    .50445 ⎬ output
      R3 =   3.57617 ⎭
```

Determine Equivalent Beam

**Program Output 13.7**

STRUCTURAL ANALYSIS

2-DIMENSIONAL STRUCTURES
Axial loading    (E*A)
Bending          (E*I)

**********************************************************************

Input data to define the problem

Number of ELEMENTS = 5
Number of NODES    = 5

Input data for each NODE:
    For Degrees of Freedom, 'active'=1,  'non-active'=0

Input data for structural analysis:      set $\delta_{0,\text{bolt}} = -0.025$ in

| Node | DOF X Y A | Applied Forces X | Y | Moment A | Coordinates X | Y | |
|------|-----------|------------------|---|----------|---------------|---|--|
| 1 | 0 1 0 | 0 | 0 | 0 | 0 | 0 | OK? (CR/N) |
| 2 | 1 1 1 | 0 | 0 | 0 | 50 | 0 | OK? (CR/N) |
| 3 | 1 1 1 | 0 | 0 | 0 | 50 | 4 | OK? (CR/N) |
| 4 | 1 1 1 | 0 | 0 | 0 | 50 | 14.1 | OK? (CR/N) |
| 5 | 1 0 0 | 0 | 0 | 0 | 50 | 25.5 | OK? (CR/N) |

Input data to define ELEMENT location, stiffness, and initial conditions
    Note multipliers for E*A, E*I, and Temp-Strain
        For 'pin-joint', negate (-) node number

| Element No. | Located Between Nodes | E*A *(10^-6) (lbs.) (N) | E*I *(10^-6) (lb-in^2) (N-m^2) | Delta- Zero (in.) (m) | Temp. Strain *(10^6) | | |
|-------------|----------------------|--------------------------|-------------------------------|------------------------|----------------------|--|--|
| 1 | 1 2 | 212 | 119 | -.025 | 0 | OK? (CR/N) | |
| 2 | 1 2 | 471 | 1532 | 0 | 0 | OK? (CR/N) | |
| 3 | 2 3 | 5000 | 05000 | 0 | 0 | OK? (CR/N) | |
| 4 | 3 4 | 500 | 7500 | 0 | 0 | OK? (CR/N) | |
| 5 | 4 5 | 500 | 26850 | 0 | 0 | OK? (CR/N) | |

**Program Output 13.8a**

Output for $\delta_{0,\text{bolt}} = -0.025$ in

There are 11 degrees of freedom
Wait for computation

### Nodal Displacements and Rotations

| Node | X | Y | Alpha |
|------|--------|---------|--------|
| 1 | .00000 | -.00000 | .00000 |
| 2 | -.00776 | .00000 | .00000 |
| 3 | -.00776 | .00000 | .00000 |
| 4 | -.00776 | .00000 | .00000 |
| 5 | -.00776 | .00000 | .00000 |

| | | Forces | | Moments |
|---------|------|----------|------|---------|
| Element | Node | Fx | Fy | Mz |
| 1 | 1 | -73098.0 | .0 | -.0 |
| 1 | 2 | 73098.0 | -.0 | .0 |
| 2 | 1 | 73098.3 | .0 | -.1 |
| 2 | 2 | -73098.3 | -.0 | .1 |
| 3 | 2 | .4 | .0 | .0 |
| 3 | 3 | -.4 | .0 | .1 |
| 4 | 3 | .1 | .0 | .9 |
| 4 | 4 | -.1 | .0 | -1.6 |
| 5 | 4 | .1 | .0 | 1.5 |
| 5 | 5 | -.1 | .0 | -2.0 |

**Program Output 13.8b**

### Applied Forces and Moments

| Node | X | Y | M | | |
|------|---|---|---|-----|--------|
| 1 | 0 | 0 | 0 | OK? | (CR/N) |
| 2 | 0 | 0 | 0 | OK? | (CR/N) |
| 3 | 0 | 0 | 0 | OK? | (CR/N) |
| 4 | 0 | 0 | 0 | OK? | (CR/N) |
| 5 | 0 | 0 | 0 | OK? | (CR/N) |

Alter $\delta_0$ to $-0.36$ in

**Program Output 13.8c** (continued on next page)

| Element | Delta-Zero | T-Strain $\#10^6$ | | |
|---|---|---|---|---|
| 1 | -.036 | 0 | OK? | (CR) |
| 2 | 0 | 0 | OK? | (CR) |
| 3 | 0 | 0 | OK? | (CR) |
| 4 | 0 | 0 | OK? | (CR) |
| 5 | 0 | 0 | OK? | (CR) |

Wait for computation

Nodal Displacements and Rotations

| Node | X | Y | Alpha | |
|---|---|---|---|---|
| 1 | .00000 | -.00000 | .00000 | |
| 2 | -.01117 | .00000 | .00000 | |
| 3 | -.01117 | .00000 | .00000 | Output results |
| 4 | -.01117 | .00000 | .00000 | |
| 5 | -.01117 | .00000 | .00000 | |

| Element | Node | Forces Fx | Fy | Moments Mz | |
|---|---|---|---|---|---|
| 1 | 1 | -105261.1 | .0 | -.0 | |
| 1 | 2 | 105261.1 | .0 | .0 | |
| 2 | 1 | 105261.6 | .0 | -.1 | |
| 2 | 2 | -105261.6 | .0 | .1 | |
| 3 | 2 | .9 | .0 | -1.8 | |
| 3 | 3 | -.9 | .0 | -2.7 | Preload is OK |
| 4 | 3 | .1 | .0 | 1.0 | |
| 4 | 4 | -.1 | .0 | -2.3 | |
| 5 | 4 | .0 | .0 | 3.0 | |
| 5 | 5 | .0 | .0 | -2.0 | |

**Program Output 13.8c** (continued)

Applied Forces and Moments

| Node | X | Y | M | | | |
|---|---|---|---|---|---|---|
| 1 | 0 | 0 | 0 | OK? | (CR/N) | |
| 2 | 0 | 0 | 0 | OK? | (CR/N) | |
| 3 | 0 | 0 | 0 | OK? | (CR/N) | Run with proper $\delta_0$ and with +100,000 lb |
| 4 | 0 | 0 | 0 | OK? | (CR/N) | |
| 5 | 100000 | 0 | 0 | OK? | (CR/N) | |

**Program Output 13.8d** (continued on next page)

| Element | Delta-Zero | T-Strain $*10^6$ |
|---------|-----------|-----------------|
| 1 | -.036 | 0   OK? (CR) |
| 2 | 0 | 0   OK? (CR) |
| 3 | 0 | 0   OK? (CR) |
| 4 | 0 | 0   OK? (CR) |
| 5 | 0 | 0   OK? (CR) |

Wait for computation

Nodal Displacements and Rotations

| Node | X | Y | Alpha |
|------|-------|-------|--------|
| 1 | .00000 | .05114 | .00000 |
| 2 | -.00385 | .00000 | -.00205 |
| 3 | .00422 | .00000 | -.00194 |
| 4 | .01927 | .00000 | -.00081 |
| 5 | .02435 | .00000 | .00000 |

| | | | Forces | Moments |
|---------|------|----------|-----|---------|
| Element | Node | Fx | Fy | Mz |
| 1 | 1 | -136300.9 | .0 | 4868.8 |
| 1 | 2 | 136300.9 | .0 | -4868.8 |
| 2 | 1 | 36300.6 | .0 | 62681.0 |
| 2 | 2 | -36300.6 | .0 | -62681.0 |
| 3 | 2 | -100000.3 | .0 | 67548.8 |
| 3 | 3 | 100000.3 | .0 | 332451.0 |
| 4 | 3 | -100000.2 | .0 | -332451.6 |
| 4 | 4 | 100000.2 | .0 | 1342454.4 |
| 5 | 4 | -100000.2 | .0 | -1342455.0 |
| 5 | 5 | 100000.2 | .0 | 2482457.0 |

**Program Output 13.8***d* (continued)

```
          Applied Forces and Moments
  Node       X        Y        M

   1         0        0        0   OK?  (CR/N)

   2         0        0        0   OK?  (CR/N)

   3         0        0        0   OK?  (CR/N)

   4         0        0        0   OK?  (CR/N)

   5      -100000     0        0   OK?  (CR/N)
```

Run with
proper $\delta_0$ and
$-100{,}000$ lb

```
  Element    Delta-Zero    T-Strain *10^6

    1          -.036         0    OK? (CR)
    2           0        0   OK? (CR)
    3           0        0   OK? (CR)
    4           0        0   OK? (CR)
    5           0        0   OK? (CR)

        Wait for computation
```

```
         Nodal Displacements and Rotations
  Node        X          Y        Alpha

    1       .00000    -.05114    .00000
    2      -.01849     .00000    .00205
    3      -.02657     .00000    .00194
    4      -.04161     .00000    .00081
    5      -.04670     .00000    .00000
```

**Program Output 13.8e** (continued on next page)

| Element | Node | Forces | | Moments |
| | | Fx | Fy | Mz |
|---|---|---|---|---|
| 1 | 1 | -74221.3 | .0 | -4868.8 |
| 1 | 2 | 74221.3 | .0 | 4868.8 |
| 2 | 1 | 174222.6 | .0 | -62681.2 |
| 2 | 2 | -174222.6 | -.0 | 62681.2 |
| 3 | 2 | 100000.9 | .0 | -67551.2 |
| 3 | 3 | -100000.9 | .0 | -332454.0 |
| 4 | 3 | 100000.4 | .0 | 332453.4 |
| 4 | 4 | -100000.4 | .0 | -1342459.2 |
| 5 | 4 | 100000.0 | .0 | 1342458.0 |
| 5 | 5 | -100000.0 | .0 | -2482463.0 |

**Program Output 13.8e** (continued)

# ONE

## FRICTION

Whenever we attempt to slide one surface over another, the motion is resisted by a friction force $F$. When the surfaces are actually sliding relative to each other, the force $F$ is more or less proportional to the normal force $N$ between the two surfaces, and the proportionality constant is the coefficient of friction $f$. Thus, when actually sliding,

$$F = fN$$

Until sliding occurs, the friction force can have any value less than $fN$ in any direction. Its magnitude and direction are determined by conditions other than the friction conditions.

Because friction is so fundamental to our lives and to the study of mechanics, we will consider it in some detail.

Friction is one of those natural phenomena that everyone knows about, but no one thoroughly understands. In many cases we do our best to minimize friction, yet we would have a difficult time without it.

Friction is responsible for the energy losses in a great many of our systems. Friction and wear go hand in hand, and can be considered responsible for the degradation of most of our machines and devices. The annual cost associated with friction is tremendous; but if we had no friction, the cost would be greater.

There are laws concerning friction that are more or less true, and there are myths concerning friction that are definitely not true.

## LAWS OF FRICTION (COULOMB'S LAW)

"The coefficient of friction is independent of the area of contact and the normal load." This is reasonably true for modest normal contact stresses. However, the above "law" has no validity at high stresses. For metals, when the contact stress exceeds perhaps 0.2 of the tensile yield stress, we cannot rely on it.

"The coefficient of friction is independent of the sliding speed." Actually, $f$ varies considerably with both speed and time. Temperature and chemical activity of the surfaces are important.

## MYTH

"Smooth surfaces have low friction." This is definitely not so—some of our smoothest surfaces have exceedingly high friction. When surfaces are so rough that they actually "engage" each other (as with gear teeth), then the friction is high, but modest roughness has little effect.

"The friction coefficient cannot exceed one (1)." Definitely not so. For exceedingly clean surfaces, the materials can "bond" together such that the friction coefficient is actually infinite.

## SOURCE OF FRICTION

The friction force that resists motion has many sources, but three predominate:

1. Geometric interaction between rough surfaces: hard projections on one body "plowing" through another, or angular surfaces sliding "uphill"
2. Gross deformation of a solid or liquid layer between the two surfaces: the viscous friction of an oil or of the plastic deformation of a layer of lead between two steel surfaces, for example
3. Physical/chemical/atomic bonding between the surfaces of the sliding bodies

For most machine applications, which form the context for this book, we generally have lubricated sliding between relatively smooth surfaces. As such, the friction force will be due to a combination of 2 and 3 above, where the desired condition is 2.

## FRICTION COEFFICIENTS

Many handbooks will quote coefficients of friction for various material combinations. However, we must realize that although the numbers represent real data, those data are not generally very reproducible. Consequently, for a real problem, we must either obtain our own relevant data or use published values with wide latitude.

Here, we will just present a few guidelines relating primarily to metals:

1. For very clean surfaces, vacuum conditions, or high temperatures, $f$ can vary from 0.3 to 1 (or more).
2. For very good lubrication with chemically active oils and greases, $f$ can vary from 0.05 to 0.2.
3. For typical contact between surfaces that are neither very clean nor very well lubricated, $f$ will range from 0.2 to 0.4
4. High-friction materials, such as tires, can exhibit values of $f$ from 0.7 upward.
5. Low-friction materials, such as Teflon or wet ice, can give values of $f$ from 0.1 down to 0.01.
6. Bearings, either rotary or linear, hydrodynamic, hydrostatic, or rolling contact, can have values of $f$ ranging from 0.01 down to essentially 0.

# TWO

## ENERGY METHODS

The concepts of work and energy are fundamental to all engineering, and there are many methods for treating problems in terms of energy. The method we will introduce here provides yet another way to determine the deflections and support forces in an elastically deformed system. The method is named after the Italian engineer Alberto Castigliano (1847–1884), and is often called "Castigliano's theorem."

One of the major reasons that energy methods are so useful is that energy is a *scalar* quantity—it has magnitude but not direction.

We will consider an elastic structure composed of a number of elastic elements. When the structure is loaded, the elements deform and store elastic energy (potential energy). The total elastic energy stored in the structure is simply the *scalar sum* of the energies stored in the individual elements.

The source of the energy stored in the structure must be the externally applied forces and moments. As we slowly load (avoiding kinetic effects) the structure to its equilibrium position, there is motion at the points of load application, and the externally applied loads "do work" on the system.

Note that the unknown external forces and moments, due to fixed supports, do *not* move and do *not* contribute to the energy.

Now, *if* the total structure is *elastic*, the total work $W$ input to the system must equal the total energy $U$ stored in the individual elements; that is, $W = U$.

For the work input to equal the energy stored, the total system must be "conservative" of energy. The elements must be purely elastic, and all joints frictionless.

Although the elements do not have to be *linearly* elastic for the above equality, the presentation below will be restricted to the linear case. The treatment of nonlinear elastic systems can be found in advanced texts.

Before we become quantitative, let us restate the problem:

> When we load any conservative, linear, elastic structure, the externally applied loads do work on the system. The work is stored as potential elastic energy in the individual elements of the structure. The total work input $W$ must equal the total energy stored $U$.

## CASTIGLIANO'S THEOREM

Consider the conservative linear elastic structure shown in the illustration below. It is composed of $j = 1, 2, \ldots, J$ elements, and is acted on by $i = 1, 2, \ldots, I$ externally applied forces. We know that the total work input by the $I$ forces must equal the total energy stored by the $J$ elements.

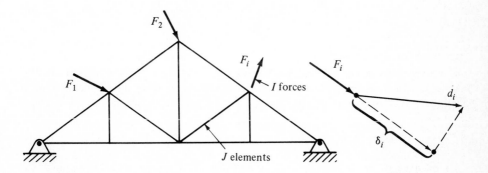

### Work Input

As shown in the illustration, the point of application of the force $F_i$ will move a distance $d_i$. However, the work done by $F_i$ is associated with the component of $d_i$ "in line" with $F_i$. Let the "in-line component" be $\delta_i$. Then the work input by $F_i$ is:

$$w_i = \int_0^{\delta_i} F_i \, d\delta_i \qquad \text{and} \qquad W = \sum_i w_i$$

### Energy Stored

Each of the $J$ elements stores energy $u_j$ because it acts as a kind of spring (linear, rotational, bending, torsional, etc.). If we know the forces acting on each element, or if we know the displacements associated with each element, we can determine the elastic energy stored. For the simple case of a linear spring:

$$u_j = \frac{F_j \delta_j}{2} = \frac{F_j^2}{2k_j} = \frac{k_j \delta_j^2}{2} \qquad \text{and} \qquad U = \sum_j u_j$$

We are going to manipulate the energy relationships by taking partial derivatives. You will recall that partial derivatives are used when a quantity of interest (the work or energy in this case) is a function of more than one variable (the forces or displacements).

If $y = f(x_1, x_2, x_3, \ldots)$, then $\partial y / \partial x_i$ is the variation in $y$ when *only* $x_i$ is varied, all other $x$'s being *constant*.

For this problem, we could write the energy as a function of either the applied forces ($F_i$), or the deformations ($\delta_i$):

$$U = f(F_1, F_2, \ldots) \qquad \text{or} \qquad U = f'(\delta_1, \delta_2, \ldots)$$

Let us consider what happens when we vary *one* of the applied forces by a small amount. The total work input will vary by $\Delta W$, which results in an equal variation in the stored energy $\Delta U$. The entire structure will deform a small amount so that the other applied forces, even though they are constant, may also do work on the system.

If, on the other hand, we consider the energy to be a function of the displacements at the applied loads, and we vary *only one* displacement by a small amount, *only* that external load moves, and *only* that external load can input work to the system.

So mathematically we vary the in-line displacement at one external load by taking the partial derivative. The work input due to the small (infinitesimal) displacement is:

$$dW = F_i \, d\delta_i = dU \qquad \text{and} \qquad \frac{\partial U}{\partial \delta_i} = F_i$$

The sketch below shows that for any linearly elastic system, because the force and deflection are proportional, the energy variation $F_i \, d\delta_i$ must equal

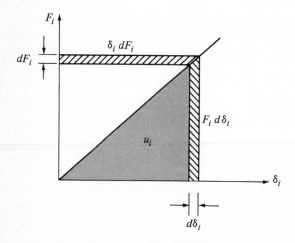

$\delta_i \, dF_i$, so that we can rewrite the above equation in two ways:

$$\frac{\partial U}{\partial \delta_i} = F_i \quad \text{and} \quad \frac{\partial U}{\partial F_i} = \delta_i$$

To summarize, if we can write the stored energy $U$ as a function of the $I$ applied loads, the partial derivative of $U$ with respect to any load must equal the in-line displacement at that load (and vice versa).

Frequently we must determine deflections where there are no loads. A very simple artifice is to apply a fictitious load at the point in question, take the derivative with respect to that fictitious load, and then set it equal to zero.

## EXAMPLES

### Springs

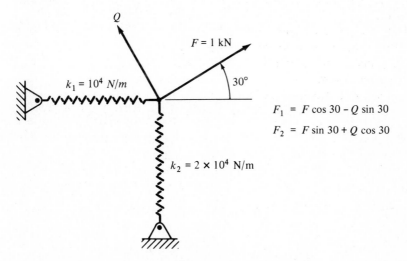

$$F_1 = F \cos 30 - Q \sin 30$$
$$F_2 = F \sin 30 + Q \cos 30$$

The force $F$ in the accompanying illustration is applied to the two-spring system; find the deflection along and perpendicular to $F$.

In order to find the perpendicular deflection, we apply the fictitious force $Q$ as shown. From equilibrium we determine the forces $F_1$ and $F_2$ in the two springs.

The energy stored in each spring is:

$$u_1 = \frac{F_1^2}{2k_1} \qquad u_2 = \frac{F_2^2}{2k_2}$$

and the total energy is:

$$U = \frac{1}{2}\left[\frac{(F \cos 30 - Q \sin 30)^2}{k_1} + \frac{(F \sin 30 + Q \cos 30)^2}{k_2}\right]$$

The deflection in line with $F$ is:

$$\delta_F = \frac{\partial U}{\partial F} = \frac{1}{2}\left[\frac{2(F\cos 30 - Q\sin 30)}{k_1}\cos 30 + \frac{2(F\sin 30 + Q\cos 30)}{k_2}\sin 30\right]$$

Setting $Q = 0$ and evaluating the equation,

$$\delta_F = 0.0875\,\text{m}$$

The deflection in line with the fictitious force $Q$ is obtained in the same way, except that the partial derivative is taken with respect to $Q$ not $F$. We obtain:

$$\delta_Q = -0.0217\,\text{m}$$

You might note that this example is identical to Example 3.6.

## Axially Loaded Members

Structures made from axially loaded members (trusses) are handled just as the spring example above. The "spring constant" for a uniform axial member is simply $AE/L$.

## Bending Members

Because the bending moment generally varies along a beam, the energy stored, per unit length, also varies. Therefore, we can consider the energy stored in a beam element of length $dx$, and integrate over the length of the beam (see the accompanying illustration):

$$du = \frac{M_b\,d\phi}{2}$$

From Chap. 8:

$$d\phi = \frac{M_b\,dL}{EI}$$

So

$$du = \frac{M_b^2\,dL}{2EI} \qquad \text{and} \qquad u = \frac{1}{2}\int_0^L \frac{M_b^2\,dL}{EI}$$

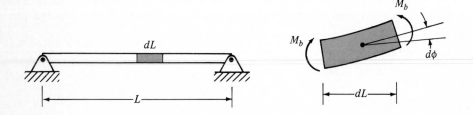

Because we will deal with partial derivatives of $u$, we can differentiate before integration and obtain (for instance):

$$\frac{\partial U}{\partial F} = \frac{\partial U}{\partial M_b} \frac{\partial M_b}{\partial F} = \int_0^L \frac{M_b}{EI} \frac{\partial M_b}{\partial F} \, dL$$

Consider the bending of a circular ring, of stiffness $EI$, and loaded by the diametral force $F$. Because of symmetry, we can study half of the ring as shown in the drawing below. The applied loads are now $F/2$ and the moment $M_0$ required for the slope to remain zero at the top.

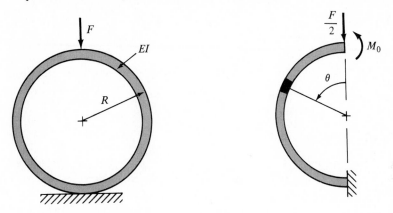

The bending moment at an angle $\theta$ is:

$$M_\theta = \frac{FR \sin \theta}{2} - M_0$$

And the total energy of the structure $(U)$ is:

$$U = \frac{1}{2EI} \int_0^\pi M_\theta^2 R \, d\theta$$

If we differentiate within the integral, we can obtain the slope (which must remain zero), and the deflection:

$$\phi = 0 = \frac{\partial U}{\partial M_0} = \frac{1}{EI} \int_0^\pi M_\theta \frac{\partial M_\theta}{\partial M_0} R \, d\theta$$

$$\delta = \frac{\partial U}{\partial F/2} = 2 \frac{\partial U}{\partial F} = \frac{2}{EI} \int_0^\pi M_\theta \frac{\partial M_\theta}{\partial F} R \, d\theta$$

Evaluating the first integral will give $M_0 = FR/\pi$. Evaluating the second integral gives $\delta = 0.488(FR^3/EI)$.

We might note that this example is identical to one of the computer examples discussed in Chap. 13.

## Complex Structures

In the ring example, the total energy was the sum (integral) of the energies in the elements. We were able to differentiate inside of the integral. Likewise, when the structure is a complex of discrete elements, and the energy is the sum of the individual energies, we can differentiate before summing. This will simplify the "bookkeeping" problems.

Given that $U = \Sigma_j\, u_j$, then:

$$\frac{\partial U}{\partial F_i} = \sum_j \frac{\partial u_j}{\partial F_i} = \sum_j \frac{\partial u_j}{\partial F_j}\frac{\partial F_j}{\partial F_i}$$

For instance for a spring-type element or truss, $u_j = F_j^2/2k_j$. Then

$$\frac{\partial u_j}{\partial F_j} = \frac{F_j}{k_j} \qquad \text{and} \qquad \frac{\partial U}{\partial F_i} = \sum_j \frac{F_j}{k_j}\frac{\partial F_j}{\partial F_i}$$

Both terms above can be precalculated prior to the various summation processes.

## THREE

## THICK-WALLED CYLINDERS (WITH PRESSURE AND ROTATION)

Earlier in this book we considered the effects of pressure and rotation on thin-walled tubes. The resulting equations were easy to develop, easy to understand, and easy to use. However, they are *not* conservative for thicker-walled cylinders. Use of the thin-wall approximation in such cases can lead to serious errors.

The equations for thick-walled cylinders have been known for many years, dating back to the work of Lamé in 1852. The solutions, derived from the theory of elasticity, are given in most elasticity texts. Here, we will not derive the equations, but will simply present the results.

Before the availability of computers, working with thick-walled cylinder systems was very tedious; now, it is very simple.

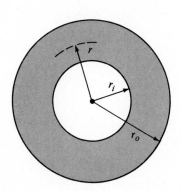

Consider the cylindrical section shown in the accompanying illustration. The cylinder can be loaded by:

Internal pressure $p_i$
External pressure $p_0$
Rotation at $\omega$

The solutions are for the case of plane stress; as such, they are correct for *thin* rotating disks (thin in the axial direction), and for pressurized cylinders of any length. For a closed-end pressure cylinder, the effects of the axial stress must be superposed on the results given below.

If we let

$$r = \text{variable radius}$$

$$r_i = \text{inner radius}$$

$$r_o = \text{outer radius}$$

then we can present the equations in terms of dimensionless variables:

$$R = \left(\frac{r}{r_i}\right)^2$$

$$S = \left(\frac{r_o}{r_i}\right)^2$$

The radial stress is $\sigma_r$, and the tangential stress is $\sigma_\theta$

## PRESSURIZED CYLINDERS

$$\sigma_r = + \frac{S(p_o - p_i)}{R(S-1)} + \frac{p_i - p_o S}{S-1}$$

$$\sigma_\theta = - \frac{S(p_0 - p_i)}{R(S-1)} + \frac{p_i - p_o S}{S-1}$$

The maximum stress is always at the inner radius ($R = 1$). If the outer pressure is zero, then

$$\sigma_{\theta_{max}} = p_i \frac{1+S}{S-1}$$

## ROTATING DISKS (THIN)

If we let:
$$A = \frac{3+\nu}{8} \qquad\qquad B = \frac{1+3\nu}{3+\nu}$$

$$\rho = \text{mass/volume} \qquad \omega = \text{radians/second}$$

then:

$$\sigma_r = \rho\omega^2 A\left(S + 1 - \frac{S}{R} - R\right)$$

$$\sigma_\theta = \rho\omega^2 A\left(S + 1 + \frac{S}{R} - BR\right)$$

$\sigma_{max}$ is at the inner radius ($R = 1$).

# FOUR

## MATERIAL PROPERTIES

This appendix gives the properties of a variety of materials taken from a variety of sources. Because data for different specimens will vary, the quantities shown have been rounded off a bit.

In general, the data are for temperatures in the vicinity of room temperature.

| Material | Elastic moduli, $10^6$ psi | | Poisson's ratio | Stress limits,† 1000 psi | | Coefficient of thermal expansion, $10^{-6}/°F$ | Weight density, $lb/in^3$ |
|---|---|---|---|---|---|---|---|
| | $E$ | $G$ | | $\sigma_{yp}$ | $\sigma_{ult}$ | | |
| Steels: Structural | 30 | 12 | 0.27 | 38 | 50 | 6.5 | 0.284 |
| C1018 Hot-Rolled | 29 | 12 | 0.3 | 48 | 69 | 6.5 | 0.284 |
| C1018 Cold-Rolled | 29 | 12 | 0.3 | 70 | 82 | 6.5 | 0.284 |
| C1045 277 Bhn | 29 | 12 | 0.3 | 90 | 120 | 6.5 | 0.284 |
| C1045 375 Bhn | 29 | 12 | 0.3 | 118 | 180 | 6.5 | 0.284 |
| AISI 4140 385 Bhn | 29 | 12 | 0.3 | 170 | 200 | 6.5 | 0.284 |
| AISI 4340 240 Bhn | 29 | 12 | 0.3 | 95 | 120 | 6.5 | 0.284 |
| AISI 4340 400 Bhn | 29 | 12 | 0.3 | 200 | 215 | 6.5 | 0.284 |
| AISI 52100 520 Bhn | 30 | 12 | 0.29 | 280 | 295 | 6.5 | 0.284 |
| Steels: Stainless | | | | | | | |
| 302 Hot-Rolled | 28 | 11 | 0.3 | 35 | 90 | 6.3 | 0.29 |
| 302 Cold-Rolled | 28 | 11 | 0.3 | 75 | 110 | 6.3 | 0.29 |
| 316 Hot-Rolled | 28 | 11 | 0.3 | 40 | 90 | 6.3 | 0.29 |
| 405 Hot-Rolled | 28 | 11 | 0.3 | 40 | 70 | 6.3 | 0.29 |
| 431 415 Bhn | 28 | 11 | 0.3 | 155 | 205 | 6.3 | 0.28 |
| AM355 | 29 | 12 | 0.3 | 181 | 216 | 6.3 | 0.28 |

| Material | Elastic moduli, $10^6$ psi | | Poisson's ratio | Stress limits,† 1000 psi | | Coefficient of thermal expansion, $10^{-6}/°F$ | Weight denisty, $lb/in^3$ |
|---|---|---|---|---|---|---|---|
| | $E$ | $G$ | | $\sigma_{yp}$ | $\sigma_{ult}$ | | |
| Aluminum: | | | | | | | |
| Pure | 10 | 3.8 | 0.32 | 15 | 0 | 13 | 0.1 |
| 2024-T4 | 10 | 3.8 | 0.32 | 45 | 70 | 13 | 0.1 |
| 6061-T4 | 10 | 3.8 | 0.32 | 21 | 35 | 13 | 0.1 |
| 7075-T6 | 10 | 3.8 | 0.32 | 70 | 85 | 13 | 0.1 |
| | | | | | | | |
| Cast iron: | | | | | | | |
| Class 25 | 12 | 4 | 0 | +25f | −100f | 6 | 0.26 |
| Class 50 | 18 | 7 | 0 | +55f | −150f | 6 | 0.26 |
| Gray A-48 | 10 | 4 | 0 | +25f | −90f | 6.7 | 0.26 |
| | | | | | | | |
| Monel: Cold Drawn | 26 | 10 | 0.3 | 85 | 100 | 7.7 | 0.32 |
| Titanium: 6Al-4V | 16 | 6 | 0.3 | 120 | 130 | 5.3 | 0.16 |
| | | | | | | | |
| Brass: | | | | | | | |
| Cartridge, 260 Cold-Rolled | 16 | 6 | 0.3 | 63 | 70 | 11 | 0.31 |
| Yellow, hard, 270 | 15 | 6 | 0.3 | 60 | 74 | 11 | 0.31 |
| Leaded, hard, 360 | 14 | 6 | 0.3 | 52 | 68 | 11 | 0.31 |
| | | | | | | | |
| Bronze, phos., 510, hard | 16 | 6 | 0.3 | 75 | 85 | 10 | 0.32 |
| Magnesium, AZ80A-T5 | 6.5 | 2.5 | 0.3 | 35 | 52 | 14 | 0.07 |
| | | | | | | | |
| Polymers: | | | | | | | |
| Epoxy | 1 | 0 | 0 | +10f | 0 | 30 | 0.06 |
| Teflon | 0.06 | 0 | 0 | 3 | 0 | 55 | 0.08 |
| Nylon 66 | 0.4 | 0.14 | 0.4 | +8f | 0 | 55 | 0.04 |
| | | | | | | | |
| Fibers: | | | | | | | |
| Glass | 11 | 0 | 0 | +500f | 0 | 0.4 | 0.08 |
| Boron | 60 | 0 | 0.2 | +500f | 0 | 3 | 0.084 |
| Carbon (Graphite) | 55 | 0 | 0 | +300f | 0 | 0 | 0 |
| Kevlar | 19 | 0 | 0 | +500f | 0 | 0 | 0.05 |
| | | | | | | | |
| Wood: (clear) | | | | | | | |
| Hickory | 2.1 | 0 | 0 | +20f | −9f | 0 | 0.028 |
| Oak, red | 1.8 | 0 | 0 | +14f | −7f | 0 | 0.024 |
| Fir, Douglas | 2 | 0 | 0 | +12f | −7f | 0 | 0.018 |
| | | | | | | | |
| Glass, silica | 10 | 4.2 | 0.2 | +10f | 0 | 0.3 | 0.08 |
| Concrete | 3 | 1.3 | 0.15 | nil | −5f | 6 | 0.085 |
| Aluminum oxide | 40 | 0 | 0 | +45f | 0 | 3.7 | 0.14 |
| Tungsten Carbide (sintered) | 45 | 0 | 0 | >+120f | 0 | 4.5 | 0.22 |

*Note*: A value of zero (0) indicates a lack of data.

† For *brittle* materials, the stress limits are the fracture stress in tension (+ · · · f), and compression (− · · · f).

# FIVE

## ANSWERS TO PROBLEMS

### Chapter 1

**1.1** Yes; acceleration forces are 33 percent of weight forces.

**1.2** $F = 0.8$ lb/in of circumference, radially outward.

**1.5** $M = Fa$ about all points.

**1.6** $F_x = -F$ $\quad F_y = 0$ $\qquad F_z = 0$
$\quad M_x = 0$ $\qquad M_y = -Fh$ $\quad M_z = Fr$

**1.7** $F_x = 0$ lb $\qquad F_y = -440$ lb $\qquad F_z = -50$ lb
$\quad M_x = -220$ ft·lb $\quad M_y = 400$ ft·lb $\quad M_z = -320$ ft·lb

**1.8** Tapered are better; extrusions are cheaper.

**1.9** (*a*) Yes; (*b*) Yes; (*c*) No; (*d*) Yes; (*e*) No; (*f*) Yes; (*g*) No.

**1.10** Tangential component of spoke tension.

**1.11** Friction between bead and rim due to pressure.

**1.13** $M = 46.95$ in·lb (5.31 N·m)

**1.14** 600 rpm and 893 in·lb (100.9 N·m)

**1.15** $x = 0.77$ in, $y = 4.1$ in

**1.16** Table for $\theta$, $\sin \theta$, $\tan \theta$

**1.17** $S = \text{sum} = 506$

### Chapter 2

**2.3** $F_{AC} = 1414(T)$ $\qquad F_{AB} = 1000(C)$ $\qquad F_{BC} = 1000(C)$

**2.4** $F_{AC} = 3\sqrt{2}F(T)$ $\qquad F_{BD} = 2\sqrt{2}F(C)$
$\quad F_{CE} = \sqrt{2}F(T)$ $\qquad F_{CD} = 2F(T)$
$\quad F_{AB} = 2F(C)$ $\qquad\; F_{DE} = F(C)$
$\quad F_{BC} = 2F(C)$

**2.5** $N_A = 150$     $N_B = 86.8$     $N_C = 86.6$     $N_D = 86.6$     $F_{z,min} = 101$

**2.6** $M_0 = 35$ N·m

**2.7** $f_{min} = 0.27$

**2.8** $\alpha_{min} = 26.6°$; 62 percent

**2.9** $M_0 = rW \tan(\alpha + \beta)$     $\tan\alpha = \dfrac{P}{2\pi r}$     $\tan\beta = f$

**2.10** $f_{min} = \dfrac{P}{2\pi r}$     Efficiency = 50 percent

**2.11** $M_0 = rW[\tan(\alpha + \beta) + f] \cong rW\left(\dfrac{P}{2\pi r} + 2f\right)$

**2.12** $M_0 \approx rW \tan(\alpha + \beta')$     $\tan\beta' = \dfrac{f}{\cos\gamma}$

    (a) $M_0 = 979$ in·lb; (b) $M_0 = 1729$ in·lb

**2.13** $f = 0$     $N_1 = 641$     $N_2 = 691$

    $f = 0.3$     $N_1 = 476$     $N_2 = 526$

**2.14** $f_{min} = 0.27$ at $B$

**2.15** $F = 2700$ N

**2.16** $f_{min} = 0.297$

**2.17** $N_o = \dfrac{N_i}{n}$     $M_o = \dfrac{M_i R}{r \tan(\alpha + \beta')}$     Efficiency = $\dfrac{\tan\alpha}{\tan(\alpha + \beta')}$

    See Prob. 2.12.

**2.18** At $A$: $F_x = 333 \to 331$; $F_y = 0$; $F_z = 500$

    At $B$: $F_x = 1793 \to 1785$; $F_y = 5841 \to 5814$; $F_z = 500$

**2.19** At $A$: $F_x$ is indeterminate; $F_y = 585$; $F_z = 50$

    At $B$: $F_x$ is indeterminate; $F_y = -585$; $F_z = -50$

**2.20** $f_{min} = 0.2$     $T_B = 9.4$ lb

**2.21** $T_A = 109.2$     $T_B = 162.3$     $f_{min} = 0.126$

**2.22** (a) $25.4°$; (b) $38.7°$; (c) 35 hp

**2.23** $\alpha = \tan^{-1}\dfrac{\rho V^2}{g}$

**2.24** $R_A = 556$     $R_B = 444$ lb

**2.25** $F = 3600$ lb; $M = 1.92 \times 10^6$ in·lb

**2.26** Approximate answers:

    $F_A = 150$     $F_B = 1200$     $F_C = 2400$     $F_D = 3600$     $F_E = 2250$

**2.27** (a) $a\gamma h + W > aH\gamma$

    (b) $F = \gamma(ah - AH) + W$

**2.28** At $A$: $F_{rad} = 94$ lb; $F_{axial} = 47$ lb

    At $B$: $F_{rad} = 77$ lb

**2.29** (a) Front wheel: $N = 88.1$ lb

       Rear wheel: $N = 61.9$ lb

    (b) $f_{min} = 0.72$

    (c) $T = 362.3$ lb

**2.30** $M_{AB} = 11.7$ in·lb

**2.31** At $C$: $F_x = 60.4$ lb; $F_y = -300$ lb

    At $D$: $F_x = -422.7$ lb; $F_y = 500$ lb

**2.32** At each end, 250-lb forces $\frac{1}{16}$ in apart (actually distributed, $q = 4000$ lb/in)

**2.34** 20.8 lb

## Chapter 3

**3.1** Plot graph $\sigma_{yp} = 50,000$ psi

**3.2** (a) $\delta = 0.0011$ in; (b) $\delta = 0.000606$ in

**3.3** $\delta = 73$ in

**3.4** $\delta = 0.0059$ in

**3.5** $\delta_x = +16.66 \times 10^{-4}$ in; $\delta_y = 0$ in

**3.6** $\delta_x = +12.5 \times 10^{-4}$ in; $\delta_y = +2.406 \times 10^{-4}$ in

**3.7** (a) $\delta_x = -48.3 \times 10^{-6}$ m; $\delta_y = 0$ m
    (b) $\delta_x = +233 \times 10^{-6}$ m; $\delta_y = -96.6 \times 10^{-6}$ m
    (c) $\delta_x = -96.6 \times 10^{-6}$ m; $\delta_y = -563 \times 10^{-6}$ m

**3.8** (a) $\delta_x = 0.0072$ in; $\delta_y = 0$ in
    (b) $\delta_x = 0.0036$ in; $\delta_y = -0.0196$ in

**3.9** 300 lb

**3.10** $\dfrac{a}{L} = \dfrac{3}{8}$

**3.11** $\delta = \dfrac{F}{7k}$ (regardless of location of $F$)

**3.12** $\delta = 2.40$ in

**3.13** $P = 12.8$ hp

**3.14** $\delta = \dfrac{FL^2}{4k_\phi}$

**3.15** $\delta = \dfrac{7FL^2}{8k_\phi}$

**3.16** $\delta = \dfrac{F\epsilon}{kL - F}$; $\delta \to \infty$ when $F = kL$

**3.17** Plot curve. $F_{max} = 13.8$ lb at $h \cong 3$ in

**3.18** $\delta = 0.54$ in

**3.19** $\delta_A = -0.00762$ m $\qquad \delta_B = -0.00381$ m $\qquad \delta_C = -0.00191$ m

**3.20** $F = 11,321$ lb $\qquad \delta_s = -0.00241$ in

**3.21** $F_{rod} = 31,573$ lb $\qquad F_{sleeve} = 18,247$ lb

**3.22** $F_{rod} = 40,256$ lb $\qquad F_{sleeve} = 9744$ lb

**3.23** $F = 166,666$ lb

**3.24** $\delta_{x'} = 0.4375 \qquad C_{x'x'} = 0.00875$
$\qquad\qquad\qquad\quad C_{y'y'} = 0.00625$
$\qquad\qquad\qquad\quad C_{x'y'} = -0.00217$

**3.25** $\alpha = -17.53°, +72.47°$

**3.26** $C_1 = 0.00000180 \qquad C_2 = 0.0000657 \qquad \alpha = 55.3°$

**3.27** $\epsilon = 2 \ln \dfrac{D_0}{D_f}$

**3.29** 23,000 psi

**3.30** 1.5 percent

## Chapter 4

**4.1** At $B$: $\delta_x = -0.0006$, $\delta_y = -0.0685$

**4.2** At $C$: $\delta_x = -0.0024$, $\delta_y = -0.273$

**4.3** At 2: $\delta_x = 0.00053$, $\delta_y = -0.0011$

**4.4** At 4: $\delta_x = 0.00014$, $\delta_y = -0.0005$

**4.5** At 3: $\delta_x = 0.167$, $\delta_y = -0.536$

**4.6** At 3: $\delta_x = 0.013$, $\delta_y = -0.447$

**4.7** At 4: $\delta_x = 0.040$, $\delta_y = -0.013$

**4.8** At 4: $\delta_x = -0.036$, $\delta_y = -0.012$

**4.9** At 4: $\delta_x = 0$, $\delta_y = -0.0217$

**4.10** At 2: $\delta_x = -0.0035$, $\delta_y = -0.0238$

**4.13** $u_1 = 0.00188$, $u_2 = -0.00070$, $u_3 = -0.00024$, $u_4 = -0.00286$, $u_5 = -0.00012$

**4.14** $u_1 = 0.00167$, $u_2 = 0$

**4.15** $u_1 = 0.00125$, $u_2 = 0.000241$

**4.16** $u_1 = 0.00024$, $u_2 = -0.0001$, $u_3 = -0.0001$, $u_4 = -0.00058$, $u_5 = -0.00005$

**4.17** $u_2 = u_5 = -0.0196$, $u_1 = u_4 = 0.00036$, $u_3 = 0.0072$

**4.18** $F_b = 11,800$    $F_s = -11,800$

**4.19** $u = -0.00536$

**4.20** $u = -0.00683$

**4.21** $F_1 = 166,666$

**4.22** At 5: $\delta_x = 11.6$, $\delta_y = -7.96$

**4.23** $F_{max} = 149$

**4.24** Program: $\delta_y = -0.00403$; "equivalent" tube: 0.00399

**4.25** $\sigma_{max} = 45,062$ psi

## Chapter 5

**5.8**

| Case | $\sigma_1$, psi | $\sigma_2$, psi | $\sigma_3$, psi | $\alpha$, deg |
|------|-----------------|-----------------|-----------------|---------------|
| (a)  | 9,000           | 0               | −1,000          | 18.4          |
| (b)  | 84,080          | 0               | −24,080         | 16.85         |
| (c)  | 30,000          | 0               | −30,000         | 0             |
| (d)  | 6,810           | 0               | −8,810          | 25.1          |
| (e)  | 890             | 0               | −26,890         | 29.9          |

**5.9**

| Case | $\sigma_1$, MPa | $\sigma_2$, MPa | $\sigma_3$ MPa | $\alpha$, deg |
|------|-----------------|-----------------|----------------|---------------|
| (a)  | 640.5           | 0               | −140.5         | 25.1          |
| (b)  | 354             | 0               | −174           | 18.7          |
| (c)  | 700             | 0               | −700           | 0             |
| (d)  | 89.7            | 0               | −249.7         | 22.5          |
| (e)  | 0               | −150            | −900           | 31.7          |

**5.10** $\tau_{max} = \dfrac{pr}{2t}$

**5.11** $\tau_{max} = \dfrac{pr}{2t}$

**5.13** $\tau_{max} = \dfrac{pr}{4t}$

**5.14** $\sigma_\theta = 20$ ksi $\qquad \tau_{max} = 10$ ksi

**5.15** $\tau_{max} = 1813$ psi

**5.16**

| Case | $\epsilon_1 \times 10^6$ | $\epsilon_2 \times 10^6$ | $\alpha$, deg |
|------|------|------|------|
| (a) | 1171 | −171 | 31.7 |
| (b) | 681 | −1881 | 19.3 |
| (c) | 400 | −600 | 18.4 |
| (d) | 357 | −1057 | 22.5 |

**5.17** $\gamma_{max} = 1500$

**5.18**

| Case | $\epsilon_1 \times 10^6$ | $\epsilon_2 \times 10^6$ | $\alpha$, deg |
|------|------|------|------|
| (a) | 1000 | 200 | 0 |
| (b) | 1062 | −462 | 11.6 |
| (c) | 1470 | −1171 | 30.2 |
| (d) | 966 | −166 | 22.5 |

**5.19**

| Case | $\epsilon_1 \times 10^6$ | $\epsilon_2 \times 10^6$ | $\alpha$, deg |
|------|------|------|------|
| (a) | 1067 | 0 | 0 |
| (b) | 1645 | −319 | 14 |
| (c) | 1000 | −1000 | 15 |
| (d) | 1379 | 287 | 23.9 |

**5.20** Hexical preferred. For 75 percent, 45° helix.

**5.21** $\alpha = 54.7°$

## Chapter 6

**6.1** $\delta = -\dfrac{FL}{AE}\left(1 - \dfrac{2\nu^2}{1-\nu}\right)$

**6.5** $\sigma_y = \nu\sigma_x = -\dfrac{\nu F}{bh}$

**6.6** $x = \dfrac{A(1-\nu) + \nu(B+C)}{(1-2\nu)(1+\nu)}$

where $A = E\epsilon_x - \alpha\,\Delta T$
$\qquad\quad B = E\epsilon_y - \alpha\,\Delta T$
$\qquad\quad C = E\epsilon_z - \alpha\,\Delta T$

And so forth.

**6.7** $\dfrac{\Delta V}{V} = 3\alpha \, \Delta T$

**6.8**

| Case | $\epsilon_1 \times 10^6$ | $\epsilon_2 \times 10^6$ | $\epsilon_z \times 10^6$ | $\alpha$, deg |
|------|------|------|------|------|
| (a) | 3200 | −50 | −1450 | 26.6 |
| (b) | 1083 | −1083 | 0 | 0 |
| (c) | 1417 | −1651 | 100 | 22.5 |
| (d) | 4290 | −2769 | −2174 | 16.1 |
| (e) | 1449 | 1449 | −4830 | 45 |

**6.9**

| Case | $\sigma_1$, psi | $\sigma_2$, psi | $\alpha$, deg | $\epsilon_z \times 10^6$ |
|------|------|------|------|------|
| (a) | 30,490 | −3,663 | 15.2 | −268 |
| (b) | 49,490 | 23,370 | 22.5 | −729 |
| (c) | −28,256 | −48,887 | 13.3 | +771 |

**6.10** $\sigma_1 = 484$ MPa $\qquad \sigma_2 = -208$ MPa $\qquad \sigma_3 = 0$ PMa $\qquad \alpha = 12.8°$

**6.11** $\sigma_x = 37,500$ psi

**6.12** $\sigma_1 = 27,189$ psi $\qquad \sigma_2 = -7904$ psi $\qquad \sigma_3 = 0$ psi $\qquad \tau_{max} = 17,546$ psi

**6.13** $\tau_{max} = 4,539$ psi

**6.15** $p \cong 320$ psi

**6.16** $p = -720$ psi (negative for external pressure)

**6.17** $p = \dfrac{2E\epsilon t}{r(1 - \nu)}$

**6.18** $p = 1176$ psi; $F = 2609$ lb

**6.19** $p = -1656$ psi; $F = -48,298$ lb

**6.20** "Equivalent" modulus $= E_f V_f + E_r(1 - V_f) \cong E_f V_f$

**6.21** "Equivalent" modulus $= \dfrac{1}{V_r/E_r + V_f/E_f} \cong \dfrac{E_r}{V_r}$

## Chapter 7

**7.1** SF = 1.33 (Tresca) or 1.52 (von Mises)

**7.2** $r = 7.44$ in, $t = 0.83$ in

**7.3** SF = 1.82 (Tresca) or 2.04 (von Mises)

**7.4** SF = 1.96 at location $a$, 1.77 at location $b$, 1.97 at location $c$, and 1.96 at location $d$

**7.5** 182 lb

**7.6** 2000 psi

**7.7** 225,000 psi (but will actually be lower)

**7.8** 30,000 psi

**7.9** 95.7°F

**7.10** +10,000 psi

**7.11** 68 minutes

**7.12** 1515 psi

**7.13** 3333 psi

**7.14** Not possible!

**7.15** $r = 3.6$ in, $t = 0.65$ in

**7.16** $r = 6.2$ in, $t = 0.65$ in

**7.17** SF = 1.31

**7.18** SF = 1.76

**7.19** 1316 lb

**7.20** SF = 1.53

**7.21** $N \simeq 34,000$

## Chapter 8

**8.1** 50 in

**8.2** $M_{max} = -FL$

**8.3** $M_{max} = -9$ kN·m

**8.4** $M_{max} = -3Fa$

**8.5** $M_{max} = +666.7$ ft·lb

**8.6** $M_{max} = -5000$ ft·lb

**8.7** $M_{max} = -\dfrac{w_0 L^2}{3}$

**8.8** $\sigma_{max} = 51,758$ psi

**8.9** 1.035 in

**8.10** 51.2 in

**8.11** (See Table 8.1)

**8.12** $\dfrac{5P}{16}$

**8.13** $\dfrac{PL^3}{192EI}$

**8.14** $M_{max} = \dfrac{6EI\delta}{L^2}$ at the ends

**8.15** 23.0 in$^4$

**8.16** $0.307 \times 10^{-6}$ m$^4$

**8.17** 0.640 in$^4$

**8.18** 617.4 mm$^4$ ($6.17 \times 10^{-12}$ m$^4$)

**8.19** 256 in$^4$

**8.20** $I_{1,2} = 4.72$, 0.60 in$^4$, $\alpha = 32.8°$

**8.21** $I_{1,2} = 93,589$, 9879 mm$^4$, $\alpha = 14.3°$

**8.22** $I_{1,2} = 11.05$, 1.448 in$^4$, $\alpha = 19.33°$

**8.23** $I_{1,2} = 3.333$, 0.833 in$^4$, $\alpha = 18.4°$

**8.24** ($a$) 2239 psi at right-hand end
   ($b$) −1343 psi at left-hand end

**8.25** $\phi = +0.00136$, $\delta = -0.009$ in

**8.26** $\delta_y = -0.059$ in, $\delta_z = +0.034$ in

**8.29** $\delta = 54$ in; *no!*

**8.30** 37.5 psi

**8.31** (*b*) is preferred because $Q$ is less

**8.32** 506 lb/in

**8.33** 550 lb/in

**8.34** $M_{max} = 62.7 \, \text{N} \cdot \text{m}, \; V_{max} = 2051 \, \text{N}$

**8.35** 32.8 psi; 6858 psi; 0.0424 in

## Chapter 9

**9.6** $\sigma_{max} = 48 \, \text{MPa}$

**9.7** $\sigma_{max} = 14.2 \, \text{MPa}$

**9.8** $\sigma_{max} = 8.65 \, \text{MPa}$

**9.9** $\sigma_{max} = 10.82 \, \text{MPa}$

**9.10** $\sigma_{max} = 7.21 \, \text{MPa}$

**9.11** $\sigma_{max} = 50.67 \, \text{MPa}$

**9.12** $\delta = 0.1605 \, \text{in}$

**9.13** $\delta = 0.0672 \, \text{in}$

**9.14** $\delta = 0.0382 \, \text{in}$

**9.15** $\delta = 0.0270 \, \text{in}$

**9.16** $\delta = 0.0165 \, \text{in}$

**9.17** $\delta = 0.0108 \, \text{in}$

**9.18** $\delta = 0.0078 \, \text{in}$

**9.19** (*a*) $\sigma_{max} = 12{,}500 \, \text{psi}$ at center of top member
(*b*) $\sigma_{max} = 922 \, \text{psi}$ at all pipe ends

**9.20** $M_{max}$ at lower left corner $= 365{,}928 \, \text{in} \cdot \text{lb}$

**9.21** Design (*a*) is better: Design (*b*) has 6.58 times the deflection and 14.6 times the stress.

**9.22** Little difference in deflection:

$$\text{Pin joint} \rightarrow 0.2568 \, \text{in}$$

$$\text{Gussets} \rightarrow 0.2517 \, \text{in}$$

Maximum stress is 50 percent greater with gussets.

**9.23** $f_{min} = 0.27$

**9.24** $F = 2700 \, \text{N}$

**9.25** $\delta = 0.00104 \, \text{in}$

## Chapter 10

**10.1** $\tau = 25{,}465 \, \text{psi}, \; \phi = 14.6$

**10.2** $\phi_{yp} = 14.3°$

**10.3** $d = 1.04 \, \text{in}, \; L = 109 \, \text{in}$

**10.4** $M_t = 2330 \, \text{N} \cdot \text{m}, \; \phi = 1.75°$

**10.5** $M_t = 59{,}774 \, \text{in} \cdot \text{lb}$

**10.6**

| Case | hp | kW |
|------|------|------|
| (a) | 16.32 | 12.17 |
| (b) | 32.64 | 24.35 |
| (c) | 81.6 | 60.87 |
| (d) | 326 | 243 |

**10.7** $L = 10$ in, $d = 0.209$ in

**10.8** First yield at $A$. SF = 2.25; $\delta_C = -0.174$ in

**10.9** $\delta = \dfrac{4nPR^3}{Gr^4}$

**10.10** SF = 1.53

**10.11** SF = 1.34

**10.12** SF = 1.11

**10.13** $r = 3.0$ in, $t = 0.1$ in (yield and buckle)

**10.14** $\delta = 0.0274$ in

**10.15** At $A$, SF = 7.22; at $C$, SF = 4.00 (yield)

**10.16** At $A$, SF = 1.68; at $C$, SF = 0.96 (fatigue)

**10.17** $\delta_y = -0.0145$; $\delta_z = 0.0050$

**10.18** $M_t = 60{,}000$ in·lb, $K_\phi = 1.28 \times 10^6$ in · lb/rad

**10.19** Same by both methods:

$$\tau = \frac{M_t r}{2\pi r^2 t} \qquad \phi = \frac{M_t L}{2\pi r^3 t G}$$

**Chapter 11**

**11.4** $\delta_x = 2.991$ in, $\delta_y = -2.713$ in, $\delta_z = -4.207$ in

**11.5** $\delta_z = 3.344$ in

**11.6** $F_{max} = 335$ lb

**11.7** Weight = 2.78 lb, 34 dof

**11.8**

| Element | Node | Safety factor |
|---------|------|---------------|
| 1 | 1 | 4.74 |
| 2 | 3 | 5.62 |
| 3 | 4 | 2.25 ("Worst" point) |
| 4 | 4 | 4.0 |
| 5, 6 | 3 | 6.47 |
| 7, 8 | 4 | 10.7 |

**11.9**

| Element | OD in | After first iteration |
|---------|-------|----------------------|
| 1 | 1.25 | |
| 2 | 0.81 | 2.2 lb |
| 3 | 1.10 | |
| 4 | 0.91 | SF of element 3 is now 2.32 |
| 5, 6 | 0.39 | not 3.0. Continue to iterate. |
| 7, 8 | 0.5 | |

## Chapter 12

**12.1** $M_{max} = Fa$

**12.2** $M_{max} = \dfrac{FL}{4}$

**12.3** $M_{max} = \dfrac{2FL}{9}$

**12.4** $M_{max} = \dfrac{Fa}{2}$

**12.5** $M_{max} = -\dfrac{wL^2}{2}$

**12.6** $M_{max} = \dfrac{3wL^2}{64}$

**12.7** $M_{max} = -\dfrac{wa^2}{2}$

**12.8** $M_{max} = \dfrac{5FL}{8}$

**12.9** $M_{max} = -\dfrac{9FL}{8}$

**12.10** $M_{max} = \dfrac{wL^2}{12}$

**12.11** $M_b = FR(1 - \cos\theta)$

**12.12**

| Point | $M_b$ | $M_t$ |
|-------|-------|-------|
| A | −2000 | 1000 |
| B | 0 | 1000 |
| C | +1000 | 0 |

**12.13** $M_b = \sqrt{a^2 + b^2}\,F$

**12.14** $M_b = PR\cos\theta$

**12.15** 0.866    $(x/L)$

**12.16** 0.207    $(x/L)$

**12.17** 0.3229    $(x/L)$

**12.18** 0.167    $(x/L)$

**12.19** $0.707 \qquad (x/L)$

**12.20** $x_1 = 0.1811L$, $x_2 = 0.7573L$

**12.21** $x = \dfrac{L}{3}$

**12.22** $\delta_{max} = \dfrac{9wL^4}{24EI}$

**12.23** $\delta_{max} = -\dfrac{0.005wL^4}{EI}$ at $\dfrac{x}{L} = 0.5785$

**12.24** $\delta_{max} = \dfrac{0.00796FL^3}{EI}$

**12.25** $\delta_{max} = \dfrac{M_0L^2}{216EI}$

**12.26** $\delta_{max} = 0.00656\dfrac{wL^4}{EI}$

**12.27** $\delta_{max} = \dfrac{FL^3}{12EI}$

**12.28** $M_{max} = \dfrac{FL}{2} - \dfrac{FL(11/48 + K)}{\frac{1}{3} + K}$, $K = \dfrac{EI}{kL^3}$

**12.29** $M_{max} = \dfrac{wL^2[\frac{1}{2} - (\frac{5}{24} + K)]}{(\frac{1}{3} + K)}$, $K = \dfrac{EI}{kL^3}$

**12.30** $\dfrac{wL^4}{384EI}$

**12.31** $\dfrac{0.00131wL^4}{EI}$

**12.32** $\dfrac{11wL^4}{120EI}$

**12.33** $s = 10.1$ mm

**12.34** $\sigma_{max} = 18{,}502$ psi; $\delta_y = -0.0377$ in, $\delta_z = 0.0400$ in

**12.35** $\delta_y = -0.0289$ in, $\delta_z = 0.4165$ in

**12.36** $\delta_y = -0.0525$ in, $\delta_z = 0.1052$ in

**12.37** Stress: wood, $-2348$ psi; aluminum, $9138$ psi. Deflection: $0.074$ in

**12.38** Deflection: $0.636$ in

**12.39** Deflection: $43.37$ in

**12.40** $M_L = 9.25\sigma_{yp}$

**12.41** $M_L = 8.125\sigma_{yp}$

**12.42** $M_L = 4.5\sigma_{yp}$

**12.43** $P_L = 4.5M_L$

**12.44** $P_L = 6M_L$

**12.45** $P_L = 9M_L$

# INDEX